# Life-Cycle Performance of Green Cementitious Composites under Complex Environmental Conditions

# Life-Cycle Performance of Green Cementitious Composites under Complex Environmental Conditions

Shan Gao
Shaobo Kang

MDPI

Basel • Beijing • Wuhan • Barcelona • Belgrade • Novi Sad • Cluj • Manchester

Shan Gao
School of Civil Engineering
Harbin Institute of Technology
Harbin
China

Shaobo Kang
School of Civil Engineering
Chongqing University
Chongqing
China

*Editorial Office*
MDPI AG
Grosspeteranlage 5
4052 Basel, Switzerland

This is a reprint of articles from the Special Issue published online in the open access journal *Materials* (ISSN 1996-1944) (available at: www.mdpi.com/journal/materials/special_issues/green_cementitious_composites_environment).

For citation purposes, cite each article independently as indicated on the article page online and using the guide below:

| Lastname, A.A.; Lastname, B.B. Article Title. *Journal Name* **Year**, *Volume Number*, Page Range. |
| --- |

**ISBN 978-3-7258-2088-7 (Hbk)**
**ISBN 978-3-7258-2087-0 (PDF)**
**https://doi.org/10.3390/books978-3-7258-2087-0**

Cover image courtesy of Shan Gao

# Contents

# About the Editors

**Shan Gao**

Shan Gao is an associate professor at the School of Civil Engineering, Harbin Institute of Technology, China. He obtained his PhD degree from Harbin Institute of Technology, China, in July 2014. His expertise includes the analysis and testing of high-performance composite structures.

**Shaobo Kang**

Dr. Kang Shao-Bo is professor at the School of Civil Engineering, Chongqing University, China. He obtained his PhD degree from Nanyang Technological University, Singapore, in February 2016. His expertise includes the analysis and testing oof building structures subjected to progressive collapse and the behaviour of high-performance fibre-reinforced concrete under various loading conditions.

*materials*

Review

# Application of Coal Gangue as a Coarse Aggregate in Green Concrete Production: A Review

**Shan Gao [1,2,*], Sumei Zhang [1] and Lanhui Guo [1]**

1 School of Civil Engineering, Harbin Institute of Technology, Harbin 150090, China; smzhang@hit.edu.cn (S.Z.); guolanhui@hit.edu.cn (L.G.)
2 Shaanxi Key Laboratory of Safety and Durability of Concrete Structures, Xijing University, Xi'an 710123, China
* Correspondence: gaoshan@hit.edu.cn

**Abstract:** Among the techniques for converting stacked coal gangue to reusable material, one of the most effective ways is to use coal gangue as a coarse aggregate in green concrete productions. The physical and chemical properties of rock and spontaneous-combustion coal gangue are generally suitable for being used as a coarse aggregate in green concrete. Coal gangue concrete is not recommended to be used in subsurface structures, as its water absorption law would be changed under a large replacement ratio. The mechanical performance of coal gangue concrete is degraded by raising the replacement ratio. Over-low and -high concrete grades are not suggested to be used as coal gangue aggregate, unless extra admixtures or specific methods are used. The influence of coal gangue on the durability of coal gangue concrete is remarkable, resulting from the porous structure of coal gangue that provides more transmission channels for air and liquid in concrete, but is beneficial for thermal insulation. The usage of coal gangue in structural concrete members is still limited. The mechanical behavior of some structural members using coal gangue concrete has been reported. Among them, concrete filled steel tubes are a preferable configuration for using coal gangue concrete, regarding both the mechanical and durability performance.

**Keywords:** coal gangue; green concrete; coarse aggregate; structural member; eco-friendly production



**Citation:** Gao, S.; Zhang, S.; Guo, L. Application of Coal Gangue as a Coarse Aggregate in Green Concrete Production: A Review. *Materials* **2021**, *14*, 6803. https://doi.org/10.3390/ma14226803

Academic Editor: Alessandro P. Fantilli

Received: 11 October 2021
Accepted: 8 November 2021
Published: 11 November 2021



## 1. Introduction

Even though many kinds of new energy and renewable fuels have been explored nowadays, coal is still the main energy for many countries in the world, especially for countries in Asia Pacific region, such as China, Indonesia, and Vietnam. In fact, the increase of global coal production in 2019 resulted from the significant production increases of China and Vietnam, while that of other regions in the world declined, as shown in Figure 1 [1]. Coal still accounts for more than 70% of energy expenditure in China [2], while the world coal expenditure fell to its lowest level in 16 years, as a result of being displaced by natural gas and renewables.

Along with the excavation and washing of coal in the production process, large amounts of rock-like coal gangue (CG) accompanying the carboniferous beds are discharged as industrial residues, accounting for around 10–20% of the coal production [2–4]. In the largest coal production country, namely China, more than 700 million tons of coal gangue was expected in 2020, with the annual rate of 300–350 million tons [4,5], not to mention that almost 5 billion tons of that have been accumulated in China [4,6] and the number for Europe is 175 million tons [2].

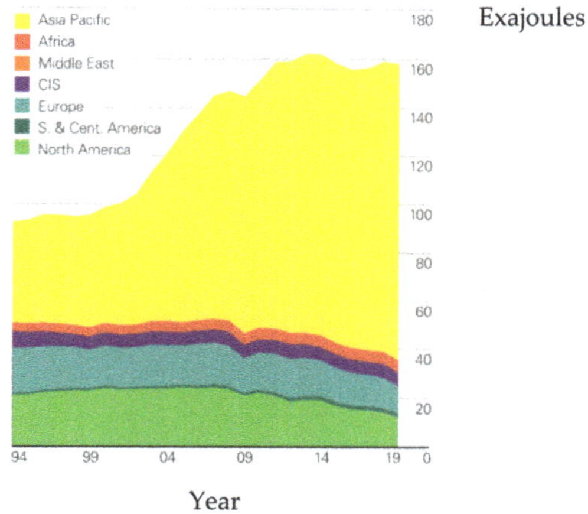


**Figure 1.** Coal consumption by region [1].

Even though the storage location of coal gangue is normally far away from downtown areas, the considerable stockpile of coal gangue still threatens human society, besides the natural ecosystem. Firstly, from the point of geotechnical engineering, rock-like coal gangue has to be stored in dumps, which obviously occupy enormous land resources that can be used for farming and construction. Furthermore, normal disordered gangue stacking tends to result in debris flow [2,4,5]. Secondly, it is well-known that the composition of coal gangue changes along with storage time and environment [7]. In other words, even though the major chemical compositions of gangue are $SiO_2$ and $Al_2O_3$ in the form of quartz and feldspar [3], certain traces of various heavy metal elements, including $Pb^{2+}$, $Zn^{2+}$, and $Cu^{2+}$, are still contained in coal gangue. After long-term weathering and soaking, the leaching of heavy metals would result in the contamination of soil, groundwater, and surface water, further harming the surrounding ecosystem and human health through bioaccumulation [2,4,6,7]. Last but not least, stocked coal gangue is also an emission resource of volatile organic chemical gas. The heat continuously accumulating inside the stocked coal gangue leads to the spontaneous combustion of coal gangue in the form of fire, even blast disasters, due to the existence of unburned coal, pyrite, and oxygen, releasing a large amount of harmful gases, including $SO_2$, $NO_x$, and CO, into the atmosphere. It is a pity that this air pollution from coal gangue is hard to control [2,4].

The gangue utilization rate of the main developed regions, such as US and Europe, is more than 90%. In contrast, that value in the largest coal gangue production country, namely China, is only about 60–70%, while the coal gangue utilization of China through combustion and calcination normally results in secondary pollution [4,6]. How to deal with the accumulation of coal gangue and improve its utilization rate has become one of the most crucial issues related to the sustainable development of the coal industry.

## 2. Utilization of Coal Gangue

For years, researchers have been working on the effective utilization of the enormous stocked coal gangue. In China, more than a half of discharged coal gangue is used for land reclamation—one third of that is used for power generation and the rest is used for producing building materials [2,4].

However, the main compositions of natural rock coal gangue are inactive $SiO_2$ and $Al_2O_3$, whose cementitious ability is rather weak. Calcination or spontaneous combustion are necessary to add CaO to coal gangue. For now, the percentage of calcined gangue combination with Portland cement is still low, normally less than 15% [3,6]. In contrast, as additive and fine aggregates, calcined gangue could also be used to prepare cementitious material, and is a promising replacement for fly ash and slag [8,9]. Meanwhile, coal gangue can be crashed and directly used as a lightweight aggregate, with a good internal curing capacity, but a high demand for water for mixing [6,10,11].

Besides power generation, the current utilization methods of coal gangue, including land reclamation and producing cementitious materials, tend to cause secondary environmental contamination and extra energy consumption, as the calcination and superfine grinding of coal gangue seem inevitable. In contrast, the utilization of coal gangue as a coarse aggregate has a relatively low energy consumption and environmental impact. However, currently, most of the studies on the utilization of coal gangue focus on producing cementitious materials and additives. The limited studies on the utilization of coal gangue as a coarse aggregate only consider the mechanical and durability performance of the concrete material. The structural application of coal gangue coarse aggregate concrete, such as in beams, columns, and walls, is rather little and is not deep enough, which also includes design methods for structural members with coal gangue concrete.

Hereafter, this paper introduces the direct utilization of coal gangue for preparing structural concrete as a replacement of coarse aggregates, aiming at sustainable, comprehensive, and efficient cleaner production. Through this review, it is hoped that more structural applications of coal gangue coarse aggregate concrete would be proposed and studied experimentally and theoretically.

## 3. Methods

The references for the review were searched for in December 2020 in Scopus and the China National Knowledge Internet database, with the consideration of the following factors [4]:

(1) The presence of terms "coal gangue" and "coarse aggregate" in the documents;
(2) The considered period: 1990–2020;
(3) The publication form of paper: "Article" or "Review";
(4) Searching area: "Engineering" and "Material".

It should be mentioned that as the focus of this review is the coal gangue utilization of coarse aggregates, most of the references regarding gangue utilization in producing cementitious materials are not mentioned. In fact, studies on gangue utilization of coarse aggregates for the preparation of concrete mainly performed by Chinese researchers are much less than that for the production of cementitious materials, as China remains the largest production country in the world and the coal consumption in other countries and regions has been dropping for decades. Referring to the application of coal gangue in concrete structural members, the relevant studies have only been performed by Chinese researchers for the same reason. The main work of this review is summarized in Figure 2.

**Figure 2.** Flowchart of the review.

## 4. Properties of Coal Gangue Aggregate

### *4.1. Physical Properties*

Natural coal gangue (namely rock coal gangue) is a rock-like mining waste that normally appears from gray to black. The color changes to whitish gray and even red after spontaneous combustion or after being calcinated, according to its iron oxide content and weathering degree [2]. According to Chinese standard JGJ-52-2006, the physical properties of rock coal gangue (RCG) and spontaneous-combustion coal gangue (SCG) vary from different origins in China, as summarized in Table 1 [12–24].

**Table 1.** Physical properties of coal gangue aggregate from different origins in China.

| Type | Region of China | Province | Apparent Density (kg/m³) | Bulk Density (kg/m³) | Water Absorption (%) | Crushing Value (%) | Void Ratio (%) | Ref. |
|---|---|---|---|---|---|---|---|---|
| SCG | Northeast | Heilongjiang | 2588 | 1289 | **3.8** | 18.9 | 50.2 | [13] |
| | | Liaoning | 2497 | - | 9.80 | 25.2 | - | [12] |
| | | Liaoning | 2630 | 1550 | 8.5 | 16.6 | **22.0** | [14] |
| | | Liaoning | **2276** | 1220 | 7.55 | 21.2 | 52.8 | [15] |
| | South | Anhui | 2630 | **976** | 9.2 | 16.6 | 57.9 | [16] |
| | | Jiangsu | 2624 | 1430 | 4.88 | 24.4 | - | [17] |
| RCG | Northeast | Liaoning | 2653 | 1489 | 3.15 | **9.9** | 49.5 | [15] |
| | North | Shanxi | **2083** | 1130 | 4.7 | 18.7 | - | [18] |
| | | Shanxi | 2689 | 1201 | **11.4** | 18.1 | - | [19] |
| | | Beijing | 2640 | - | 1.8 | 22.6 | - | [20] |
| | | Henan | 2510 | 1320 | 1.7 | 19.1 | 47.5 | [21] |
| | | Henan | 2540 | 1264 | 3.9 | 17.9 | 51.5 | [22] |
| | Northwest | Shaanxi | **2090** | - | 4.92 | 19.0 | - | [23] |
| | South | Jiangsu | 2712 | - | 1.7 | 16.8 | - | [24] |
| | | Jiangsu | 2620 | 1612 | 3.9 | 18.6 | | [17] |

The apparent density of gangue is approximately 2600 kg/m$^3$, which is similar to natural stone, while its bulk density is approximately 1200 kg/m$^3$, which belongs to lightweight aggregate. The water absorption and crushing value of gangue (approximately 5% and 20% respectively) are relatively larger than those of natural stone (approximately 2% and 15%, respectively) due to its higher carbon content. Although calcination would reduce the carbon content in RCG, its water absorption and crushing value would also be further enlarged. Even though some crashing values of CG in Table 1 are close to or even smaller than the conventional crashing values of natural rock, these values are still larger than those of the natural rock collected from the local in the corresponding references.

As the replacement of natural coarse aggregate, both RCG and SCG seem to be a little fragile to be used in concrete, which may likely result in the fracture in coal gangue aggregate itself, besides the fracture in the interfacial transition zone (ITZ).

### *4.2. Chemical Properties*

According to its chemical properties, coal gangue could be categorized into four types, namely claystone gangue ($SiO_2$: 40–70% and $Al_2O_3$: 15–30%), sandstone gangue ($SiO_2$ > 70%), aluminous stone gangue ($Al_2O_3$ > 40%), and calcareous stone gangue (CaO > 30%) [25]. Table 2 lists the mineralogical composition of coal gangue in various countries [26]. Typical coal gangue contains 50–70% clay minerals, 20–30% quartz, and 10–20% other minerals and carbon impurity. Compared with Europe and South America, the content of kaolinite in the coal gangue of China and USSR is richer, showing more potential for reuse, as kaolinite is apt to be activated. Instead, the content of illite in the coal gangue of Europe is much richer than that in USSR and China.

**Table 2.** Mineralogical composition of coal gangue in various countries/% [26].

| Minerals | Czechoslovakia | Germany | Spain | Britain | USSR | China |
|---|---|---|---|---|---|---|
| Illite | 10–45 | 41–66 | 20–60 | 10–31 | 5–30 | 10–30 |
| Kaolinite | 24–45 | 4–25 | 3–30 | 10–40 | 1–60 | 10–67 |
| Chlorite | 0–15 | 1–3 | 0–7 | 2–7 | – | 2–11 |
| Quartz | 10–50 | 13–27 | 5–57 | 15–25 | – | 15–35 |
| Iron ore | 0–25 | 0.5–5 | – | 2-10 | 0.2–8 | 2–10 |
| Organic matters | 0–25 | 5–10 | 4–30 | 5–25 | 8–40 | 5–25 |

In general, as coal gangue is actually a transition substance between rock and coal, its chemical composition is similar to that of natural coarse aggregate, which also confirms the feasibility of being a coarse aggregate in concrete production.

As listed in Table 3, the content of $SiO_2$ is always maximum among the chemical composition in coal gangue, regardless of the origins, while the content of $Al_2O_3$ is normally half that of $SiO_2$. The major chemical compositions of coal gangue, namely $SiO_2$ and $Al_2O_3$, tend to increase after spontaneous-combustion or being calcinated, but still vary from different origins. As the content of ($SiO_2$ + $Al_2O_3$ + $Fe_2O_3$) in gangue normally accounts for more than 70%, coal gangue could be treated as a pozzolanic material, regardless of its origin [2].

**Table 3.** Chemical composition of CG (mass fraction)/%.

| Type | Region | Province | $SiO_2$ | $Al_2O_3$ | $Fe_2O_3$ | CaO | MgO | $Na_2O$ | $K_2O$ | $TiO_2$ | $P_2O_5$ | $SO_3$ | Loss | Ref. |
|---|---|---|---|---|---|---|---|---|---|---|---|---|---|---|
| RCG | Northeast of China | Heilongjiang | 61.0 | 23.6 | 6.70 | 1.18 | 0.52 | - | - | - | - | - | 2.50 | [11] |
| | | Heilongjiang | **34.1** | 26.0 | 0.49 | 0.67 | 0.61 | - | 0.16 | - | - | 0.28 | **32.8** | [27] |
| | | Liaoning | 50.3 | 26.3 | 6.11 | 7.74 | 2.00 | 1.10 | 3.28 | 1.13 | 0.15 | 0.93 | - | [15] |
| | North of China | Beijing | 49.9 | 24.4 | 6.42 | 0.82 | 1.59 | 1.46 | 2.06 | 0.88 | - | 0.12 | 11.8 | [3] |
| | | Shandong | 48.8 | 19.0 | 4.47 | 2.03 | 2.29 | 1.43 | 0.19 | - | - | - | 16.8 | [3] |
| | | Henan | 59.9 | 20.7 | 6.70 | 2.00 | 1.80 | 0.65 | 2.40 | - | - | 1.53 | - | [7] |
| | | Shanxi | 35.1 | 16.8 | **27.3** | 3.82 | 1.60 | - | - | - | - | 3.00 | 0.83 | [19] |
| | | Hebei | 48.3 | 23.1 | 4.30 | 4.10 | 1.70 | 0.10 | 1.50 | 0.80 | 0.10 | 1.00 | 14.7 | [28] |
| | Northwest of China | Shaanxi | 49.5 | 33.3 | 7.60 | 6.09 | 0.97 | 0.52 | 0.94 | 0.83 | - | - | - | [23] |
| | South of China | Jiangsu | 55.5 | 18.2 | 5.42 | 3.38 | 1.23 | 0.64 | 1.67 | - | - | 0.64 | 13.3 | [3] |
| | | Jiangsu | 59. 8 | 29.4 | 1.44 | 0.68 | 0.51 | 0.08 | 1.76 | - | - | - | - | [8] |
| | Poland | - | 58.9 | 20.5 | 6.63 | 0.35 | 1.93 | 0.54 | 3.19 | 1.05 | 0.06 | 0.17 | 5.50 | [5] |
| | Italy | No. 1 | 43.7 | 21.4 | 5.57 | 0.89 | 0.77 | 0.11 | 0.16 | 1.05 | 0.16 | 1.02 | 25.2 | [29] |
| | | No. 2 | 46.8 | 17.2 | 7.67 | 7.60 | 0.99 | 0.15 | 2.40 | 0.78 | 0.16 | 0.34 | 15.8 | |
| | | No. 3 | 57.2 | 18.7 | 6.25 | 1.86 | 1.42 | 0.46 | 3.59 | 0.95 | 0.14 | 0.08 | 9.28 | |
| SCG | Northeast of China | Liaoning | 55.6 | 21.0 | 6.57 | 3.65 | 2.50 | 1.90 | 4.10 | 0.84 | 0.24 | 0.82 | - | [15] |
| | North of China | Shanxi | 56.6 | 36.8 | 1.95 | 0.62 | 0.22 | 0.42 | - | 2.10 | - | - | - | [9] |
| | Northwest of China | Shaanxi | 55.2 | 31.1 | 2.94 | 1.31 | 0.75 | 0.12 | 1.13 | 1.12 | 0.07 | - | 5.94 | [30] |
| | Italy | No. 1 | 56.4 | 26.3 | 6.42 | 1.06 | 1.07 | 4.02 | 0.17 | 1.21 | 0.20 | 0.65 | 2.38 | [29] |
| | | No. 2 | 52.3 | 19.5 | 8.35 | 8.41 | 1.21 | 2.65 | 0.20 | 0.85 | 0.17 | 0.49 | 5.61 | |
| | | No. 3 | 61.9 | 20.2 | 6.77 | 2.00 | 1.57 | 3.78 | 0.53 | 1.01 | 0.15 | 0.11 | 1.81 | |

### 4.3. Trace Metal

The heavy metal content of coal gangue and the permissible values in soils and surface lands are listed in Table 4. Similar to the chemical composition, the heavy metals in gangue also vary according to the origins. The Cr content of coal gangue in Italy is much higher than that in China and Poland, while the As content of coal gangue in China is a little higher than the permissible value. Besides this, the heavy metal content of gangue is much lower than the permission level of the Polish Regulation of the Ministry of Environment of 9 September 2002 for soil quality standards and earth quality standards. In addition, the short-term leachability of the heavy metals is relatively weak [5]. In this case, in terms of environmental protection, coal gangue could also be used as a coarse aggregate in green concrete production. However, more data regarding the heavy metal content in coal gangue should be collected from testing in order to propose a more accurate requirement regarding the heavy metal content of discharged coal gangue. In addition, the content of heavy metal in discharged coal gangue should be tested before being used in a concrete mixture so as to make sure that its utilization would not pose an environmental, health, and safety concern.

**Table 4.** Content of heavy metals in coal gangue/mg kg$^{-1}$.

| Type | Region | Province | As | Ba | Cd | Co | Cr | Cu | Hg | Mo | Ni | Pb | Sn | Zn | Ref. |
|---|---|---|---|---|---|---|---|---|---|---|---|---|---|---|---|
| RCG | China | Jiangsu | 27.3 | - | 0.58 | - | 42.3 | 24.3 | 0.24 | - | 20.5 | 4.32 | - | 39.5 | [31] |
| SCG | Italy | No. 1 | - | - | - | - | **190** | - | - | - | 56 | 11 | - | 27 | [29] |
| | | No. 2 | - | - | - | - | **228** | - | - | - | 43 | 5 | - | 46 | |
| | | No. 3 | - | - | - | - | **308** | - | - | - | 57 | 10 | - | 37 | |
| RCG | Poland | - | <2 | 84.5 | 2.9 | 14.6 | 61.9 | 48.3 | 0.04 | <2 | 41.2 | 20 | 3 | 89.5 | [5] |
| Permission content in Poland | | | 20 | 200 | 4 | 20 | 150 | 150 | 2 | 10 | 100 | 100 | 20 | 300 | [5] |

## 5. Properties of Coal Gangue Concrete

Even though using crashed and sieved coal gangue as a coarse aggregate is the most convenient and energy-saving method to utilize coal gangue, the feasibility of this method still depends on the properties of coal gangue. The performance of coal gangue concrete needs to meet the standard for preparing the concrete in construction. Calcination is beneficial for improving the interfacial bond between coal gangue and cement mortar.

### *5.1. Workability*

#### 5.1.1. Water Absorption

It is foreseeable that the water absorption of coal gangue concrete would be increased by raising the coal gangue replacement ratio, as its water absorption is higher than that of natural gravel. Figure 3a shows the permeability coefficient of the concrete under 25% and 30% replacement ratios. The increase of the pore phase conductivity by increasing the coal gangue replacement is supposed to result in an increase of the permeability coefficient [31]. Furthermore, the water absorption law tested according to ASTM C1585 was remarkably changed, as shown in Figure 3b. The inflection point for the water absorption speed in the curves became more and more obvious with the increase of the replacement ratio. Under a large replacement ratio, coal gangue aggregate in concrete became almost saturated in a short time, and then the unsaturated mortar and natural aggregates continued to absorb water, with a water absorption speed much lower than that of the coal gangue. Comparing the results in [31,32], as shown in Figure 3b, it seems a coal gangue replacement ratio lower than 30% would not significantly affect the water absorption of the concrete. In general, coal gangue concrete is not recommended for use in subsurface structures; besides, it is obvious that a large water absorption would reduce the strength of the coal gangue concrete.

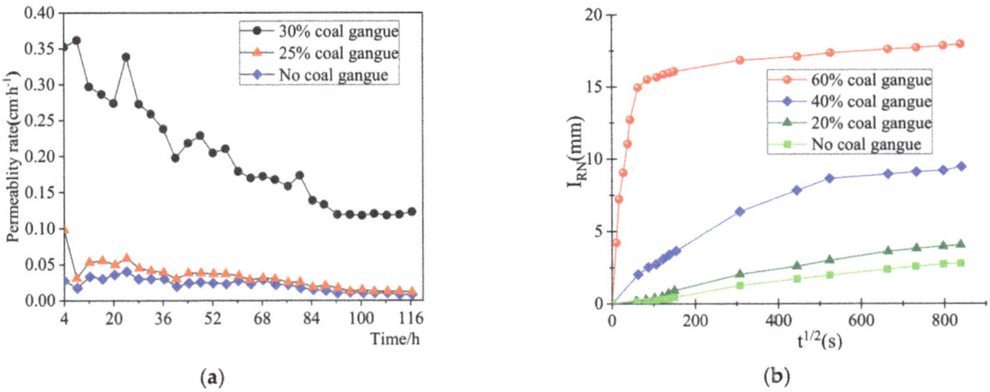


**Figure 3.** Effect of rock coal gangue replacement on the water absorption of coal gangue concrete. (**a**) Permeability coefficient [31]; (**b**) Absorption depth [32].

5.1.2. Slump

As coal gangue (CG) shows a larger water absorption capacity than natural aggregate, the workability of coal gangue concrete would obviously be decreased by raising the CG replacement ratio according to Chinese standard GB 50164-2011, as shown in Figure 4. Under a 50% CG replacement ratio, the slump of the coal gangue concrete would be reduced by about 10%, while the number would be 20% under 100% CG replacement ratio. As the properties of CG vary from the origin, the difference between RCG and SCG shown in Figure 5 is subtle [33,34]. To enhance its workability, adding fly ash and superplasticizer in coal gangue concrete are both effective and simple methods [35,36]. Additional water consumption for coal gangue could also be a choice to enhance the workability, but would result in a strength reduction of coal gangue concrete resulting from the extra water [37]. It is surprising from Figure 4 that the pre-soaking method is not as effective as it is supposed to be, which may be also due to the properties of coal gangue varying from the origin.

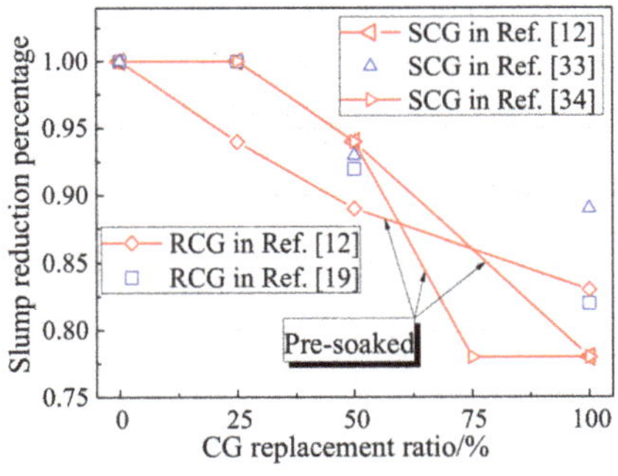


**Figure 4.** Effect of coal gangue replacement on the slump of coal gangue concrete.

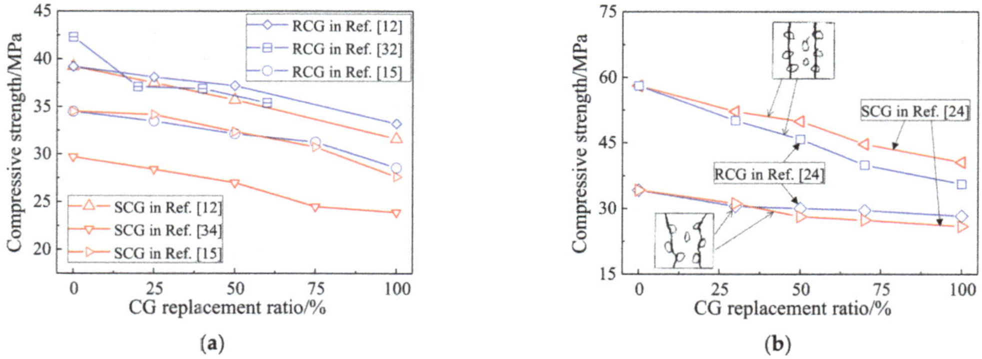


**Figure 5.** Effect of coal gangue replacement on the compressive strength of coal gangue concrete. (**a**) Compressive strength; (**b**) Failure pattern.

## 5.2. Mechanical Performance

### 5.2.1. Compressive Strength

It is assumed that the porous structure of CG results in a strength reduction of coal gangue concrete. As shown in Figure 5a, the compressive strength of coal gangue concrete tested according to Chinese standard GB 50010-2010 decreases approximately linearly with the rise of the replacement ratio, regardless of concrete strength and the type of coal gangue. Under 100% CG replacement ratio, the compressive strength of coal gangue concrete decreased by around 15–20% [12,15,32,34]. Figure 5b indicates that the failure pattern of coal gangue concrete is different according to the concrete grade [24]. Under a low concrete grade, the failure of coal gangue concrete occurs in the mortar, where the coal gangue shows little effect. In contrast, under a high concrete grade, the failure of coal gangue concrete happens in both the mortar and coal gangue aggregate. In this case, the types of coal gangue aggregate contribute to the compressive strength of coal gangue concrete. As mentioned above, calcinating CG would activate the cementitious composition in it and improve the strength of the mortar, but would degrade the strength of the coal gangue itself. In addition, an over-low concrete grade, for example lower than C20, is not suggested for use as a coal gangue aggregate, as the mortar is not enough to fully cover the coarse aggregate, which may lead to a significant reduction in the mechanical properties of coal gangue concrete [38].

Similar to ordinary concrete, adding steel fiber could reinforce the mechanical properties of coal gangue concrete. The compressive strength of coal gangue concrete with 2% steel fiber would be improved by 16.7% [39]. Furthermore, calcianting RCG [40] and presoaking SCG [17] would also achieve the same effect.

As presented in Figure 6, even though a general descending trend in the elastic modulus of coal gangue concrete was found when raising the CG replacement ratio, the comparsion between the RCG concrete and SCG concrete was contrary between [12] and [15]. In other words, the elastic modulus of RCG concrete was larger than that of the SCG concrete [12], whiet the former was smaller than the latter in [15], except for the 100% replacement ratio. Besides the discreetness of coal gangue properties, the presoaking method, which is supposed to effectvely improve the mechanical behaviors of SCG concrete, rather than those of RCG concrete, was used in [15].

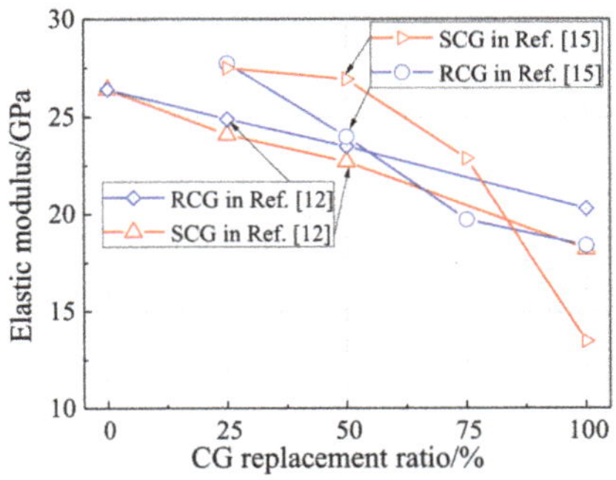


**Figure 6.** Effect of coal gangue replacement on the elastic modulus of coal gangue concrete.

Figure 7 presents the stress−strain relationship of the concrete with different contents of RCG and SCG [15]. It indicates that increasing both the RCG and SCG replacement ratios would increase the peak strain of the coal gangue concrete. The drop beyond the peak in the curves of RCG concrete was sharper than that of the SCG concrete, due to the porous structure of SCG. These phenomena imply that the ductility of SCG concrete is better than that of RCG concrete. Based on the tested stress−strain relationship of coal gangue concrete and the model in Chinese standard GB 50010-2010, a dimensionless model ($x = \varepsilon/\varepsilon_c$, $y = \sigma/\sigma_c$) for both RCG and SCG concrete is proposed for the first time, as follows [15]. The proposed model should consider the aggregate replacement ratio and CG types. However, more data regarding the stress−strain relationship model of CG concrete are still needed to validate the model, as it was empirically obtained and the properties of CG vary according to the origin.

$$y = \begin{cases} ax + (3-2a)x^2 + (a-2)x^3, 0 \leq x \leq 1 \\ x/(b(x-1)^2 + x), x > 1 \end{cases} \quad (1)$$

where $a$ is ($-0.0112r + 1.844$) and ($7 \times 10r^{-5} - 0.0117r^2 + 1.6866$) for RCG and SCG, respectively, and $b$ is ($0.1052r + 1.824$) and ($-0.0118r + 2.38$) for RCG and SCG, respectively.

#### 5.2.2. Splitting Tensile Strength

As shown in Figure 8, generally, the splitting tensile strength of coal gangue concrete decreases as the CG replacement ratio increases. However, due to the discreetness of thecoal gangue properties, a minimum value was observed in the results of RCG concrete in [15] under 50% CG replacement ratio. By using CG as the coarse aggregate, more splitting tensile damage was found in the coarse aggregate, namely coal gangue, rather than in the bond between the coarse aggregate and the matrix.

### 5.3. Durability Performance

#### 5.3.1. Shrinkage

The cracks resulting from shrinkage affect the durability of concrete, especially for concrete with CG whose water absorption characteristic is different to the natural coarse aggregate. Figure 9a shows the effect of concrete grade on the coal gangue concrete shrinkage tested according to the GB/T 50082-2009 standard and ASTM C157. It is obvious that higher-grade coal gangue concrete shows larger shrinkage. It is also interesting that the early shrinkage of coal gangue concrete is smaller than that of normal concrete, which results from the inner curing effect of coal gangue with a large water absorption [41]. As shown in Figure 9b, the increase of shrinkage due to the increase of curing temperature would be enlarged by raising the CG replacement ratio [42]. As expected, the concrete using an SCG aggregate and large replacement ratio showed large shrinkage, as shown in Figure 9c [12]. However, a totally opposite trend, where increasing the replacement ratio decreased the shrinkage of concrete was still observed, as seen in Figure 9d [43]. This finding may result from the larger volume of coal gangue used in the concrete mixture under the same mass replacement ratio of the coarse aggregate, which would reduce the shrinkage of the mortar matrix.

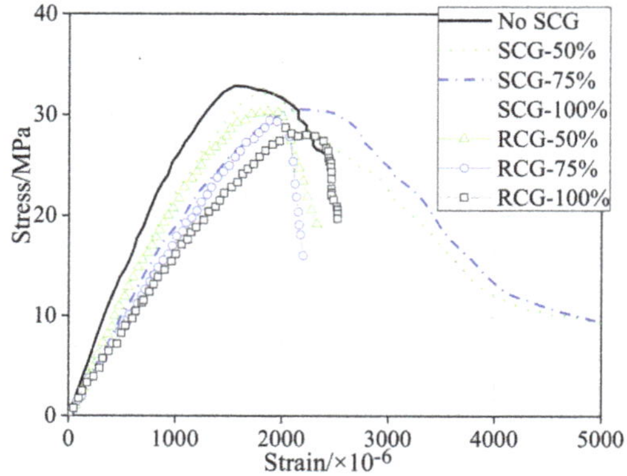


**Figure 7.** Stress–strain curves of the coal gangue concrete [15].

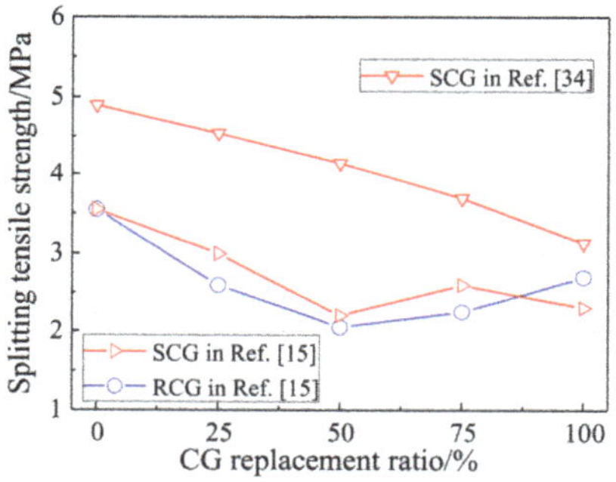


**Figure 8.** Effect of coal gangue replacement on splitting tensile strength of coal gangue concrete.

5.3.2. Freeze–Thaw Resistance

The failure of concrete under freezing–thawing cycles results from the frost–heave stress related to the capillary water absorption properties [44]. Due to the large frost–heave stress resulting from the porous structure of CG, the mechanical performance of coal gangue concrete was significantly affected by freezing–thawing cycles from the test complying with Chinese standard GB/T 50082-2009, which was enlarged by raising the CG replacement ratio [32,45]. The freezing–thawing mass loss of coal gangue concrete would also be enlarged by increasing the water–cement ratio [46].

Based on the tested results, a dimensionless model ($x = \varepsilon/\varepsilon_c, y = \sigma/\sigma_c$) of the stress–strain relationship is proposed for coal gangue concrete after freezing–thawing cycles [23]. The model modified from the model in Chinese standard GB 50010-2010 considers the effects of the CG replacement ratio and the cycle number of freezing–thawing, rather than the CG type. Therefore, similar to the model proposed in [15], even though this model fit well with the test results, it was also empirically obtained, which means that more data are needed in order to validate it.

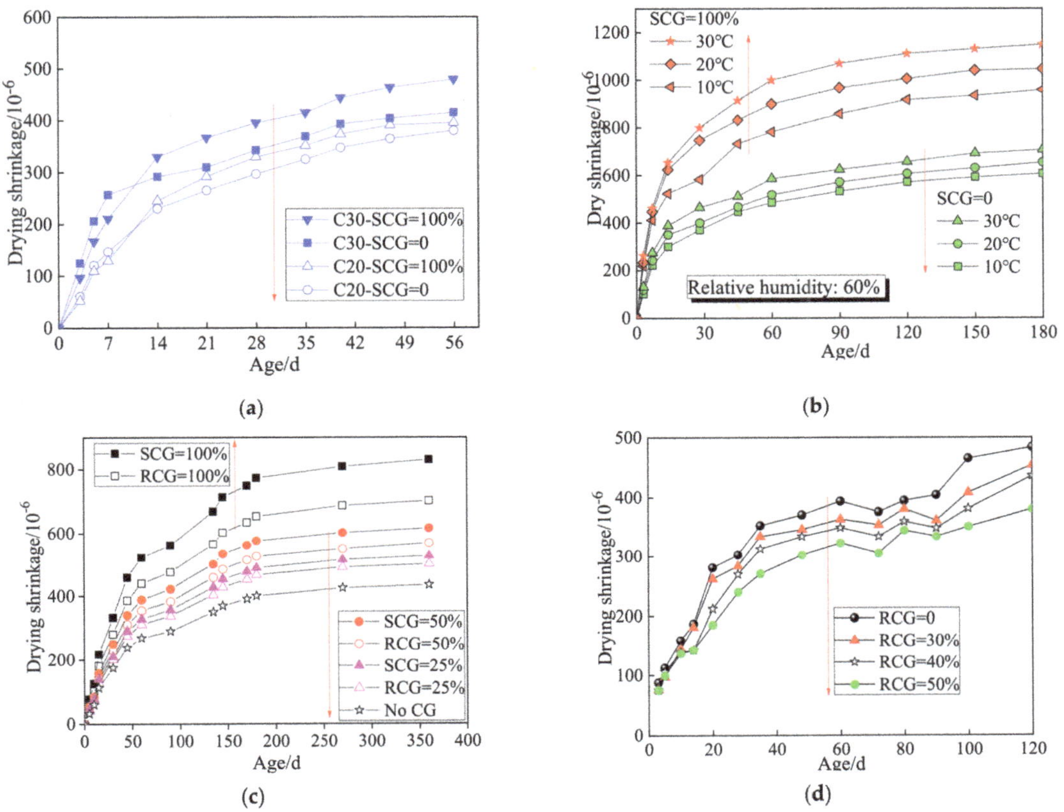


**Figure 9.** Effect of coal gangue replacement on the shrinkage of coal gangue concrete. (**a**) Concrete grade [41]; (**b**) Curing condition [42]; (**c**) Coal gangue type [12]; (**d**) Replacement ratio (RCG) [43].

$$y = \begin{cases} (1.92 - \mu N)x + (1.08 + \mu N)x^2 - (0.08 + \mu N)x^3, 0 \le x < 1 \\ x/((1.3 + \eta N)(x-1)^2 + x), x \ge 1 \end{cases} \quad (2)$$

where $\mu$ is $(0.0025e^{0.0552rCG})$, $\eta$ is $(0.0035 + 0.0011r_{CG})$, $N$ is the cycle number of freezing–thawing, and $r_{CG}$ is the CG replacement ratio.

The admixture of fly ash and silica fume [47], organic anti-freezing agent [48], and polypropylene fiber [49] are added into the coal gangue concrete to enhance its freeze–thawing resistance. Up to 1% of organic anti-freezing agent, namely ethylene glycol, could improve the freeze–thawing resistance of coal gangue concrete by 10–30%, which is believed to be one of the most effective methods.

#### 5.3.3. Carbonization Resistance

The transmission of $CO_2$ in concrete is mainly affected by the porosity ratio and connected channels. The coal gangue aggregate generally has more pores than natural gravel. The air containing $CO_2$ enters into the concrete more easily, resulting in a poor durability of the concrete [50]. Therefore, it is obvious that increasing the coal gangue replacement ratio would reduce the carbonization resistance of coal gangue concrete (seen in Figure 10, showing the results from the test complying with Chinese standard GB/T 50082-2009) [51].

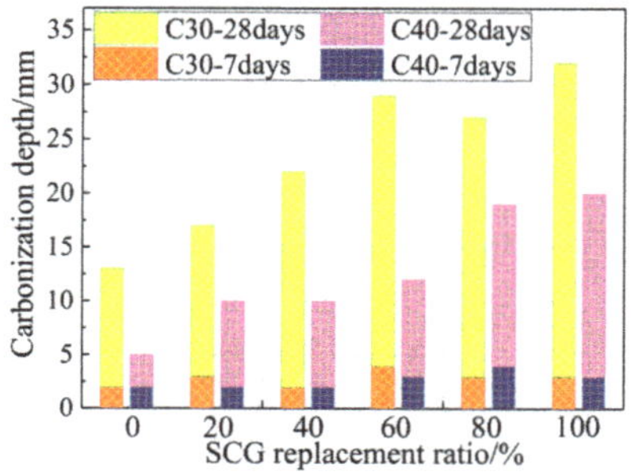


**Figure 10.** Carbonization depth−SCG replacement ratio relationship.

Figure 10 also implies a finding that the carbonization depth of SCG concrete is negatively related to its strength, which was also found in other relavent tests [52,53]. A formula describing the relationship between carbonization depth $d_c$ and strength $f_{CG}$ of SCG concrete after 28 days of curing is proposed as follows [53]. This empirical model is not only related to one parameter, namely CG strength, but is also valid in the scope of concrete strength.

$$d_c = -0.31f_{CG} + 29.6, (18.7 \leq f_c \leq 51.2) \tag{3}$$

Compared with RCG, the concrete using 700 °C-calcinated coal gangue has a better carbonization resistance than that of RCG concrete. C-S-H hydrogel, which is produced by the second hydration reaction among the $SiO_2$, $Al_2O_3$, and cement hydration product $Ca(OH)_2$ produced after the calcination of the gangue aggregate, compacts the pores and reduces the $CO_2$ transmission channel in concrete. A formula for predicting the carbonization depth of RCG concrete regarding age ($t$), water−cement ratio ($m_w/m_c$), and coal gangue replacement ratio ($r_{CG}$) is proposed [54]. Compared with Equation (3) from [53], this empirical formula considers the effects of three different parameters, instead of just the concrete strength. However, as mentioned by the authors themselves, this formula is only suitable for RCG, rather than SCG.

$$d_c = 0.098\sqrt{t} + 50.66m_w/m_c + 2.015r_{CG} - 17 \tag{4}$$

5.3.4. Chloride Resistance

More capillary pores in SCG are the main reason for the poor chloride resistance of coal gangue concrete. Therefore, in the coal gangue concrete with a low water−cement ratio, the gel pores and capillary pores are filled with hydrated gel, increasing the compactness of the concrete larger, decreasing the electric flux, and enhancing the chloride ion penetration resistance [55–57].

To improve the chloride resistance of coal gangue concrete, calcination [24], adding steel fiber [39], and adding the mineral admixture and superplasticizer [58] are effective ways that could reduce the micro cracks of mortar matrix and change the ion permeation path. As shown in Figure 11, where the results are from the test complying with ASTM C 1202-05, adding fly ash would reduce the strength of the coal gangue concrete, but increase its chloride resistance (namely decreasing the chloride migration coefficient), regardless of the water−cement ratio [57].

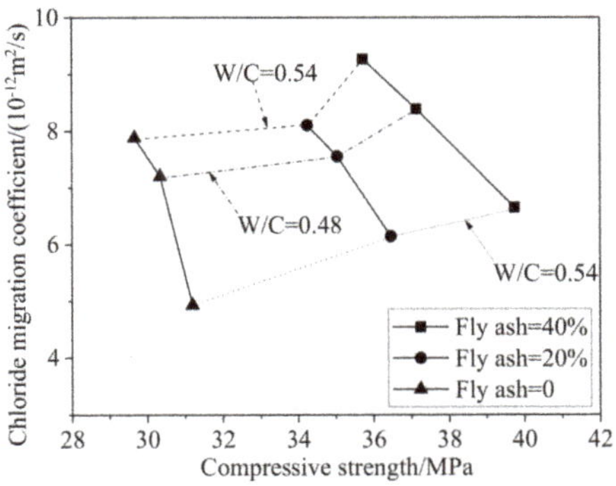


**Figure 11.** Relationship between chloride migration coefficient and strength of coal gangue concrete [57].

*5.4. Thermal Characteristic*

As mentioned above, increasing the CG content would normally degrade the mechanical and durability performance of the coal gangue concrete. In contrast, the porous structure of CG is beneficial for thermal insulation. Therefore, a novel type of thermal insulation coal gangue concrete using coal gangue is proposed [19]. Besides coal gangue, glazed hollow bead (GHB), nano-silica (NS), and ultra-fine slag (UFS) are also used to improve the performance of the novel concrete. As shown in Figure 12, increasing the CG content of the concrete would significantly reduce its thermal conductivity coefficient, even though the GHB content marks the maximum influence on the thermal conductivity coefficient of the concrete, while the NS and UFS content show subtle effects.

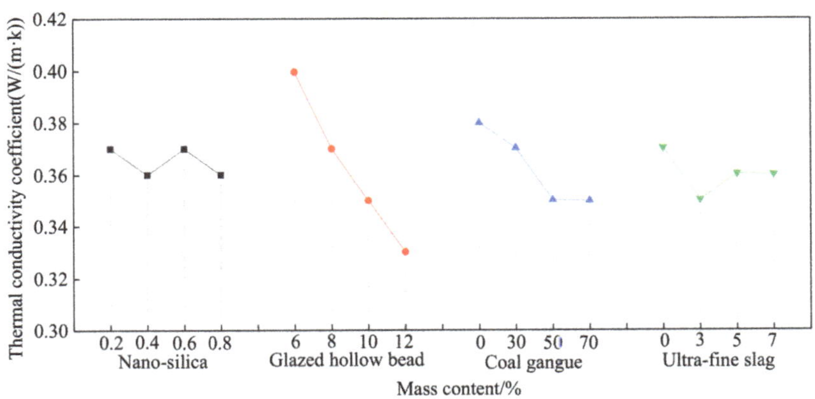


**Figure 12.** Thermal conductivity coefficient of the novel coal gangue concrete [19].

## 6. Application of Coal Gangue Concrete

### *6.1. Reinforced Coal Gangue Concrete*

#### 6.1.1. Beam

In [59–61], the mechanical performance of SCG concrete and the flexural performance of reinforced coal gangue concrete beam are both studied. As shown in Figure 13, the reduction of the ultimate moment for reinforced coal gangue concrete beams is a little larger than that of the coal gangue concrete itself, while the cracking moment of reinforced coal gangue concrete seems unrelated to the coal gangue concrete replacement ratio, which may result from the various properties of coal gangue. In general, the ultimate moment reduction of reinforced coal gangue concrete beam is below 10%, which seems acceptable in practical engineering. Furthermore, [59–61] also indicate that the current design method in

Chinese standards could be directly used to predict the flexural resistance of the reinforced coal gangue concrete beam by using the corresponding compressive strength of the coal gangue concrete.

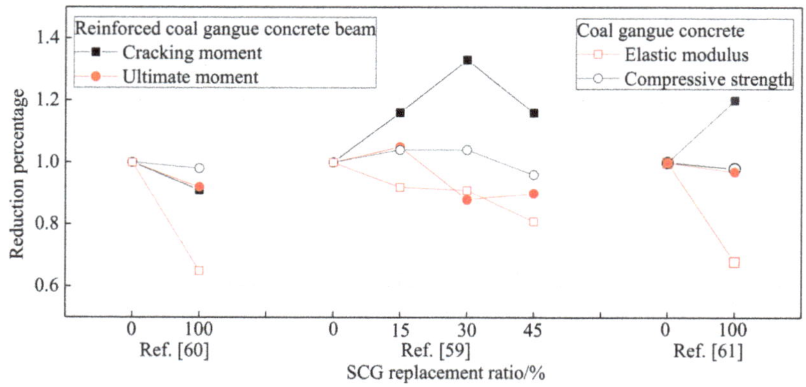


**Figure 13.** Flexural performance reduction of the reinforced coal gangue concrete beam.

As shown in Figure 14 [62], the compressive strength reduction of coal gangue concrete is remarkable, which results in a remarkable reduction in the shear resistance of the reinforce coal gangue concrete beam. Regardless of the CG replacement ratio, the beams show a shear–compression failure pattern. The shear resistance of reinforced coal gangue concrete beams could also be predicted by the current design method using Chinese standards.

6.1.2. Column

The reinforced concrete columns using RCG show similar failure patterns to the normal reinforced concrete columns under axial and eccentric compression [63], except that the cracking load of the columns is reduced by increasing the CG replacement ratio. The effects of the eccentric ratio and CG replacement ratio on the resistance reduction of the columns are shown in Figure 15. The compressive resistance of reinforced coal gangue concrete columns could be predicted by the current design method in Chinese standards. In contrast, a large coal gangue replacement and axial compression ratio is not suggested in the columns under seismic loads [64].

6.1.3. Wall

As shown in Figure 16, the failure pattern of coal gangue concrete shear wall under cyclic loads is similar to that of a normal concrete shear wall, even though the failure of coal gangue concrete shear wall tends to be slightly more severe when raising the CG replacement ratio [65,66]. About a 10% reduction in the horizontal resistance of the wall was caused when using a 100% SCG replacement ratio, while the ultimate horizontal displacement of the wall was reduced by 3%. By considering the degradation of shear resistance under cyclic loads, namely a 30% reduction, the current design method in Chinese standards was modified for the shear resistance of coal gangue concrete wall [67].

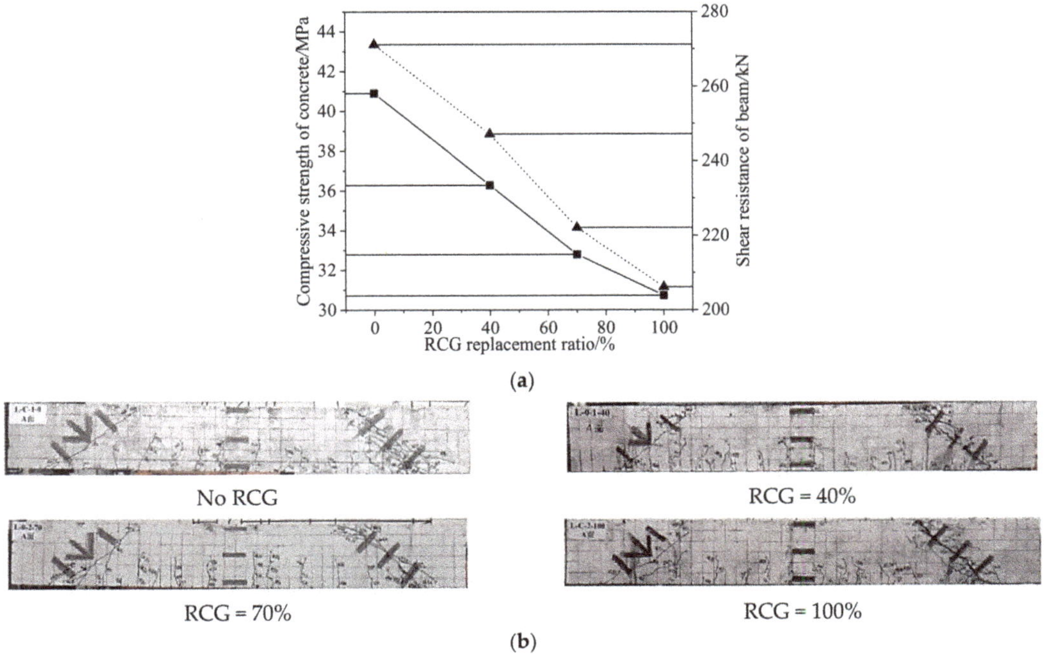


**Figure 14.** Shear performance of a reinforced coal gangue concrete beam [62]. (**a**) Shear resistance reduction; (**b**) Shear−compression failure patterns.

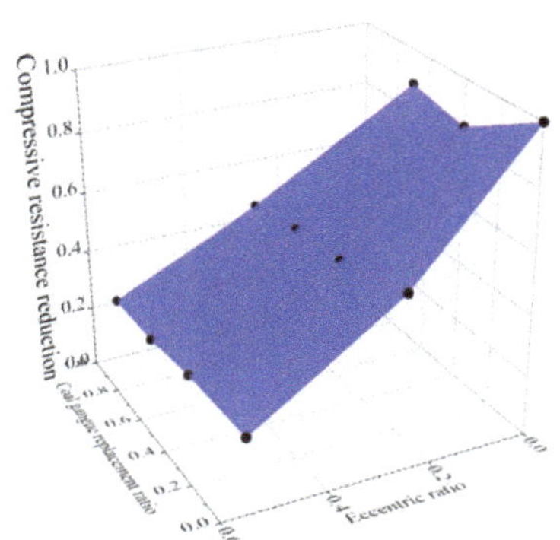


**Figure 15.** Compressive performance of reinforced coal gangue concrete columns.

*6.2. Coal Gangue Concrete Filled Steel Tube*

As mentioned above, the mechanical and durability performance of concrete is degraded when using coal gangue concrete. Therefore, concrete filled steel tubes (CFSTs) are an applicative configuration for using coal gangue concrete, as the external tube not only enhances the mechanical behavior of the internal core coal gangue concrete, but also protects it from the effect of severe environment [68–70].

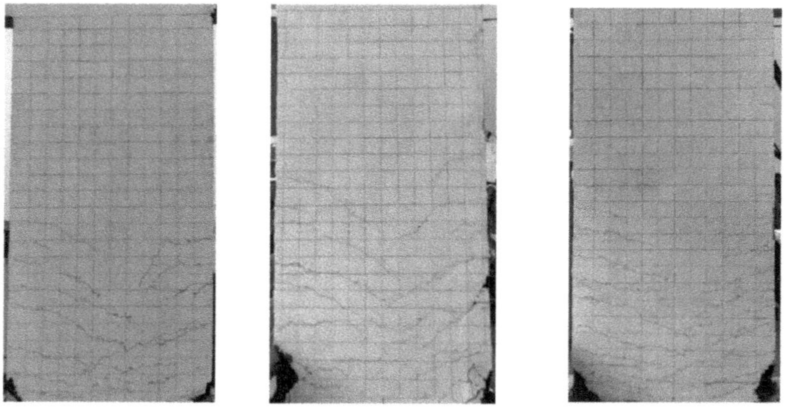

No SCG ($f_c$ = 29.2 MPa)  SCG = 50% ($f_c$ = 27.7 MPa)  SCG = 100% ($f_c$ = 26.99 MPa)

**Figure 16.** Failure patterns of coal gangue concrete wall [65].

Similar to reinforced coal gangue concrete stubs, the failure pattern of CFST stubs with coal gangue is also barely influenced by the CG replacement ratio, regardless of the concrete grade, as shown in Figure 17a. The axial strength of CFST with coal gangue is linearly related to the RCG replacement ratio [71]. Under a 100% replacement ratio, the axial capacity of CFST with coal gangue is reduced by 38%, while the compressive strength of coal gangue concrete is reduced by 50%. The current design method in Chinese standard GB50936-2014 for normal CFST was modified regarding the effect of CG replacement ratio, as follows [72]. The corresponding error of the modified method was in the range of $\pm$10%. Even though this semi-empirical prediction method is proposed here for the first time, it is only valid for the CG from specific origins, as with the above-mentioned formulas regarding CG concrete.

$$N_{up}^{\mathrm{CGCFST}}(r) = (A_c + A_s)(1.212 + (\frac{0.176 f_y}{213} + 0.974)\,\xi + (\frac{-0.104 k_{sr}^{\mathrm{CGC}} f_c}{14.4} + 0.031)\,\xi^2) k_{sr}^{\mathrm{CGC}} f_c \quad (5)$$

where $k_{sr}^{\mathrm{CGC}}$ is the strength reduction factor of coal gangue concrete, $k_{sr}^{\mathrm{CGC}} = 1 - 0.909 r_{\mathrm{CG}} + 0.425 {r_{\mathrm{CG}}}^2$.

### *6.3. Steel—Coal Gangue Concrete Composite Structure*

#### 6.3.1. Wall

Some researchers have proposed setting an internal steel plate into reinforced coal gangue concrete walls to enhance their seismic capacity [73,74], as shown in Figure 18. Even though this configuration is supposed to be beneficial, and both experimental and numerical studies have been conducted in [73,74], the effect of the CG replacement ratio has not been represented in these studies, which mostly focus on the effect of the internal steel plate, rather than the coal gangue concrete. In addition, the simplified theoretical prediction for the shear resistance and the seismic model of this coal gangue concrete shear wall with internal steel plate is also not presented.

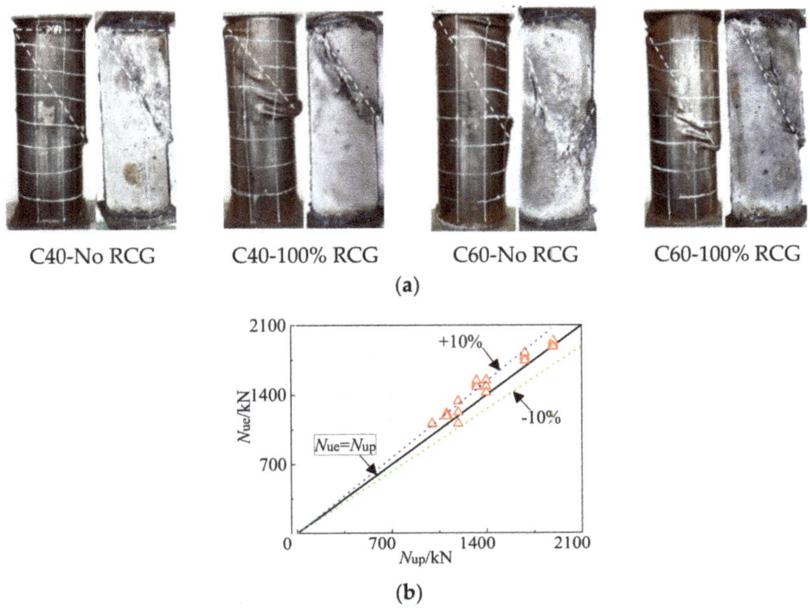


**Figure 17.** Compressive performance of coal gangue concrete filled steel tubes [72]. (**a**) Failure pattern; (**b**) Axial strength prediction.

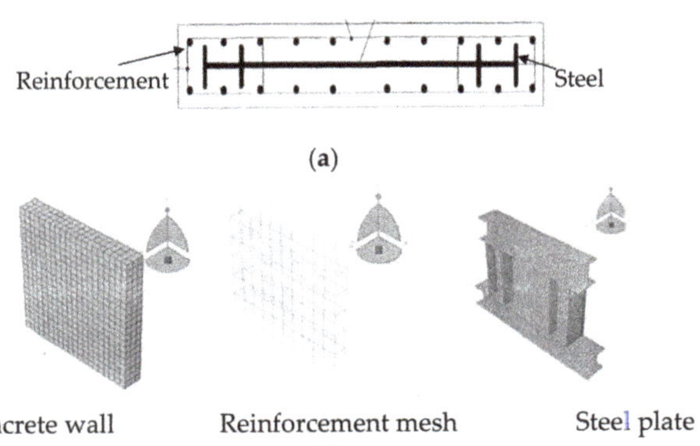


**Figure 18.** Coal gangue concrete shear wall with an internal steel plate [73,74]. (**a**) Profile of the tested composite wall; (**b**) Finite element model of the composite wall.

### 6.3.2. Slab

As a light-weight aggregate (bulk density less than 1200 $kg/m^3$ [75]), coal gangue is also preferable in concrete slabs, especially with profiled steel sheeting, namely composite slabs, as shown in Figure 19. However, the performance of the slab with coal gangue concrete and profiled steel sheeting is still limited. Some preliminary studies confirm the feasibility of using coal gangue concrete in composite slab with profiled steel sheeting, even though the deflection of the slab with coal gangue concrete is about 10% larger than that with normal concrete [76,77].

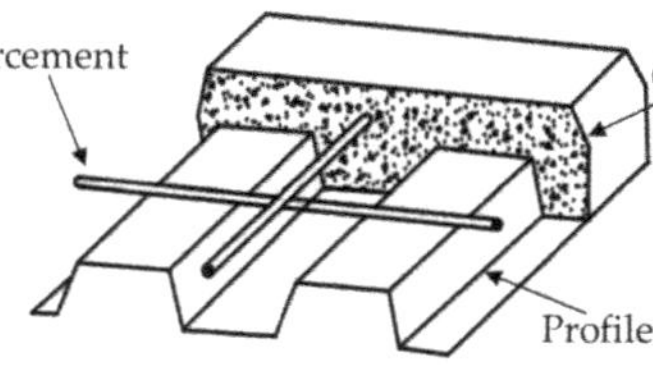


**Figure 19.** Composite slab with coal gangue concrete and profiled steel sheeting.

## 7. Discussion and Research Gap

Based on the above analysis, generally, the properties of coal gangue coarse aggregates are feasible for substituting natural coarse aggregates in the preparation of green concrete. The mechanical performance, durability performance, and even the thermal characteristic of the coal gangue coarse aggregate concrete have been investigated. Even though the properties of coal gangue vary according to origins, the usage of coal gangue coarse aggregate would decrease the mechanical and durability performance, but improve the thermal characteristic of the concrete. The mechanical behaviour of coal gangue concrete is generally acceptable in practical applications, while its durability is relatively poor. The proposed stress–strain models of coal gangue concrete could be used in the numerical and theoretical prediction for its mechanical performance [77,78]. However, these models are empirically-based, and are highly related to the properties of the local coal gangue.

In contrast, the application of coal gangue coarse concrete in structural members is relatively limited. The mechanical performance of reinforced concrete, CFST, and steel–concrete composite structures with coal gangue concrete have been investigated considering the structural members of beams, columns, and walls. In particular, most of the formulas for the load-bearing capacity prediction of these members do not even consider the influence of the CG replacement ratio [79,80].

It is worthy of attention that CFST is an applicative configuration for using coal gangue concrete, even without a thorough study on its durability performance, since the external tube could protect core coal gangue concrete from the effect of severe environments. The prediction method for the axial resistance of CFST with coal gangue by directly considering the influence of the CG replacement ratio is relatively accurate and simple for practical application.

It has been found that structural concrete-based coal gangue has several issues that, need to be resolved before its usage in the structural engineering. The following research gaps have been identified.

(1) The physical properties of RCG and SCG should be investigated comparatively in particular, since they are quite crucial to the properties of coal gangue concrete. In other words, the physical change of coal gangue as a coarse aggregate in concrete mixture before and after spontaneous/artificial combustion should be compared and identified quantitatively.

(2) Even though the main content in the chemical composition of RCG and SCG is actually stable, heavy metal and organic matter leaching of RCG and SCG should be studied comparatively. More specific permission regarding heavy metal and organic matter leaching in coal gangue should be issued by governments to avoid posing environmental, health, and safety concerns.

(3) The workability of coal gangue concrete significantly varies from the gangue origins, which makes the improvement method ineffective in some cases. This phenomenon may be solved if the physical properties of coal gangue are assessed more accurately. A proper index to assess the workability of coal gangue concrete for practical use should also be proposed based on more experimental data.

(4) Even though the degradation effect of coal gangue aggregate on the mechanical properties of coal gangue concrete is obvious, the empirical stress–strain model should be validated against more data. In particular, the elastic modulus and tensile

strength of coal gangue concrete have not been described through theoretical models, not to mention bending and creep performance and damage accumulation process.

(5) The studies on the durability performance of coal gangue concrete are directly related to the properties of the coal gangue aggregate. Therefore, the specific properties and replacement ratio of coal gangue should be considered in the prediction formulas regarding the durability performance of coal gangue concrete. More tests on the durability performance of coal gangue concrete are still needed, such as the drying and watering cycle, high temperature, abrasion resistance, and sulfate resistance.

(6) Besides the experimental and theoretical studies on the structural members with coal gangue concrete are rather limited, the durability performance of those has never been studied. Specifically, the interface bonding behavior between the steel component and coal gangue coarse aggregate concrete under the severe environment and long-term loading should be studied in detail. More design methods from formal standards should be modified by considering the properties of coal gangue for practical applications.

## 8. Conclusions

This paper has reviewed the utilization of coal gangue (CG) in green concrete as a coarse aggregate. The mechanical and durability performance of coal gangue concrete and the corresponding structural members have analyzed from the reported literature. The main conclusions are as follows:

(1) The physical and chemical properties of rock and spontaneous-combustion coal gangue are generally suitable for being used as a coarse aggregate in green concrete, even though these properties vary from the origin. The short-term heavy metal leachability of coal gangue is also relatively weak. However, more specific permission regarding heavy metal and organic matter leaching in coal gangue should be issued by governments to avoid posing environmental, health, and safety concerns.

(2) Coal gangue concrete is not recommended to be used in subsurface structures, since its water absorption law would be changed by increasing the CG replacement ratio. The slump of coal gangue concrete with a 100% CG content would be reduced by about 10–20%. Adding fly ash, superplasticizer, and additional water are effective and simple ways to enhance the workability of coal gangue concrete. A proper index to assess the workability of coal gangue concrete for practical use should also be proposed based on more experimental data.

(3) The mechanical performance, including compressive strength, elastic modulus, and splitting tensile strength of coal gangue concrete is degraded by raising the CG replacement ratio. This degradation caused by the porous structure of CG varies significantly from the CG origin. Over-low and -high concrete grades are not suggested to be used as coal gangue aggregates, unless extra admixture or specific methods are used. The models for describing the elastic modulus, tensile strength, bending strength, and creep performance of coal gangue concrete are still needed for practical applications.

(4) The porous structure of coal gangue provides more transmission channels for air and liquid in concrete. Therefore, the influence of CG on the durability of coal gangue concrete is more remarkable than on its mechanical performance, which could be reduced by adding super-fine admixture and calcination. In contrast, the porous structure of CG is beneficial in the preparation of thermal insulation concrete. The specific properties and replacement ratio of coal gangue should be considered in the prediction formulas regarding the durability performance of coal gangue concrete.

(5) The application of coal gangue concrete in structural members is still limited. Even though the static and seismic behavior of some structural members using coal gangue concrete has been investigated, the durability of these structural members has never been reported. Specifically, the interface bonding behavior between steel component and coal gangue coarse aggregate concrete under the severe environment and long-term loading should be studied in detail. Among them, concrete filled steel tubes

are a preferable configuration for using coal gangue concrete, regarding both the mechanical and durability performance.

**Author Contributions:** Conceptualization, writing-original draft preparation, S.G.; methodology, S.Z.; writing-review and editing, L.G. All authors have read and agreed to the published version of the manuscript.

**Funding:** This research was funded by the Youth Innovation Team of Shaanxi Universities (21JP138).

**Institutional Review Board Statement:** Not applicable.

**Informed Consent Statement:** Not applicable.**Data Availability:** The raw/processed data required to reproduce these findings cannot be shared at this time as the data also forms part of an ongoing study.

**Conflicts of Interest:** The authors declare no conflict of interest.

## References

1. BP. *BP Statistical Review of World Energy*; BP: London, UK, 2019.
2. Zhang, Y.L.; Ling, T.C. Reactivity activation of waste coal gangue and its impact on the properties of cement-based materials—A review. *Constr. Build. Mater.* **2020**, *234*, 117424. [CrossRef]
3. Li, D.; Song, X.; Gong, C.; Pan, Z. Research on cementitious behavior and mechanism of pozzolanic cement with coal gangue. *Cem. Concr. Res.* **2006**, *36*, 1752–1759. [CrossRef]
4. Li, J.Y.; Wang, J.M. Comprehensive utilization and environmental risks of coal gangue: A review. *J. Clean. Prod.* **2019**, *239*, 117946. [CrossRef]
5. Jabłońska, B.; Kityk, A.V.; Busch, M.; Huber, P. The structural and surface properties of natural and modified coal gangue. *J. Environ. Manag.* **2017**, *190*, 80–90. [CrossRef] [PubMed]
6. Moghadam, M.J.; Ajalloeian, R.; Hajiannia, A. Preparation and application of alkali-activated materials based on waste glass and coal gangue: A review. *Constr. Build. Mater.* **2019**, *221*, 84–98. [CrossRef]
7. Huang, Y.L.; Li, J.M.; Song, T.Q.; Sun, Q.; Kong, G.; Wang, F. Microstructure of coal gangue and precipitation of heavy metal elements. *J. Spectrosc.* **2017**, 3128549. [CrossRef]
8. Dong, Z.; Xia, J.; Fan, C.; Cao, J. Activity of calcined coal gangue fine aggregate and its effect on the mechanical behavior of cement mortar. *Constr. Build. Mater.* **2015**, *100*, 6–69. [CrossRef]
9. Cheng, Y.; Hongqiang, M.; Hongyu, C.; Jiaxin, W.; Jing, S.; Zonghui, L.; Mingkai, Y. Preparation and characterization of coal gangue geopolymers. *Constr. Build. Mater.* **2018**, *187*, 318–326. [CrossRef]
10. Ma, X.W.; Liu, J.H.; Shi, C.J. A review on the use of LWA as an internal curing agent of high performance cement-based materials. *Constr. Build. Mater.* **2019**, *218*, 385–393. [CrossRef]
11. Cong, X.; Lu, S.; Yao, Y.; Wang, Z. Fabrication and characterization of self-ignition coal gangue autoclaved aerated concrete. *Mater. Des.* **2016**, *97*, 155–162. [CrossRef]
12. Wang, Q.; Li, Z.; Zhang, Y.; Zhang, H.; Zhou, M.; Fang, Y. Influence of coarse coal gangue aggregates on elastic modulus and drying shrinkage behaviour of concrete. *J. Build. Eng.* **2020**, *32*, 101748. [CrossRef]
13. Cheng, W.H. *The Influence of Argillaceous Shale Coal Gangue on Normal Temperature and High Temperature Performance of Concrete*; Harbin Institute of Technology: Harbin, China, 2015.
14. Wu, X.F.; Zhou, M.; Cui, Z.L. The experiment research on concrete strength influence of self combustion coal gangue coarse aggregate. *Ind. Constr.* **2009**, *39*, 81–85.
15. Zhou, M.; Dou, Y.; Zhang, Y.; Zhang, Y.; Zhang, B. Effects of the variety and content of coal gangue coarse aggregate on the mechanical properties of concrete. *Constr. Build. Mater.* **2019**, *220*, 386–395. [CrossRef]
16. Zhu, Z.Z. *The Development of Coal Gangue Concrete and Its Performance Test*; Anhui University of Science and Technology: Huainan, China, 2014.
17. Duan, X.M. *Study on the Micro-Structure and Physical-Mechanical Performance of Concrete with Coal Gangue as Aggregate*; China University of Mining and Technology: Xuzhou, China, 2014.
18. Miao, H.Y.; Zhang, Z.P.; Gong, L.; Wang, P. College of Architecture and Civil Engineering, Taiyuan University of Technology. Experimental study on coal gangue thermal insulation glazed hollow bead concrete. *Concrete* **2017**, *2*, 152–160.
19. Zhang, Y.; Ma, G.; Liu, Y.; Li, Z. Mix design for thermal insulation concrete using waste coal gangue as aggregate. *Mater. Res. Innov.* **2015**, *19* (Suppl. S5), 878–884. [CrossRef]
20. Zhang, J.X.; Chen, W.L.; Yang, R.J. Experimental study on basic properties of coal gangue aggregate. *J. Build. Mater.* **2010**, *13*, 739–743.
21. Bai, C.N.; Li, L.H.; Sheng, Y.; Long, G.C.; Ma, K.L. The effects of coal gangue as coarse aggregate on mechanical property of concrete. *Sichuan Build. Mater.* **2018**, *44*, 1–3.
22. Fang, Q.C. Study on preparation and properties of shale gangue concrete. *Sichuan Build. Sci.* **2017**, *43*, 107–111.
23. Guan, X.; Qiu, J.; Song, H.; Qin, Q.; Zhang, C. Stress-strain behaviour and acoustic emission characteristic of gangue concrete under axial compression in frost environment. *Constr. Build. Mater.* **2019**, *220*, 476–488. [CrossRef]

24. Ma, H.Q.; Yi, C.; Zhu, H.G.; Dong, Z.C.; Chen, H.Y.; Wang, J.X.; Li, D.Y. Compressive strength and durability of coal gangue aggregate concrete. *Mater. Rep. B* **2018**, *32*, 2390–2395.
25. Wang, A.G.; Zhu, Y.Y.; Xu, H.Y.; Liu, K.W.; Jing, Y.; Sun, D.S. Research progress on coal gangue aggregate for concrete. *Bull. Chin. Ceram. Soc.* **2019**, *38*, 2076–2086.
26. Guo, Y.X.; Zhang, Y.Y.; Cheng, F.Q. Industrial development and prospect about comprehensive utilization of coal gangue. *CIESC J.* **2014**, *65*, 2443–2453.
27. Wang, C.; Ni, W.; Zhang, S.-Q.; Wang, S.; Gai, G.-S.; Wang, W.-K. Preparation and properties of autoclaved aerated concrete using coal gangue and iron ore tailings. *Constr. Build. Mater.* **2016**, *104*, 109–115. [CrossRef]
28. Koshy, N.; Dondrob, K.; Hu, L.; Wen, Q.; Meegoda, J.N. Synthesis and characterization of geopolymers derived from coal gangue, fly ash and red mud. *Constr. Build. Mater.* **2019**, *206*, 287–296. [CrossRef]
29. Frías, M.; de Rojas, M.I.S.; García, R.; Valdés, A.J.; Medina, C. Effect of activated coal mining wastes on the properties of blended cement. *Cem. Concr. Compos.* **2012**, *34*, 678–683. [CrossRef]
30. Sun, Z.H.; Liu, K.P.; Wang, M.Q.; Wang, Z.M.; Yang, F.F. Study on activation properties of spontaneous combustion coal gangue. *J. Build. Mater.* **2013**, *16*, 497–502.
31. Wang, J.; Qin, Q.; Hu, S.; Wu, K. A concrete material with waste coal gangue and fly ash used for farmland drainage in high groundwater level areas. *J. Clean. Prod.* **2016**, *112*, 631–638. [CrossRef]
32. Guan, X.; Chen, J.; Qiu, J.; Gao, Y.; Gao, J. Damage evaluation method based on ultrasound technique for gangue concrete under freezing-thawing cycles. *Constr. Build. Mater.* **2020**, *246*, 118437. [CrossRef]
33. Chen, Y.W.; Sun, Q.; Wang, N.; Niu, W.Y. Experimental study on the spontaneous combustion coal gangue concrete. *Concrete* **2014**, *10*, 63–65.
34. Zhang, Y.; Wang, Q.; Zhou, M.; Fang, Y.; Zhang, Z. Mechanical properties of concrete with coarse spontaneous combustion gangue aggreagate (SCGA): Experimental investigation and prediction methodology. *Constr. Build. Mater.* **2020**, *255*, 119337. [CrossRef]
35. Zhu, K. Study on performance of lightweight aggregate concrete with fly ash and gangue. *Railw. Eng.* **2014**, *3*, 118–121.
36. Li, Y.J.; Yan, X.P.; Guo, R.Q.; Zhou, M. Study on the gradual loss in slump of the spontaneous combustion coal gangue full-lightweight concrete mixture. *Bull. Chin. Ceram. Soc.* **2013**, *32*, 727–731.
37. Zhou, M.; Wang, Q.; Mu, S. The effects on workability of concrete mixture of spontaneous combustion coal gangue coarse aggregate characteristics. *Non-Met. Mines* **2013**, *36*, 8–11.
38. Chen, B.P. Strength of gangue concrete. *Ind. Constr.* **1994**, *7*, 29–32.
39. Wang, Z.S.; Zhao, N. Properties of steel fiber reinforced coal gangue coarse aggregate concrete. *Wuhan Univ. J. Nat. Sci.* **2014**, *19*, 262–268. [CrossRef]
40. Yang, Q.B.; Lu, M.X.; Luo, Y.B. Effects of surface-activated coal gangue aggregates on properties of cement-based materials. *J. Wuhan Univ. Technol.-Mater. Sci. Ed.* **2013**, *28*, 1118–1121. [CrossRef]
41. Cui, Z.L.; Li, J. Effect of coarse aggregate with different water absorption on strength and drying shrinkage of concrete. *Bull. Chin. Ceram. Soc.* **2016**, *35*, 2396–2399.
42. Cui, Z.L.; Hao, J.L.; Chen, L.; Lan, Y. Experimental study on strength and drying shrinkage crack of spontaneous combustion gangue concrete. *Non-Met. Mines* **2015**, *38*, 76–78.
43. Wang, Q.; Zhao, Z.Q.; Liu, Y.Q.; Liu, S. Research on shrinkage performance of coal gangue concrete. *Concrete* **2015**, *10*, 68–70.
44. Qiu, J.S.; Zheng, J.J.; Guan, X.; Gu, S.; Pan, D.; Zhang, C. Capillary water absorption properties of coal gangue concrete under freezing-thawing circumstance. *J. Build. Mater.* **2017**, *20*, 881–885.
45. Zhang, J.X.; Chen, W.L.; Jin, S.S.; Chen, C.-Z.; Yang, R.-J. Investigation on durability of coal gangue aggregate concrete. *J. Beijing Univ. Technol.* **2011**, *37*, 116–125.
46. Li, Y.J.; Han, J.J.; Xing, Y.; Yan, X.P.; Zhang, X. Research on the freezing and thawing cycle test of coal gangue aggregate concrete. *Concrete* **2013**, *12*, 100–103.
47. Zhou, M.; Zhao, H.M.; Wang, R.; Chen, C.; Qu, H.-L. Effect of admixture on impermeability and frost resistance of spontaneous combustion gangue sand-lightweight concrete. *Bull. Chin. Ceram. Soc.* **2015**, *34*, 131–137.
48. Zhang, X.D.; Li, Q.W.; Song, Y. Study of anti-freezing agent on frost resistance of coal gangue concrete mixing with water reducers. *Non-Met. Mines.* **2016**, *39*, 48–51.
49. Zhu, K.; Yuan, S.L.; Ma, X.W. Study of the road performance of fiber reinforced coal gangue light-weight aggregate concrete. *China Concr. Cem. Prod.* **2015**, *6*, 58–61.
50. Sun, J.Y.; Lu, L.G. Coupled effect of axially distributed load and carbonization on permeability of concrete. *Constr. Build. Mater.* **2015**, *79*, 9–13. [CrossRef]
51. Chen, Y.W.; Ding, Z.Y.; Zhu, J.S.; Sun, X.W. Experimental study on the spontaneous combustion coal gangue concrete. *Concrete* **2015**, *5*, 87–89.
52. Li, Q.W.; Zhang, X.D.; Li, G.X.; Li, Y.; Zhang, J. Study on carbonation depth of spontaneous combustion coal gangue lightweight aggregate concrete. *Chin. J. Environ. Eng.* **2016**, *10*, 2616–2620.
53. Li, Y.J.; Xing, Y.; Han, J.J.; Cao, S. Experimental research on gas permeability and carbonation resistance of coal gangue aggregates. *Non-Met. Mines* **2016**, *39*, 17–20.

54. Yi, C.; Ma, H.Q.; Zhu, H.G.; Dong, Z.C.; Su, Z.J.; Zhang, Y.T.; Chu, Z. Investigation on anti-carbonation performance of coal gangue coarse aggregate concrete. *J. Build. Mater.* **2017**, *20*, 787–793.
55. Zhou, M.; Tian, B.Y.; Wang, Q.; Mou, S.; Pu, B.C. Experimental study on the influence of spontaneous combustion gangue coarse aggregate on sand lightweight concrete performance. *Bull. Chin. Ceram. Soc.* **2013**, *32*, 2231–2237.
56. Wang, Q.; Liu, S.; Wang, J.B.; Li, M.; Liu, Y. Research on resistance to chloride ion of the coal gangue coarse aggregate concrete. *Concrete* **2016**, *8*, 36–38.
57. Li, Y.J.; Xing, Y. Experiment study on chloride ion penetration of coal gangue concrete. *Non-Met. Mines* **2016**, *39*, 11–13.
58. Zhou, M.; Li, G.N.; Zhang, Q.; Cui, H. Study on application of spontaneous combustion coal gangue aggregate in ready-mixed concrete. *J. Build. Mater.* **2015**, *18*, 830–835.
59. Feng, J.X. *Coal Gangue Type of Self-Compacting Concrete Beam Bending Performance Test Research*; Yianbian University: Yianbian, China, 2017.
60. Wang, Q.H.; Li, Z.; Zhou, M.; Zhang, Y.Z. Effects of spontaneous-combustion coal gangue aggregate replacement ratio on flexural behavior of reinforced concrete beams. *J. Build. Struct.* **2020**, *41*, 64–74.
61. Li, G.N. *Study of Flexural Behavior on Spontaneous Combustion Gangue Aggregate Reinforced Concrete Beam*; Liaoning Technical University: Liaoning, China, 2015.
62. Bai, G.L.; Zhu, C.; Wang, J.W.; Li, X.X.; Liu, H.; Lu, G.K. Experimental study on shear behavior of coal gangue concrete beams. *J. Build. Struct.* **2020**, *41*, 49–55.
63. Liu, H.Q. *Experimental Study on Bearing Capacity of Coal Gangue Concrete Column*; Xi'an University of Architecture and Technology: Xi'an, China, 2019.
64. Li, Y.J.; Yan, X.P.; Zhang, X.; Geng, X.; Zhang, J. Experimental research on seismic behavior of reinforced concrete column composed of coal gangue aggregate. *J. China Coal Soc.* **2013**, *38*, 1006–1011.
65. Wang, S.X. *Experimental Study on Seismic Performance of Coal Gangue Concrete Shear Wall under Horizontal Reciprocating Load*; Liaoning Technical University: Liaoning, China, 2020.
66. Zhuang, Y.Z. Mechanical properties of self-combusted gangue reinforced concrete shear walls under low-cyclic load. *J. Zhejiang Univ.* **2000**, *34*, 325–330.
67. Tan, X.M.; Guo, M.H.; Wang, F.M. Earthquake-resistant behavior of low-rise self-combusted gangue reinforced concrete shear wall. *J. South China Univ. Technol. Nat. Sci.* **1999**, *27*, 67–73.
68. Li, G.C.; Liu, Z.Y.; Yang, L.Z. The constitutive equation and strength criterion on core concrete of gangue concrete filled steel tube. *J. Northeast. Univ. Nat. Sci.* **2002**, *23*, 64–66.
69. Li, G.C.; Xu, Z.; Zhang, C.Y. Finite element analysis of gangue concrete-filled steel tubular beam-columns. *J. Shenyang Jianzhu Univ. Nat. Sci.* **2010**, *26*, 1052–1057.
70. Zhang, N.; Zheng, C.; Zhao, Z.; Yang, B. Experimental investigations on compression behavior of GFRP-coal gangue concrete-steel tubular columns. *J. Build. Mater.* **2020**, *9*, 1–10.
71. Zhang, H.B. *Experimental Study on Bearing Capacity of Self Compacting Concrete-Filled Steel Pipe Short Column under Axial Compression*; Yanbian University: Yanbian, China, 2018.
72. Gao, S.; Zhao, G.; Guo, L.; Zhou, L.; Yuan, K. Utilization of coal gangue as coarse aggregates in structural concrete. *Constr. Build. Mater.* **2020**, *268*, 121212. [CrossRef]
73. Peng, Q. *Experimental Study on the Mechanical Properties of the Coal Gangue Concrete Shear Wall with Internal Steel Plate*; Shenyang Jianzhu University: Shenyang, China, 2015.
74. Liu, H. *Study on Seismic Performance of a Steel Plate Gangue Concrete Composite Shear Wall*; Shenyang Jianzhu University: Shenyang, China, 2015.
75. GB/T 17431.1-2010. *Lightweight Aggregates and Its Test Methods-Part 1: Lightweight Aggregates*; Standardization Administration of China: Beijing, China, 2010.
76. Li, G.C.; Chang, C.; Qu, Z.S. Mechanics function of composite slab with gangue concrete and profiled sheeting. *J. Liaoning Tech. Univ.* **2003**, *22*, 61–63.
77. Li, Z.H. *Finite Element Analysis of Profiled Steel Sheet-Coal Gangue Lightweight Aggregate Concert Composite Slabs*; Taiyuan University of Technology: Taiyuan, China, 2008.
78. Javed, M.F.; Amin, M.N.; Shah, M.I.; Khan, K.; Iftikhar, B.; Farooq, F.; Aslam, F.; Alyousef, R.; Alabduljabbar, H. Applications of gene expression programming and regression techniques for estimating compressive strength of bagasse ash based concrete. *Crystals* **2020**, *10*, 737. [CrossRef]
79. Rehman, S.K.U.; Kumarova, S.; Memon, S.A.; Javed, M.F.; Jameel, M. A review of microscal, rheological, mechanical, thermoelectrical and piezoresistive properties of graphene based cement composite. *Nanomaterials* **2020**, *10*, 2076.
80. Rehman, S.K.U.; Ibrahim, Z.; Memon, S.A.; Javed, M.F.; Khushnood, R.A. A sustainable graphene based cement composite. *Sustainability* **2017**, *9*, 1229. [CrossRef]

*materials*

MDPI

*Review*

# The Durability of Recycled Fine Aggregate Concrete: A Review

**Changming Bu [1,2], Lei Liu [1,2], Xinyu Lu [1,2], Dongxu Zhu [1,2], Yi Sun [1,2,*], Linwen Yu [3], Yuhui OuYang [1,2], Xuemei Cao [1,2] and Qike Wei [4]**

1 School of Civil Engineering and Architecture, Chongqing University of Science & Technology, Chongqing 401331, China; buchangming@126.com (C.B.); 2020206111@cqust.edu.cn (L.L.); 2020206077@cqust.edu.cn (X.L.); 2020206029@cqust.edu.cn (D.Z.); 2020206038@cqust.edu.cn (Y.O.); 2020206051@cqust.edu.cn (X.C.)

2 Chongqing Key Laboratory of Energy Engineering Mechanics & Disaster Prevention and Mitigation, Chongqing 401331, China

3 College of Materials Science and Engineering, Chongqing University, Chongqing 400044, China; linwen.yu@cqu.edu.cn

4 China Metallurgical Construction Engineering Group Construction Ltd., Chongqing 400084, China; weiqike@cmccltd.com

* Correspondence: sunyi@cqust.edu.cn; Tel.: +1-35-9416-9610

**Abstract:** With the rapid development of urbanization, many new buildings are erected, and old ones are demolished and/or recycled. Thus, the reuse of building materials and improvements in reuse efficiency have become hot research topics. In recent years, scholars around the world have worked on improving recycle aggregates in concrete and broadening the scope of applications of recycled concrete. This paper reviews the findings of research on the effects of recycled fine aggregates (RFAs) on the permeability, drying shrinkage, carbonation, chloride ion penetration, acid resistance, and freeze–thaw resistance of concrete. The results show that the content of old mortar and the quality of recycled concrete are closely related to the durability of prepared RFA concrete. For example, the drying shrinkage value with a 100% RFA replacement rate is twice that of normal concrete, and the depth of carbonation increases by approximately 110%. Moreover, the durability of RFA concrete decreases as the RFA replacement rate and the water–cement ratio improve. Fortunately, the use of zeolite materials such as fly ash, silica fume, and meta kaolin as surface coatings for RFAs or as external admixtures for RFA concrete had a positive effect on durability. Furthermore, the proper mixing methods and/or recycled aggregates with optimized moisture content can further improve the durability of RFA concrete.

**Keywords:** recycled fine aggregate concrete; impermeability; drying shrinkage; chloride penetration resistance; carbonation resistance; acid resistance; resistance to freeze–thaw cycles



**Citation:** Bu, C.; Liu, L.; Lu, X.; Zhu, D.; Sun, Y.; Yu, L.; OuYang, Y.; Cao, X.; Wei, Q. The Durability of Recycled Fine Aggregate Concrete: A Review. *Materials* **2022**, *15*, 1110. https://doi.org/10.3390/ma15031110

Academic Editor: Dario De Domenico

Received: 30 December 2021
Accepted: 28 January 2022
Published: 31 January 2022



## 1. Introduction

With the rapid economic and social development of the 21st century, the urbanization process has produced a large amount of construction waste. According to statistics, China produces more than 3 billion tons of construction waste each year, and waste concrete accounts for 50–60% of all construction waste [1,2]. Such a large amount of waste concrete is usually placed in landfills and roadbed backfills and not effectively utilized, which is not only a waste of resources but also greatly damages the environment and goes against the path of sustainable development [3,4]. Therefore, researchers at home and abroad have conducted prospective research on the application of waste concrete in construction [5–13], e.g., making recycled aggregates and geopolymers from waste concrete [14–16]. The geopolymer is a new environmentally friendly material that can replace cement in construction. Some people believe that it is the same as, or subordinate to, alkali-activated materials [17]. However, this is wrong, as the two are essentially different processes and produce different chemical products. An alkali-activated material is a monomeric precipitate, whereas a geopolymer is a true polymer [18]. Furthermore, ground polymerization processes require alkaline and acidic environments, whereas alkali-activated materials require mainly high alkalinity environments.

Currently, the mechanical properties and durability of recycled aggregate concrete (RAC) have been studied in depth. A comprehensive review of the mechanical properties of RAC has been conducted, and the results show that degradation of the mechanical properties of RAC is mainly related to the high water absorption of the aggregates and the weak areas formed by the mortar adhering to the old interface [19,20]. The durability of RAC has also been investigated, and the results show that the higher the recycled aggregate (RA) content, the mortar content attached to the RA, and the W/C (water-to-ash ratio), the worse the durability of RAC [21].

Recycled aggregates can be divided into recycled coarse aggregate (RCA) and recycled fine aggregate (RFA). Compared to recycled coarse aggregate concrete (RCAc), recycled fine aggregate concrete (RFAc) has been less studied, but RFA is also important in replacing sand in the preparation of concrete. The application of RFA has a positive impact on the environment [22]. However, the application of RFA is limited in various ways due to its many drawbacks. It was found that the high water absorption of RFA reduces the workability of the prepared fresh concrete. In coarse aggregate concrete, an increase in the water-to-ash ratio (W/C) or the addition of additives or mineral additives is required to offset the lower workability of the RAC [23]. Regarding fresh state characteristics of recycled aggregate concrete, to achieve the same target slump, both the effective and apparent w/c ratios should increase as the amount of RCA incorporated increases. Fine RCA concrete mixes require more water than coarse RCA concrete mixes [24]. The RFA provides more and finer powder material and a more complex interfacial transition zone (ITZ) [25]. Since the old mortar and cement paste attached to RFA severely affect the performance of RFAc, removing them can effectively increase the quality of RFA. A variety of old mortar removal techniques have been studied by domestic and foreign scholars. The mechanical treatment method is simple and easy to operate [26]; however, it tends to crack the RA and has a negative impact on the workability and compressive strength of the concrete. The heat treatment method dehydrates the old mortar adhering to the surface of the RA and renders it loose and porous [26,27]; however, high-temperature heating is highly energy-intensive. The acid presoaking method effectively breaks down the bond between the hydration products of the cement, thus removing the old mortar adhering to the RA surface [28]; however, this method increases the cost of concrete production. In the ultrasonic cleaning method, the cleaned RAC has a 7% increase in compressive strength compared to the uncleaned RAC [29]. Microwave-assisted beneficiation can create higher thermal stresses on the bonding surface of the old mortar and aggregate, thus causing delamination, and this method is less energy-intensive than conventional heating and machine grinding [30]. Some studies have broken away from the tradition and invented the process of crushing all waste concrete to obtain fully recycled fine aggregates [31]. The fully recycled fine aggregate obtained by this process has excellent physical properties and can replace 100% of the river sand in concrete [32].

In this paper, the durability of RFAc is reviewed in terms of impermeability, drying shrinkage, carbonation resistance, chloride penetration resistance, acid resistance, and freeze–thaw resistance. Understanding the effect of RFA on the durability of concrete can provide a theoretical basis for the future large-scale application of RFA as a substitute for sand in practical projects.

## 2. Impermeability

Impermeability is a characteristic of the durability of concrete and is closely related to other durability characteristics. For example, water-soluble chloride ions, $CO_2$, and sulfate can enter concrete, or other external factors can cause damage to the concrete, making the service life of concrete greatly affected and thus becoming a global concern [33–35]. The impermeability of concrete is mainly determined by its own microscopic properties such as pore size and interfacial transition zone. The impermeability of RFAc is mainly influenced by the replacement rate of RFA, W/C, and mineral admixtures. It was found that the addition of admixtures and additives to concrete can improve the microscopic effect and achieve improved impermeability [36–38].

The substitution rate of RFA is closely related to its physical property index. The permeability of the concrete prepared by replacing RFA with different physical properties is also different. Compared with natural fine aggregates, the physical properties of RFA are poor. The higher the substitution rate, the worse the performance of the prepared RFAc. The poor grading of RFA particles with a large number of voids on the surface and micro cracks attached to the old mortar renders the prepared RCAc less dense, so the level of RCAc impermeability decreases with the increase in the RCA substitution rate [39,40]. In self-compacting concrete (SSC), water absorption increases as the RFA substitution rate increases, which is caused by the high porosity and high water absorption of RFA [41,42]. In a sulfate environment, RFAc exhibited better permeability resistance than normal concrete. The water absorption of RFAc exposed to a sulfate environment decreased by an average of 25% compared to normal concrete [43].

The water–cement ratio is an important parameter in the preparation of concrete, which will affect the permeability resistance of concrete. A water–cement ratio of 0.65 has a lower impermeability rating than RFAc with a water–cement ratio of 0.55, indicating that the higher the water–cement ratio, the worse the impermeability of RFAc, because the higher the water–cement ratio, the slower the hydration of the RFAc cement and the more pores it produces, which leads to the poor compactness of RFAc [39,40,44].

Due to its volcanic ash nature, fly ash (FA) has been widely used as a replacement for silicate cement in engineering construction [45]. The addition of fine fly ash to concrete can promote hydration reactions, zeolite reactions, accumulation effects, and nucleation effects [46]. Thus, FA can participate in the early hydration reaction and increase the hydration products to fill the pores and increase the density of the concrete [47]. In self-compacting concrete mixed with 100% RFA (SSC) and 10% (FA), the permeability of SSC prepared with different water–cement ratios was found to be significantly lower compared to the control group without FA [44]. The physical and chemical properties of RFA are defective, which weakens the transition zone at the interface of the prepared RAC. In order to improve the ITZ of RAC, several researchers have developed different concrete mixing methods. Figure 1 shows the double mixing method (DM) [48] and the triple mixing method (TM) [49] as well as the optimized triple mixing method (OTM). The difference between the OTM and the other two methods is that the water-reducing agent (SP) is added in a different order. In the OTM, SP is added together with other gelling materials to promote additional gelling. The dispersion of this material further increases the content of additional gelling material in the ITZ, which can better exploit the zeolite effect at a later stage, such that the use of the OTM can result in a permeability rating of 8 for RAC [50].

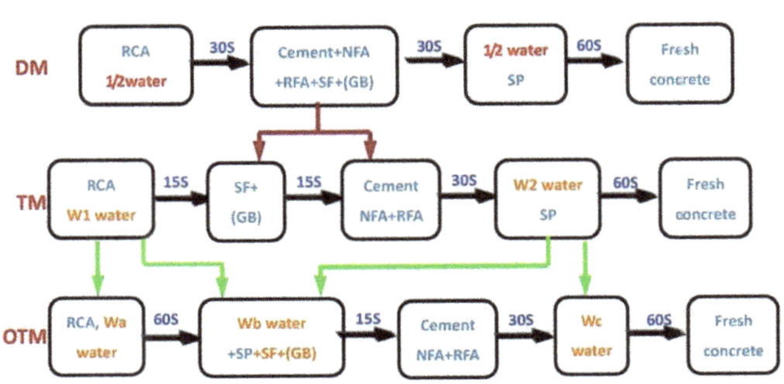


**Figure 1.** Comparison of mixing methods for the preparation of RAC [50] (Wa equals 60–80% of the product of the weight of RA multiplied by its water absorption, Wb equals the weight percentage of residual water of SCM in the total gelling material, and Wc = mixed water − Wb − Wa).

The water absorption rate of RFA is much higher than that of natural fine aggregate. RFA absorbs water when freshly mixed with concrete, which has an effect on the mechanical properties and durability of concrete, so the water content of RFA before mixing should not be neglected. Therefore, the effect of RFA in three states of air-drying (AD), oven-drying (OD), and saturated surface drying (SSD) on the permeability of the prepared RFAc was investigated, and it was found that the permeability of these states at the same RFA replacement rate was ordered as follows: SSD > AD > OD. This finding was different from that of RCA. The permeability resistance of concrete prepared from RCA with different moisture contents was ordered as follows: SSD < AD < OD [51]. The reason for this may be that the particle size of RFA is smaller than RCA, but the specific surface area is larger, so the water can become more confined [52].

The addition of RFA has a serious effect on the permeability of concrete. An increase in the RFA replacement rate and the W/C ratio will lead to a decrease in the permeability resistance. However, the use of saturated surface-dried RFA, optimized mixing methods, and the addition of fly ash can improve the impermeability of RFAc. In addition to fly ash, silica fume, metakaolin, and other mineral admixtures also have zeolite properties, so the impermeability of RFAc mixed with silica fume and higher mineral admixtures is also beneficial. Table 1 summarizes the effects of different factors on the impermeability of RFAc.

**Table 1.** The influence of different factors on the impermeability of RFAc.

| Factor | Change | Influence |
|---|---|---|
| RFA % | Increase | Negative |
| W/C ratio | Increase | Negative |
| RFA water content | Increase | Positive |
| Optimized triple mixing method (OTM) | - | Positive |
| Mineral admixture | - | Positive |

## 3. Drying Shrinkage

Drying shrinkage requirements are already an important part of project specifications [53]. The evaporation of water from the interior of concrete in a dry environment results in a change in concrete dimensions with time. The drying shrinkage process starts with the evaporation of free water from the pores, followed by the evaporation of adsorbed water from the very small capillary pores; the evaporation of free water is not related to drying shrinkage, but the evaporation of adsorbed water leads to the shrinkage and internal generation of tensile stresses to the extent that drying shrinkage cracks appear in the concrete [54]. As for the coarse aggregate concrete, it was found that the "electric arc furnace slag" improves the strength and modulus of elasticity and leads to shrinkage reduction of concrete [55,56].

The drying shrinkage value increases as the RFA substitution rate increases, and when the substitution rate is 100%, the drying shrinkage value of RFAc is twice as high as that of normal concrete [57]. The drying shrinkage value of RFAc increases with the substitution rate and decreases as the interim period increases. One study found that the drying shrinkage value of RFAc with a 30% substitution rate was 14.0% lower than that of normal concrete and 8.2% higher than that of normal concrete with a 100% substitution rate at 90 days of the prophase [58] (Figure 2). It has also been found that the relationship between drying shrinkage values and the RFA substitution rate is divided by 30%; drying shrinkage values after a substitution rate of 30% increase as the substitution rate increases, and the substitution rate before a substitution rate of 30% has no significant effect on drying shrinkage values [59].

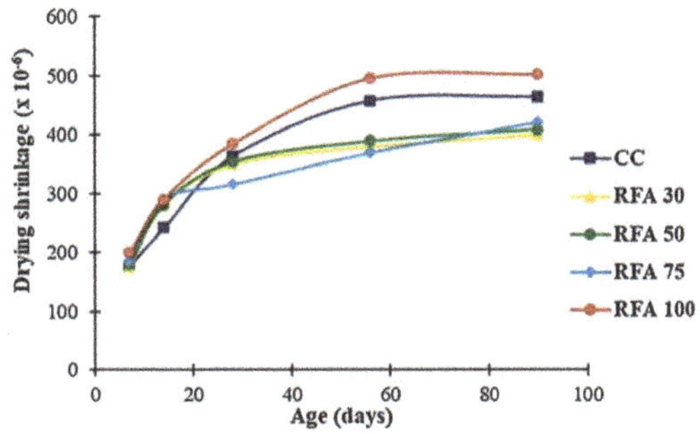


**Figure 2.** Dry shrinkage of RS (river sand) and RFA concrete [58].

Different sources of RFA also have an important effect on RFAc, and the study of RFAc prepared from different sources of RFA has great significance for the application of waste concrete in practical engineering. RFA from recycled laboratory-produced concrete and RFA from recycled precast concrete, both of which have similar strengths, were studied for drying shrinkage values of RFAc under both recycling routes and were found to be similar [60]. However, in another study, which also investigated RFA from two different sources, RFA1 from flat-plate concrete and RFA2 from composite beam specimen concrete, drying shrinkage tests on RFAc prepared from both RFAs at 0%, 50, and 100% substitution rates were conducted, and it was found that, regardless of the substitution rate of RFA, the drying shrinkage values of RFAc prepared from RFA2 were lower than that of RFAc prepared by RFA1. The differences in drying shrinkage values exhibited by RFAc prepared by the two sources are mainly related to the physical properties of RFA: (1) the density of RFA2 is greater than that of RFA1 (RFA1: 2222 kg/m$^3$, RFA2: 2326 kg/m$^3$), indicating that RFA1 contains a higher amount of old cement paste than RFA2; and (2) the water absorption is lower than that of RFA1 (RFA1: 11.2%, RFA2: 8.1%). These two differences in physical properties indicate that the water content of RFAc prepared by RFA1 is higher than that of RFAc prepared by RFA2 and that the high water content has a low binding effect on concrete [61] (Figure 3).

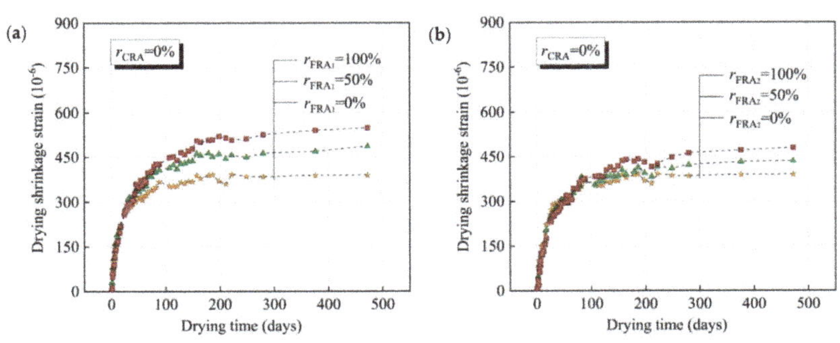


**Figure 3.** Development of dry shrinkage of concrete with time using different sources of FRA [61]: (**a**) concrete with FRA1, rCRA = 0%; (**b**) concrete with FRA2, rCRA = 0%.

The drying shrinkage of RFAc with a 20% RFA replacement rate and an effective water–cement ratio of 0.43 was similar to that of the control with a water–cement ratio of 0.45 at the same amount of RFAc mixing water, but the drying shrinkage of RFAc with a 30% RFA replacement rate and an effective water–cement ratio of 0.41 was lower than that of the control [62]. SP is a very frequently used admixture, and its addition reduces the amount of mixed water in the workable fresh concrete mixture, reducing the water–cement ratio for the same amount of cement and thus improving the performance of RFA [63–65]. The effect of different SPs on the drying shrinkage of RFAc prepared with different replacement rates of RFA was investigated, mainly comparing RFAc without SP, with conventional plasticizer (SP1), and with high-performance and high-efficiency plasticizer (SP2), and it was found that the drying shrinkage values of RFAc without SP increased to 28 and 57% at 7 and 91 days, respectively; the drying shrinkage values of RFAc with SP1 increased at 7 days to 44%, while the drying shrinkage value at 91 days decreased to 2%; the drying shrinkage value of RFAc7d with SP2 increased to 38%, while the drying shrinkage value at 91 days decreased to 30%, which shows that the addition of SP was effective in reducing the drying shrinkage value of RFAc, and the effectiveness of SP2 was higher than that of SP1 [66].

Silica fume (SF), produced during the smelting process of the ferrosilicon industry, has a silica content of up to 75% and is an effective volcanic ash material [67]. Due to the large amount of water absorption of the old mortar attached to the RA surface, it can lead to an increase in the drying shrinkage value of SSC prepared with RFA. It was found that the drying shrinkage value of the ternary RFA SSC prepared with finely ground blast furnace slag (GGBFS) and SF instead of cement was greatly reduced, on the one hand because the two external admixtures promoted hydration and on the other hand because the ultrafine material filled the pores inside the concrete [44].

Different mixing methods also have an effect on the performance of concrete. RFA is one of the important reasons for the performance of concrete due to its excessive water absorption; however, the use of moisture in the two-part addition, i.e., the two-stage mixing method (TSMA), can effectively prevent the excessive water absorption of RFA [68]. The performance of concrete with different RFA replacement rates prepared by the normal mixing method and TSMA was investigated, and it was found that the drying shrinkage values of all RFAc using TSMA were lower than those of the concrete prepared by the normal mixing method, mainly because of the higher compactness and better ITZ of RFAc prepared by TSMA [69].

In summary, the drying shrinkage values of the concrete prepared by the RFA substitution of sand became larger with the increase of the RFA substitution rate. In order to reduce the effect of RFA on the drying shrinkage value of concrete, the drying shrinkage value can be effectively reduced by adding admixture (SP) and mineral admixture, and the drying shrinkage value of RFAc with TSMA is also smaller than that of RFAc with a normal mixture. Drying shrinkage is a long-term performance feature of concrete, which is of great significance to concrete in the process of use, but RFAc is significantly affected by RFA, so it is necessary to consider an optimal replacement rate of RFA when preparing RFAc. Therefore, an optimal replacement rate of RFA needs to be considered when preparing RAFc, and the effect of RFA on the drying shrinkage of concrete should be reduced by other means.

## 4. Carbonation Resistance

The amount of $CO_2$ in the air is about 0.03%. The pores present in concrete allow $CO_2$ to diffuse within it and react with hydration products to form a carbonation process. Carbonation is mainly a chemical reaction between carbon dioxide ($CO_2$) and calcium hydroxide (CH), hydrated calcium silicate (C-S-H), or bentonite (AFt) [70]. Determining the carbonation resistance of RFAc is fundamental to the study of its durability, since concrete is exposed to natural environments that induce carbonation corrosion for long periods of time, and the rate of carbonation is directly related to the lifetime of concrete. The carbonation resistance of RFAc is mainly related to the minimum particle size of RFA, the substitution rate, the W/C ratio, and the mineral external admixtures.

The smaller the RFA particle size, the higher the amount of old mortar adhered to the surface, so the carbonation depth of RFAc increases instead as the minimum particle size of the RFA decreases [71]. $CO_2$ curing is one of the means of treating RFA, and $CO_2$ curing can fill the pores and cracks on the surface of the aggregate and improve the performance of the old and new ITZ [72]. Carbonation studies were conducted on concrete prepared from $CO_2$-cured recycled coarse and fine aggregates; it was found that the treated RFAc had improved carbonation resistance compared to the untreated one, with the maximum carbonation resistance depth of the former being about 42.5% lower than that of the latter (Figure 4), and it also showed the same trend under freeze–thaw cycling conditions [73].

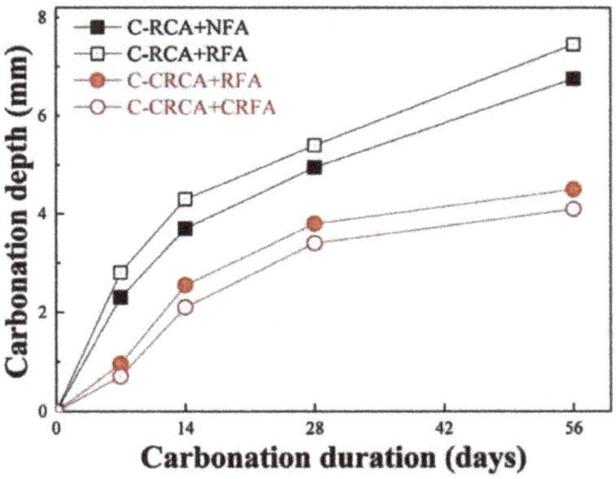


**Figure 4.** Carbonation behavior of CRA concrete exposed to carbonation [73].

When the replacement rate of RFA in concrete is 30 and 100%, the carbonation depth of concrete $CO_2$ increases by about 40 and 110%, respectively, indicating that the resistance of RFAc to carbonation decreases as the replacement rate of RFA increases, which is caused by the internal capillarity of RFAc [71,74,75]. In self-compacting concrete (SSC), the depth of carbonation also increases with the increase in the substitution rate [76].

The lower carbonation resistance of RFAc compared to the control is not only due to the high porosity of RFA but also related to the increase in the effective W/C, which improves the diffusion of $CO_2$ [77]. Fly ash is beneficial for the anticarbonation properties of RFAc; when fly ash replaced cement in the range of 10–30%, a replacement rate of 20% was found to be most beneficial for the anticarbonation properties of RFAc [71]. Some studies, however, have found that the carbonation depth of RFAc increased with increasing fly ash when the RFA substitution rate was below 30% [78]. In self-compacting concrete (SSC) incorporating RFA, it was found that partial age soils (MK) were beneficial to the anticarbonation properties, and SSC with a 50% RFA substitution rate and incorporating MK showed a 20 and 18% decrease in carbonation depth after 4 and 12 weeks of accelerated carbonation exposure, respectively, while SSC with a 100% RFA substitution rate and incorporating MK showed the same carbonation depth after 4 and 12 weeks of accelerated carbonation exposure as the control group. The reason for this may be the high reactivity of the added MK and volcanic ash products such as calcium hydroxide, which caused a large consumption of early hydrated calcium (CH) and thereby reduced the carbonation depth [76].

The carbonation depth increased with time, and the RFAc with SP2 (SikaPlast898) addition had a lower carbonation depth than both without plasticizer and with SP1 (Sikament 400 plus) RFAc at each prophase, mainly because the addition of SP2 decreased the effective w/b of RFAc. Thus, the internal porosity was lower, and the gas permeability was lower [79]. The resistance to carbonation of recycled concrete prepared by one generation of RFA (RAC1) and recycled concrete prepared by two generations of RFA (RAC2) was investigated at four bending stress levels (40, 70, 100, and 120%), and it was found that the carbonation depth increased with stress, except for the 70% stress level, and the carbonation depth of RAC2 was higher than that of RAC1 [75].

The particle size of RFA affects the carbonation resistance of RFAc, and the increase of the RFA substitution rate and W/C can negatively affect the carbonation resistance of RFAc. The anticarbonation performance of RFAc can be improved by the $CO_2$ curing of RFA and the addition of mineral ex-dopant. $CO_2$ curing is one of the methods that can be used to improve the quality of RFA, and mineral ex-dopant coating and microbial carbonate precipitation treatment can also improve the quality of RFA and thus the anticarbonation performance of RFAc. Table 2 summarizes the influence of different factors on the anti-carbonation performance of RFAc.

**Table 2.** The influence of different factors on the anti-carbonation performance of RFAc.

| Factor | Change | Influence |
|---|---|---|
| RFA minimum particle size | Decrease | Negative |
| RFA% | Increase | Negative |
| W/C ratio | Increase | Negative |
| $CO_2$ pretreatment RFA | - | Positive |
| Mineral admixture | - | Positive |

## 5. Chloride Penetration Resistance

Buildings in harsh marine environments are subjected to long-term chloride ion corrosion, leading to the degradation of reinforced concrete structures, which can seriously affect the service life of buildings [80,81]. There are two main effects of chloride ions on buildings: (1) the physical salt attack of chloride ions causes cracking and scaling on building surfaces [82]; and (2) chloride ions corrode the reinforcement inside the building, penetrate the protective layer of concrete, reach the reinforcement, and gradually accumulate, causing the demineralization of the reinforcement, which corrodes the prebuilt reinforcement [83]. RFA has various poorer physical properties compared to natural sand, so the chloride ion permeability resistance of RFAc is more importantly affected by the RFA substitution rate, curing conditions, $CO_2$ curing, and mineral admixtures.

The chloride ion diffusion coefficient and electric flux of RFAc are much higher than those of control groups and show a linear growth trend with the increase of the RFA substitution rate. Reasons include the following: (1) a large amount of old mortar attached to the surface of RFA leads to high porosity, which provides a channel for chloride ion transmission; (2) the old mortar on the surface of RFA absorbs water and reaches the saturation surface, which prevents the transition area of the old and new mortar interfaces from becoming dense, causing it to form chloride ion transmission channels; and (3) additional water cannot be completely absorbed by RFA, resulting in a water–cement ratio that is higher than the effective water–cement ratio and thus a low compactness in the RFAc [40,84]. The total charge of SSC increased significantly with an increasing RFA substitution rate, but the resistivity and chloride resistance of SSC at a 25% substitution rate did not change significantly compared to a control [41]. The RAC prepared by considering coarse and fine recycled concrete aggregates in plain concrete, because of the high porosity of the aggregate itself, was accompanied by a high diffusion coefficient and increased conductivity, so the conductivity of the prepared RAC increased with increases in the substitution rate of the coarse and fine recycled concrete aggregates [85,86].

Curing is an important stage in the early stages of concrete. Reactive powder concrete (RPC) was cured using two curing conditions, normal curing and thermal curing, and the resistivity of RPC was found to be higher under thermal curing conditions, which is due to the improved volcanic ash reaction of RPC under thermal curing conditions, producing a denser CHS gel that fills the pores [87].

The properties of RFAs obtained by different treatments are different, and the prepared RFAc exhibits different properties. The resistivity of RFAc was studied using two types of RFA, (1) recycled fine aggregate and recycled coarse aggregate mixed together (R1) and (2) fully recycled fine aggregate (R2), and it was found that the resistivity of RFAc mixed with R2 was higher than that mixed with R1 due to the improved physical properties and the lower water absorption of R2 [86]. $CO_2$-cured RFA and untreated RFA were compared to study the effect of both on the chloride ion permeability of RFAc, and it was found that the chloride ion permeability of RFAc prepared from $CO_2$-cured RFA was higher at all substitution rates (Figure 5), and the reason for the reduced chloride ion resistance of $CO_2$-cured RFAc was as follows: When RFA was carbonated, chloride ions were attached as Friedl's salt on the cement paste and then released into the pore solution, leading to an increase in chloride ion concentration. As carbonation proceeded, the chloride ion concentration in the carbonated zone increased, while the chloride ion concentration gradient in the noncarbonated zone also increased, so chloride ions entered the noncarbonated zone, leading to an increase in chloride ion concentration in the noncarbonated zone [88].

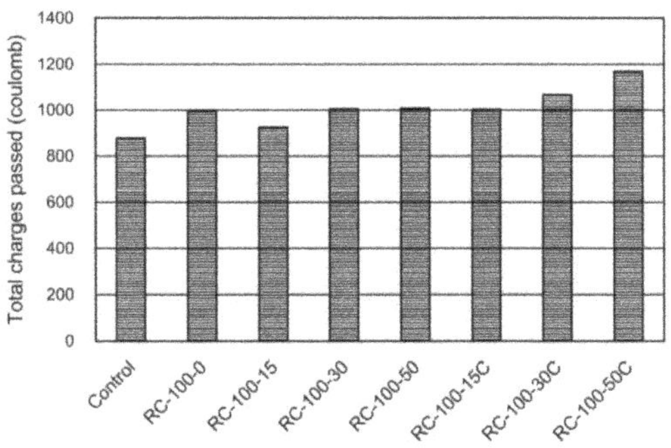


**Figure 5.** Chloride ion penetration at 56 days [88].

The chloride ion resistance of the concrete added to RFA was generally reduced, and it was found that the chloride ion resistance of RFAc could be improved by adding an external admixture. The addition of 10% silica fume (SF) to self-compacting RFA concrete had a lower chloride ion permeability than the control group for all prophases, and this was because SF acts as a microfiller, increasing the density of the SSC interface transition zone [89]. A high amount of clay produces a volcanic ash material called metakaolin (MK) when calcined, which is used as an additional cementitious material (SCM) in concrete because of its volcanic ash nature. The addition of MK to mortar and concrete has been found to improve mechanical and durability properties [90,91]. The chloride ion permeability of regenerated SSC with the addition of 10% MK decreased in all properiods, where C100F25M with a 25% minimum RFA substitution decreased by 64% relative to the control group for 28 days, and C100F100M with a 100% maximum RFA substitution decreased by 52% relative to the control group for 28 days, and this was because of the volcanic ash effect of MK and the additional hydration products, which resulted in a denser microstructure of the regenerated SSC [42]. Rice husk ash (RHA), obtained after combustion treatment, can be about 90% $SiO_2$ and has high volcanic ash properties [92]. RHA is used as a cement replacement for concrete because of its high $SiO_2$ content and volcanic ash effect [93–95]. It has been found that RHA can reduce chloride ion penetration in concrete [20]. The durability of RFAc was greatly affected by the addition of recycled aggregates, in which the resistance to chloride ion penetration was significantly reduced. Therefore, finely ground rice husk ash (GRHA) was used as a cementitious material to prepare RAC with a 20–50% replacement of cement, and it was found that the coarse and fine aggregates were completely replaced by RA in RAC at 90 and 180 days. The depth of chloride ion penetration decreased with the increase of GRHA admixture, as shown in Figure 6. This was mainly because the CSH generated by the reaction between GRHA and volcanic ash filled the pores and because the $SiO_2$ and CSH in GRHA impeded the movement of chloride ions [96]. The durability performance of high-performance recycled aggregate concrete (HPRAC) prepared with ground blast furnace slag (GGBFS) and recycled aggregates is also of interest, so the effect of fly ash (FA) and MgO on the durability performance of HPRAC was investigated, with an RCA replacement rate of 40%, with 30% RFA in HPRAC, with 0, 15, 30, and 45% FA dosing to replace GGBFS, and at 15% FA doping, with 2.5, 5, and 7.5% MgO. It was found that the chloride permeability of HPRAC was reduced with the increase of doping after 15% FA doping. This is because the appropriate amount of FA substitution for GGBFS can fill the pores to promote hydration reactions, but when the FA doping is too high, FA exhibits particle stability but will reduce the hydration rate; in contrast, GGBFS has the best impermeability performance with the addition of 5% MgO chloride ions in a 15% FA admixture, and this is because MgO promotes the AAS hydration reaction, which fills the pores of concrete and thus increases the density of the structure [43,97].

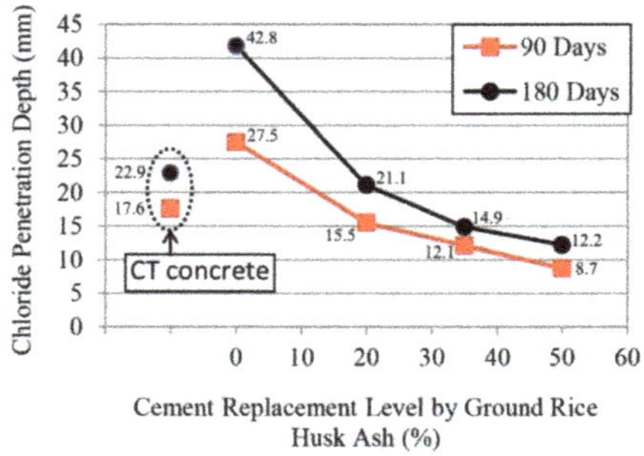


**Figure 6.** Relationship between the depth of chloride ion penetration in concrete and the level of rice husk ash replacement [96].

Surface coating treatment is one of the methods used to enhance the performance of RFA. The RFA was coated with fly ash, slag, and PVA (polyvinyl alcohol), and 25% of the natural sand was replaced with the coated RFA. It was found that the resistivity of the slag-coated RFA was maximal, regardless of the w/b, and increased with the thickness of the coating [25].

The addition of SP can improve the durability performance of concrete, and the addition of SP in RFAc can increase the density of the concrete, so the durability performance of RFAc can be improved. However, it was found that there was no significant difference in chloride ion penetration between RFAc with SP and those without [98].

The resistance of RFAc to chloride ion permeation was decreased with the increase of the RFA substitution rate. The treatment of RFA with $CO_2$ was detrimental to the chloride ion permeation resistance of RFAc, but the coating-treated RFA was positive against chloride ion permeation, and the addition of mineral admixtures enhanced the chloride ion permeation resistance of RFAc. It is impossible to avoid the reduction of the chloride ion permeation resistance of concrete by adding RFA, so we should consider upgrading the quality of RFA itself or adding mineral admixtures and SP to mitigate the reduction in its performance. Table 3 summarizes the effects of different factors on the resistance of RFAc to chloride ion penetration.

**Table 3.** The influence of different factors on the resistance of RFAc to chloride ion penetration.

| Factor | Change | Influence |
|---|---|---|
| RFA% | Increase | Negative |
| Thermal curing | - | Negative |
| $CO_2$ pretreatment RFA | - | Negative |
| Surface coating treatment RFA | - | Positive |
| Mineral admixture | - | Positive |

## 6. Acid Resistance

Industrial production causes a large amount of sulfur dioxide emissions, as well as an increasing number of nitrogen oxides from vehicle exhaust, which are the main components that contribute to the production of acid rain ($NO_x$) and sulfur dioxide ($SO_2$) [99]. The concrete structure of buildings exposed to acid rain for long periods of time is affected by the deterioration of concrete and the rusting of reinforcements. The deterioration of concrete by acid is the reaction of sulfuric acid with calcium hydroxide to produce gypsum, as shown in Equations (1)–(3), and gypsum formation expands concrete and accelerates corrosion. RFAc containing 100% RFA is less durable than natural sand concrete because of such properties as the internal capillarity of RFAc [100].

$$Ca(OH)_2 + H_2SO_4 \rightarrow CaSO_4 \cdot 2H_2O \quad (1)$$

$$CaO \cdot Al_2O_3 \cdot 12H_2O + 3(CaSO_4 \cdot 2H_2O) + 14H_2O \rightarrow CaO \cdot Al_2O_3 \cdot 3CaSO_4 \cdot 32H_2O \quad (2)$$

$$CaO \cdot SiO_2 \cdot 2H_2O + H_2SO_4 \rightarrow CaSO_4 + Si(OH)_4 + H_2O \quad (3)$$

The acid resistance of concrete depends mainly on its percentage reduction in mass and compressive strength in acid solutions. The reduction in mass loss and compressive strength in a 3% sulfuric acid solution was discussed by adding RFA with different substitution rates to the activated concrete, which was considered from two sources, one from normal concrete and the other from activated powder concrete. It was found that the mass loss and compressive strength of the activated powder concrete prepared from both sources of RFA subjected to sulfuric acid attack increased as the substitution rate of RFA increased, but both had good acid resistance [87]. This is mainly due to the reaction of sulfuric acid with calcium hydroxide through the porous activated powder concrete to produce gypsum, which expands calcium salts in the pores in aqueous solutions and finally leads to a decrease in the mass and compressive strength of the activated powder concrete [101]. The acid resistance of high-performance concrete (HPC) in sulfate solutions and sulfuric acid solutions was studied by adding 20% RCA and RFA, and it was found that HPC had good acid resistance in sulfate solutions, probably due to the addition of SF, but a severe loss of strength in sulfuric acid solutions [102]. In concentrated hydrochloric acid and concentrated sulfuric acid, the corresponding mass loss of plain concrete with added RFA increased with the soaking time and the substitution rate, and the mass loss of RFAc with a 30% substitution rate after 90 days was lower than that of the control, while the mass loss of RFAc with a 100% substitution rate was higher than that of the control. This was because the RFA with a 100% substitution rate was finer and contained more pores and old cement mortar [58]. Similar findings were found in another study on the effect of RFAs on the acid resistance of alkali-activated slag concrete (AASC), a new environmentally friendly cementitious material, where AASC with 0, 50, and 100% RFA replacement rates were exposed to a 3% sulfuric acid solution for 180 days, and it was found that the quality of AASC decreased as the replacement rate increased, but even after 180 days of exposure, the maximum mass loss of AASC was less than 20% (as shown in the figure), which was mainly due to the presence of GGBS in AASC [103] (Figure 7). The acid resistance of SSC prepared with coarse and fine recycled aggregates in 5% sulfate was better than the control. Meanwhile, the effect of natural volcanic ash (PZ) on the recycled SSC was investigated, and it was found that the addition of PZ improved the acid resistance of the recycled SSC, with 20% PZ dosing being optimal [104].

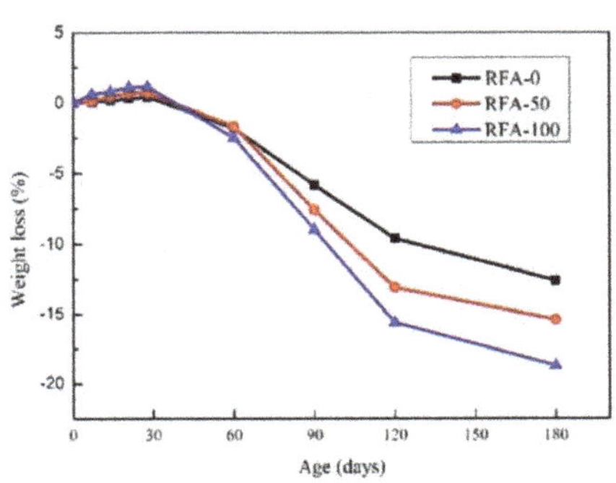


**Figure 7.** Weight loss of alkali–activated slag concrete [103].

Different types of concrete supplemented with RFA showed mass loss and compressive strength reduction in acid solutions. The more significant the mass loss and compressive strength reduction was, the more the replacement rate of RFA increased. The incorporation of natural volcanic ash into SSC incorporated with RFA improves the acid resistance. An appropriate RFA substitution rate should be considered for RFAc in acidic environments, and a certain amount of natural volcanic ash can be considered to improve the acid resistance.

## 7. Resistance to Freeze–Thaw Cycles

In the UK, freeze–thaw cycles have been identified as one of the most significant deleterious factors affecting the life of concrete structures [11,105,106]. The water inside the concrete under the effect of freeze–thaw cycles turns into micro-ice bodies, with its volume expanding by up to 9% [107]. When there is not enough space for volume expansion in concrete, hydraulic and tensile stresses are generated in the pores, which leads to further increases in pore size [108]. The increased pore space, in turn, draws in more water, which causes greater tensile stresses when it freezes again, eventually leading to concrete deterioration. This shows that pore space and water are closely related to freeze–thaw resistance. The freeze–thaw resistance of RAC is considered to be poor due to the presence of old mortar and micro cracks adhering to the RA surface, resulting in a high water absorption rate [21].

The freeze–thaw resistance of RFAc was investigated by considering three groups of RFAc: plain RFAc with a water–cement ratio of 0.53 (C); high-strength concrete (HC) without SP and with an aerator (AEA), with a water–cement ratio of 0.35; and high-strength concrete (HCAE) with SP added but without an AEA with a water–cement ratio of 0.35, at replacement rates of 0, 20, 50, and 100%. Three hundred freeze–thaw cycles were tested for the three groups of RFAc, and it was found that the mass loss of C was much higher than that of HC and HCAE (Figure 8). This result indicates that the freeze–thaw resistance of RFAc is not related to the type of aggregate but is determined by the water–cement ratio, and the addition of an AEA has little effect on the freeze–thaw resistance of RFAc, as the water–cement ratios of HC and HCAEA were the same but had similar mass loss. Therefore, reducing the water–ash ratio is an effective way to improve the freeze–thaw resistance [109]. In another study, it was found that the mass loss values of RFAc in under 300 freeze–thaw cycles decreased as the water–cement ratio decreased from 0.7 to 0.5 [110].

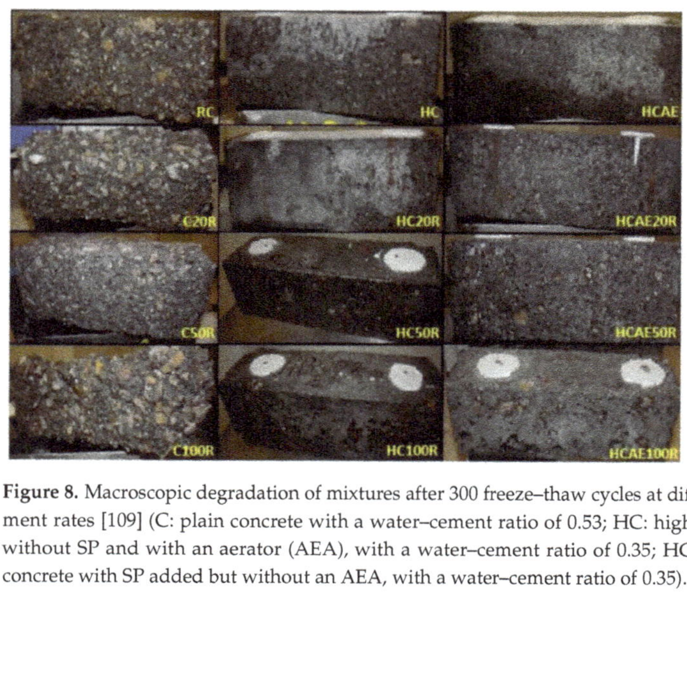


**Figure 8.** Macroscopic degradation of mixtures after 300 freeze–thaw cycles at different RFA replacement rates [109] (C: plain concrete with a water–cement ratio of 0.53; HC: high-strength concrete without SP and with an aerator (AEA), with a water–cement ratio of 0.35; HCAE: high-strength concrete with SP added but without an AEA, with a water–cement ratio of 0.35).

Internal curing is the internal water supply to concrete to achieve maintenance. Internal curing can reduce water consumption, improve the homogeneity of concrete, and reduce the cracking of concrete surfaces. Studies dedicated to internal curing found that internal curing may have an effect on the durability of concrete structures [111]. RFA has a much higher water absorption than natural aggregates and can be used as a source of internal curing. Applying RFA to the internal curing of concrete was found to be beneficial for properties such as cracking delay, freeze–thaw resistance, and the dynamic modulus of elasticity, and using 50% RFA was recommended for internal curing because the best results were achieved at that substitution rate [112]. The different moisture concentrations of RFA may also have an effect on internal curing, so RFA prepared at 0, 50, and 100% saturation were investigated for RFAc. The performance was investigated, and it was found that the best freeze–thaw resistance was achieved when RFAc was prepared at 50 and 100% saturation for internal conservation [110].

Different substitution rates of RFA also have an important effect on the freeze–thaw resistance of RFAc. By studying the performance of SSC with added RFA, it was found that the instability and the weak ITZ of SSC increased with the increase of the RFA substitution rate, and the compressive strength of 30% of the SSC specimens with added RFA after 28 and 56 days of freeze–thaw cycles decreased to less than 350 MPa [113].

In summary, the freeze–thaw resistance of RFAc is significantly affected by the substitution rate of RFA, but reducing the water–cement ratio can enhance that resistance. In addition, using RFA as a source for internal maintenance and using RFA with a suitable moisture content both have positive effects on this freeze–thaw resistance. RFA itself suffers from a high water absorption, but using it as a source for internal maintenance can be reasonably applied to enhance the resistance of RFAc to freeze–thaw cycles. Table 4 shows the influence of different factors on the freeze–thaw resistance of RFAc.

**Table 4.** The influence of different factors on the freeze–thaw resistance of RFAc.

| Factor | Change | Influence |
|---|---|---|
| RFA% | Increase | Negative |
| W/C ratio | Decrease | Positive |
| RFA as an internal conservation source | - | Positive |
| RFA moisture content: 50%/100% | - | Positive |

## 8. Conclusions

This paper focuses on the durability of RFAc. The efficiency of RFA with appropriate water content, pretreated RFA, concrete mixing methods, and mineral admixtures on the durability of RFAc is summarized. Some important points are highlighted here:

(1) The durability of RFAc is closely related to the quality of the RFA, which is mainly influenced by the content of the old mortar, specifically, the higher the content of the old mortar, the higher the porosity and water absorption. The average water absorption of natural sand is about 1.1%, while the average water absorption of RFA can reach about 8.4%.

(2) The impermeability, drying shrinkage, resistance to chloride ion penetration, carbonation, acid resistance, and freeze–thaw cycle resistance of RFAc decrease as the RFA replacement rate improves. For example, the drying shrinkage value at a 100% RFA replacement rate is twice that of normal concrete, and the depth of carbonation increases by approximately 110%.

(3) The impermeability, carbonation resistance, and freeze–thaw cycle resistance of RFAc decrease as the effective water–cement ratio increases. For example, when the water-to-ash ratio is reduced from 0.7 to 0.5, the mass loss of RFAc under freeze–thaw cycles is reduced. Meanwhile, the drying shrinkage is less affected by the change in the water–cement ratio.

(4) RFA with appropriate moisture benefits the permeation resistance and freeze–thaw cycle resistance of RFAc.
(5) The treatment of RFA with $CO_2$ has a positive effect on the anticarbonation performance of RFAc. However, it is unfavorable to the resistance to chloride ion penetration.
(6) Advanced mixing methods focused on improving the ITZ have been reported, and they can partially compensate for the poor performance of recycled concrete fine aggregates. Thus, the relative durability of RFAc can be optimized.
(7) Existing studies report an optimized triple mixing method, where the SP is added together with additional gelling material. This addition promotes the dispersion of that gelling material, which, in turn, leads to a higher content of additional gel material in the ITZ, a better volcanic ash effect at a later stage, and a reduction in the porosity of the RFAc, thus allowing for an impermeability rating of 8 for the RAC using this method. In addition, because a water-reducing agent is added early on, together with the additional gelling material in this method, a thin layer of additional gelling material can be formed, even when the mixture is insufficient.

## 9. Outlook

The replacement rate of RFA has a serious impact on the durability of concrete. Therefore, when the RFA replacement rate is 100%, 10% FA can be added to improve the impermeability. In terms of drying shrinkage values performs, the best RFA replacement rate of the concrete is 30%; as for how the resistance to chloride penetration performs, the best RFA replacement rate is 25%. The durability of RFAc is closely related to the quality of RFA. In addition to the methods mentioned in this paper to improve the quality of RFA, there are also thermal–chemical, thermal–physical, and microbial precipitation methods; thus, the durability of concrete prepared from RFA by different treatment methods is also worth studying in future research. In addition, other work will be focused on the properties of acid resistance and freeze–thaw cycle resistance. The performance of concrete in harsh environments should also be considered.

**Author Contributions:** Conceptualization, validation, and project administration: C.B.; writing (original draft preparation) and formal analysis: L.L.; formal analysis and data curation: X.L.; writing (review and editing) and visualization: D.Z.; investigation and methodology: Y.O.; writing (review and editing) and methodology: X.C.; software and funding acquisition: Y.S.; methodology and supervision: L.Y.; resources and funding acquisition: Q.W. All authors have read and agreed to the published version of the manuscript.

**Funding:** This research was funded by the Project of Natural Science Foundation of Chongqing Municipality (grant number: cstc2021jcyj-msxmX0444); the Project of Chongqing Construction Science and Technology Plan (grant number: 2021 No. 1-6); the Project of Chongqing Bureau of Human Resources and Social Security (grant number: cx2020008); the Open Fund of Chongqing Key Laboratory of Energy Engineering Mechanics & Disaster Prevention and Reduction (grant number: EEMDPM2021103); the Cooperation Project of Ministry of Education "Chunhui Planning" (grant number: Z2015147); the Fund of State Key Laboratory of Bridge Engineering Structural Dynamics; Key Laboratory of Bridge Earthquake Resistance Technology, Ministry of Communications, PRC; the second author acknowledges financial support from the Postgraduate Science and Technology Innovation Project in Chongqing University of Science and Technology (No. YKJCX2120601).

**Acknowledgments:** Thanks to all the teachers and classmates who worked hard during the experiment.

**Conflicts of Interest:** The authors declare that they have no known competing financial interests or personal relationships that could have influenced the work reported in this paper.

## References

1. Hu, M.; Yang, M. Analysis of China's construction waste resourcing policy based on policy tools. *Constr. Econ.* **2019**, *40*, 22–26.
2. Chen, X.Y.; Cheng, Z.Y.; Zhan, X.; Wu, Q. Experimental study and numerical simulation of mechanical properties of recycled concrete with nano-$SiO_2$-rubber powder. *Mater. Guide* **2021**, *35*, 23235–23240, 23245.
3. Ma, M.; Tam, V.W.; Le, K.N.; Li, W. Challenges in current construction and demolition waste recycling: A China study. *Waste Manag.* **2020**, *118*, 610–625. [CrossRef] [PubMed]

4. Liu, W.; Wang, Z.; Gao, S. A study on foreign legislation on resource utilization of construction waste. *Law Soc.* **2018**, *26*, 206–208.
5. Mohammed, M.S.; ElKady, H.; Abdel-Gawwad, H.A. Utilization of construction and demolition waste and synthetic aggregates. *J. Build. Eng.* **2021**, *43*, 103207. [CrossRef]
6. Pu, Y.; Li, L.; Wang, Q.; Shi, X.; Luan, C.; Zhang, G.; Fu, L.; Abomohra, A.E.F. Accelerated carbonation technology for enhanced treatment of recycled concrete aggregates: A state-of-the-art review. *Constr. Build. Mater.* **2021**, *282*, 122671. [CrossRef]
7. Mariaková, D.; Anna Mocová, K.; Fořtová, K.; Ryparová, P.; Pešta, J.; Pavlů, T. Ecotoxicity and Essential Properties of Fine-Recycled Aggregate. *Materials* **2021**, *14*, 463. [CrossRef]
8. Feng, C.; Cui, B.; Ge, H.; Huang, Y.; Zhang, W.; Zhu, J. Reinforcement of recycled aggregate by Microbial-Induced mineralization and deposition of calcium carbonate—Influencing factors, mechanism and effect of reinforcement. *Crystals* **2021**, *11*, 887. [CrossRef]
9. Fang, X.; Zhan, B.; Poon, C.S. Enhancement of recycled aggregates and concrete by combined treatment of spraying $Ca^{2+}$ rich wastewater and flow-through carbonation. *Constr. Build. Mater.* **2021**, *277*, 122202. [CrossRef]
10. Gebremariam, A.T.; Vahidi, A.; Di Maio, F.; Moreno-Juez, J.; Vegas-Ramiro, I.; Łagosz, A.; Mróz, R.; Rem, P. Comprehensive study on the most sustainable concrete design made of recycled concrete, glass and mineral wool from C&D wastes. *Constr. Build. Mater.* **2021**, *273*, 121697.
11. Farzampour, A. Temperature and humidity effects on behavior of grouts. *Adv. Concr. Constr.* **2017**, *5*, 659.
12. Chalangaran, N.; Farzampour, A.; Paslar, N. Nano silica and metakaolin effects on the behavior of concrete containing rubber crumbs. *CivilEng* **2020**, *1*, 17. [CrossRef]
13. Mansouri, I.; Shahheidari, F.S.; Hashemi, S.M.A.; Farzampour, A. Investigation of steel fiber effects on concrete abrasion resistance. *Adv. Concr. Constr.* **2020**, *9*, 367–374.
14. Pacheco, J.; De Brito, J. Recycled Aggregates Produced from Construction and Demolition Waste for Structural Concrete: Constituents, Properties and Production. *Materials* **2021**, *14*, 5748. [CrossRef]
15. Reis, G.S.D.; Quattrone, M.; Ambrós, W.M.; Grigore Cazacliu, B.; Hoffmann Sampaio, C. Current Applications of Recycled Aggregates from Construction and Demolition: A Review. *Materials* **2021**, *14*, 1700. [CrossRef]
16. Zhu, Z.; Gu, S.E.; Tang, Z.; Song, L. Experimental Study on Recycling of Waste Concrete Based on Geopolymer Technology. In *The International Congress on Environmental Geotechnics*; Springer: Singapore, 2018; pp. 336–343.
17. van Deventer, J.S.; Provis, J.L.; Duxson, P.; Brice, D.G. Chemical research and climate change as drivers in the commercial adoption of alkali activated materials. *Waste Biomass Valorization* **2010**, *1*, 145–155. [CrossRef]
18. Davidovits, J. Why alkali-activated materials (AAM) are not geopolymers. *Publ. Tech. Pap.* **2018**. [CrossRef]
19. Shi, C.; Cao, Z.; Xie, Z. Research progress on the mechanical properties of recycled concrete. *Mater. Guide* **2016**, *30*, 96–103, 126.
20. Chindaprasirt, P.; Rukzon, S. Strength and chloride resistance of the blended Portland cement mortar containing rice husk ash and ground river sand. *Mater. Struct.* **2015**, *48*, 3771–3777. [CrossRef]
21. Guo, H.; Shi, C.; Guan, X.; Zhu, J.; Ding, Y.; Ling, T.C.; Zhang, H.; Wang, Y. Durability of recycled aggregate concrete—A review. *Cem. Concr. Compos.* **2018**, *89*, 251–259. [CrossRef]
22. Estanqueiro, B.; Dinis Silvestre, J.; de Brito, J.; Duarte Pinheiro, M. Environmental life cycle assessment of coarse natural and recycled aggregates for concrete. *Eur. J. Environ. Civ. Eng.* **2018**, *22*, 429–449. [CrossRef]
23. Limbachiya, M.C.; Leelawat, T.; Dhir, R.K. RCA concrete: A study of properties in the fresh state, strength development and durability. In *Sustainable Construction: Use of Recycled Concrete Aggregate: Proceedings of the International Symposium Organised by the Concrete Technology Unit, University of Dundee and Held at the Department of Trade and Industry Conference Centre, London, UK, 11–12 November 1998*; Thomas Telford Publishing: London, UK, 1998; pp. 227–237.
24. Kurda, R.; de Brito, J.; Silvestre, J.D. Influence of recycled aggregates and high contents of fly ash on concrete fresh properties. *Cem. Concr. Compos.* **2017**, *84*, 198–213. [CrossRef]
25. Zhang, H.; Ji, T.; Zeng, X.; Yang, Z.; Lin, X.; Liang, Y. Mechanical behavior of ultra-high performance concrete (UHPC) using recycled fine aggregate cured under different conditions and the mechanism based on integrated microstructural parameters. *Constr. Build. Mater.* **2018**, *192*, 489–507. [CrossRef]
26. Pandurangan, K.; Dayanithy, A.; Prakash, S.O. Influence of treatment methods on the bond strength of recycled aggregate concrete. *Constr. Build. Mater.* **2016**, *120*, 212–221. [CrossRef]
27. Juan MS, D.; Gutiérrez, P.A. Study on the influence of attached mortar content on the properties of recycled concrete aggregate. *Constr. Build. Mater.* **2008**, *23*, 872–877. [CrossRef]
28. Ismail, S.; Ramli, M. Engineering properties of treated recycled concrete aggregate (RCA) for structural applications. *Constr. Build. Mater.* **2013**, *44*, 464–476. [CrossRef]
29. Katz, A. Treatments for the improvement of recycled aggregate. *J. Mater. Civ. Eng.* **2004**, *16*, 597–603. [CrossRef]
30. Akbarnezhad, A.; Ong, K.C.G.; Zhang, M.H.; Tam, C.T.; Foo, T.W.J. Microwave-assisted beneficiation of recycled concrete aggregates. *Constr. Build. Mater.* **2011**, *25*, 3469–3479. [CrossRef]
31. Yang, Y.; Liang, S.; Mo, H.; Chen, Y.W. Treatment and Recycling Method of Construction Waste. CN101099974, 9 January 2008.
32. Yang, Y.; Su, Y.; Li, Z.; Chen, C.P.; Chen, B.X.; Guo, W.E.; Wang, H.C. Pilot-scale technology research on fully recycled fine aggregate made from waste concrete. *Concrete* **2020**, *10*, 80–84.
33. Barbucci, A.; Delucchi, M.; Cerisola, G. Organic coatings for concrete protection: Liquid water and water vapour permeabilities. *Prog. Org. Coat.* **1997**, *30*, 293–297. [CrossRef]

34. Wang, Y.; Chen, G.; Wan, B.; Lin, H.; Zhang, J. Behavior of innovative circular ice filled steel tubular stub columns under axial compression. *Constr. Build. Mater.* **2018**, *171*, 680–689. [CrossRef]
35. Brenna, A.; Bolzoni, F.; Beretta, S.; Ormellese, M. Long-term chloride-induced corrosion monitoring of reinforced concrete coated with commercial polymer-modified mortar and polymeric coatings. *Constr. Build. Mater.* **2013**, *48*, 734–744. [CrossRef]
36. Zhang, Y.; Kong, X. Influences of superplasticizer, polymer latexes and asphalt emulsions on the pore structure and impermeability of hardened cementitious materials. *Constr. Build. Mater.* **2014**, *53*, 392–402. [CrossRef]
37. Güneyisi, E.; Gesoğlu, M.; Booya, E.; Mermerdaş, K. Strength and permeability properties of self-compacting concrete with cold bonded fly ash lightweight aggregate. *Constr. Build. Mater.* **2015**, *74*, 17–24. [CrossRef]
38. Duan, P.; Shui, Z.; Chen, W.; Shen, C. Efficiency of mineral admixtures in concrete: Microstructure, compressive strength and stability of hydrate phases. *Appl. Clay Sci.* **2013**, *83*, 115–121. [CrossRef]
39. Sun, Y.; Li, D.; Xu, Y.; Huang, H. Experimental study on mechanical properties and durability of recycled fine aggregate concrete in actual production. *Build. Struct.* **2019**, *49*, 81–84.
40. Mardani-Aghabaglou, A.; Tuyan, M.; Ramyar, K. Mechanical and durability performance of concrete incorporating fine recycled concrete and glass aggregates. *Mater. Struct.* **2015**, *48*, 2629–2640. [CrossRef]
41. Sasanipour, H.; Aslani, F. Durability properties evaluation of self-compacting concrete prepared with waste fine and coarse recycled concrete aggregates. *Constr. Build. Mater.* **2020**, *236*, 117540. [CrossRef]
42. Kapoor, K.; Singh, S.; Singh, B. Permeability of self-compacting concrete made with recycled concrete aggregates and metakaolin. *J. Sustain. Cem.-Based Mater.* **2017**, *6*, 293–313. [CrossRef]
43. Mostofinejad, D.; Hosseini, S.M.; Nosouhian, F.; Ozbakkaloglu, T.; Tehrani, B.N. Durability of concrete containing recycled concrete coarse and fine aggregates and milled waste glass in magnesium sulfate environment. *J. Build. Eng.* **2020**, *29*, 101182. [CrossRef]
44. Gesoglu, M.; Güneyisi, E.; Öz, H.Ö.; Yasemin, M.T.; Taha3, I. Durability and shrinkage characteristics of self-compacting concretes containing recycled coarse and/or fine aggregates. *Adv. Mater. Sci. Eng.* **2015**, *2015*, 278296. [CrossRef]
45. Hemalatha, T.; Ramaswamy, A. A review on fly ash characteristics–Towards promoting high volume utilization in developing sustainable concrete. *J. Clean. Prod.* **2017**, *147*, 546–559. [CrossRef]
46. Chindaprasirt, P.; Jaturapitakkul, C.; Sinsiri, T. Effect of fly ash fineness on microstructure of blended cement paste. *Constr. Build. Mater.* **2007**, *21*, 1534–1541. [CrossRef]
47. Zhang, M.-H.; Li, L.; Paramasivam, P. Shrinkage of high-strength lightweight aggregate concrete exposed to dry environment. *Mater. J.* **2005**, *102*, 86–92.
48. Tam, V.W. Comparing the implementation of concrete recycling in the Australian and Japanese construction industries. *J. Clean. Prod.* **2009**, *17*, 688–702. [CrossRef]
49. Kong, D.; Lei, T.; Zheng, J.; Ma, C.; Jiang, J.; Jiang, J. Effect and mechanism of surface-coating pozzalanics materials around aggregate on properties and ITZ microstructure of recycled aggregate concrete. *Constr. Build. Mater.* **2010**, *24*, 701–708. [CrossRef]
50. Zhang, W.; Wang, S.; Zhao, P.; Lu, L.; Cheng, X. Effect of the optimized triple mixing method on the ITZ microstructure and performance of recycled aggregate concrete. *Constr. Build. Mater.* **2019**, *203*, 601–607. [CrossRef]
51. Mefteh, H.; Kebaïli, O.; Oucief, H.; Berredjem, L.; Arabi, N. Influence of moisture conditioning of recycled aggregates on the properties of fresh and hardened concrete. *J. Clean. Prod.* **2013**, *54*, 282–288. [CrossRef]
52. Zhuang, Y.; Liang, Y.; Nabizadeh, A.; Ng, Z.; Ji, T. Influence of the moisture state of recycled fine aggregate on the impermeability of concrete. *Mater. Test.* **2019**, *61*, 991–998. [CrossRef]
53. Videla, C.; Aguilar, C. An updated look at drying shrinkage of Portland and blended Portland cement concretes. *Mag. Concr. Res.* **2006**, *58*, 459–476. [CrossRef]
54. Mokarem, D.W.; Weyers, R.E.; Lane, D.S. Development of a shrinkage performance specifications and prediction model analysis for supplemental cementitious material concrete mixtures. *Cem. Concr. Res.* **2005**, *35*, 918–925. [CrossRef]
55. De Domenico, D.; Faleschini, F.; Pellegrino, C.; Ricciardi, G. Structural behavior of RC beams containing EAF slag as recycled aggregate: Numerical versus experimental results. *Constr. Build. Mater.* **2018**, *171*, 321–337. [CrossRef]
56. Arribas, I.; Santamaria, A.; Ruiz, E.; Ortega-Lopez, V.; Manso, J.M. Electric arc furnace slag and its use in hydraulic concrete. *Constr. Build. Mater.* **2015**, *90*, 68–79. [CrossRef]
57. Jang, S.-J.; Yun, H.-D. Mechanical properties of ready-mixed concrete incorporating fine recycled aggregate. *Mag. Concr. Res.* **2015**, *67*, 621–632. [CrossRef]
58. Kirthika, S.; Singh, S. Durability studies on recycled fine aggregate concrete. *Constr. Build. Mater.* **2020**, *250*, 118850. [CrossRef]
59. Guo, Z.; Chen, C.; Lehman, D.E.; Xiao, W.; Zheng, S.; Fan, B. Mechanical and durability behaviours of concrete made with recycled coarse and fine aggregates. *Eur. J. Environ. Civ. Eng.* **2020**, *24*, 171–189. [CrossRef]
60. Pedro, D.D.; De Brito, J.; Evangelista, L. Structural concrete with simultaneous incorporation of fine and coarse recycled concrete aggregates: Mechanical, durability and long-term properties. *Constr. Build. Mater.* **2017**, *154*, 294–309. [CrossRef]
61. Zhang, H.; Wang, Y.; Lehman, D.E.; Geng, Y.; Kuder, K. Time-dependent drying shrinkage model for concrete with coarse and fine recycled aggregate. *Cem. Concr. Compos.* **2020**, *105*, 103426. [CrossRef]
62. Zega, C.J.; Di Maio, Á.A. Use of recycled fine aggregate in concretes with durable requirements. *Waste Manag.* **2011**, *31*, 2336–2340. [CrossRef]

63. Pereira, P.; Evangelista, L.; Brito, J.D. The effect of superplasticisers on the workability and compressive strength of concrete made with fine recycled concrete aggregates. *Constr. Build. Mater.* **2011**, *28*, 722–729. [CrossRef]
64. Pereira, P.; Evangelista, L.; De Brito, J. The effect of superplasticizers on the mechanical performance of concrete made with fine recycled concrete aggregates. *Cem. Concr. Compos.* **2012**, *34*, 1044–1052. [CrossRef]
65. Barbudo, A.; De Brito, J.; Evangelista, L.; Bravo, M.; Agrela, F. Influence of water-reducing admixtures on the mechanical performance of recycled concrete. *J. Clean. Prod.* **2013**, *59*, 93–98. [CrossRef]
66. Cartuxo, F.; De Brito, J.; Evangelista, L.; Jiménez, J.R.; Ledesma, E.F. Rheological behaviour of concrete made with fine recycled concrete aggregates—Influence of the superplasticizer. *Constr. Build. Mater.* **2015**, *89*, 36–47. [CrossRef]
67. Khan, M.I.; Siddique, R. Utilization of silica fume in concrete: Review of durability properties. *Resour. Conserv. Recycl.* **2011**, *57*, 30–35. [CrossRef]
68. Tam, V.W.; Gao, X.; Tam, C.M. Microstructural analysis of recycled aggregate concrete produced from two-stage mixing approach. *Cem. Concr. Res.* **2005**, *35*, 1195–1203. [CrossRef]
69. Sivamani, J.; Renganathan, N.T. Effect of fine recycled aggregate on the strength and durability properties of concrete modified through two-stage mixing approach. *Environ. Sci. Pollut. Res.* **2021**. [CrossRef]
70. Zhang, J.; Wang, Y.; Xu, M.; Zhao, Q. Effect of carbon dioxide corrosion on compressive strength of oilwell cement. *J. Chin. Ceram. Soc.* **2009**, *37*, 642–647.
71. Geng, J.; Sun, J. Characteristics of the carbonation resistance of recycled fine aggregate concrete. *Constr. Build. Mater.* **2013**, *49*, 814–820. [CrossRef]
72. Kaliyavaradhan, S.K.; Ling, T.-C. Potential of $CO_2$ sequestration through construction and demolition (C&D) waste—An overview. *J. CO2 Util.* **2017**, *20*, 234–242.
73. Liang, C.; Lu, N.; Ma, H.; Ma, Z.; Duan, Z. Carbonation behavior of recycled concrete with $CO_2$ -curing recycled aggregate under various environments. *J. CO2 Util.* **2020**, *39*, 101185. [CrossRef]
74. Evangelista, L.; De Brito, J. Durability performance of concrete made with fine recycled concrete aggregates. *Cem. Concr. Compos.* **2010**, *32*, 9–14. [CrossRef]
75. Zhu, P.; Chen, K.; Hu, K. Carbonation Behavior of Repeated Recycled Fine Aggregate Concrete under Bending Load. *KSCE J. Civ. Eng.* **2019**, *23*, 729–736. [CrossRef]
76. Singh, N.; Singh, S. Validation of carbonation behavior of self compacting concrete made with recycled aggregates using microstructural and crystallization investigations. *Eur. J. Environ. Civ. Eng.* **2020**, *24*, 2187–2210. [CrossRef]
77. Evangelista, L.; De Brito, J. Durability of crushed fine recycled aggregate concrete assessed by permeability-related properties. *Mag. Concr. Res.* **2019**, *71*, 1142–1150. [CrossRef]
78. Sim, J.; Park, C. Compressive strength and resistance to chloride ion penetration and carbonation of recycled aggregate concrete with varying amount of fly ash and fine recycled aggregate. *Waste Manag.* **2011**, *31*, 2352–2360. [CrossRef]
79. Cartuxo, F.; De Brito, J.; Evangelista, L.; Jiménez, J.R.; Enrique, F.L. Increased durability of concrete made with fine recycled concrete aggregates using superplasticizers. *Materials* **2016**, *9*, 98. [CrossRef]
80. Angst, U.; Elsener, B.; Larsen, C.K.; Vennesland, Ø. Critical chloride content in reinforced concrete—A review. *Cem. Concr. Res.* **2009**, *39*, 1122–1138. [CrossRef]
81. Shi, X.; Xie, N.; Fortune, K.; Gong, J. Durability of steel reinforced concrete in chloride environments: An overview. *Constr. Build. Mater.* **2012**, *30*, 125–138. [CrossRef]
82. Butler, W.B.; Janssen, D.J.; Schell, H.C.; Cabrera, J.G. *Guide to Durable Concrete*; ACI: Montreal, QC, Canada, 2001.
83. Bazant, Z.P. Physical model for steel corrosion in concrete sea structures-theory. *ASCE J. Struct. Div.* **1979**, *105*, 1137–1153. [CrossRef]
84. Zhenhua, D.; Shaodan, H.; Zhisheng, P.; Shanshan, J.; Jianzhuang, X. Rheology of recycled fine aggregate concrete and its influence on strength and durability. *J. Build. Struct.* **2020**, *S2*, 420–426.
85. Velay-Lizancos, M.; Martinez-Lage, I.; Azenha, M.; Granja, J.; Vazquez-Burgo, P. Concrete with fine and coarse recycled aggregates: E-modulus evolution, compressive strength and non-destructive testing at early ages. *Constr. Build. Mater.* **2018**, *193*, 323–331. [CrossRef]
86. Fan, C.-C.; Huang, R.; Hwang, H.; Chao, S.-J. Properties of concrete incorporating fine recycled aggregates from crushed concrete wastes. *Constr. Build. Mater.* **2016**, *112*, 708–715. [CrossRef]
87. Salahuddin, H.; Qureshi, L.A.; Nawaz, A.; Razad, S.S. Effect of recycled fine aggregates on performance of Reactive Powder Concrete. *Constr. Build. Mater.* **2020**, *243*, 118223. [CrossRef]
88. Chinzorigt, G.; Lim, M.K.; Yu, M.; Lee, H.; Enkbold, O.; Choi, D. Strength, shrinkage and creep and durability aspects of concrete including $CO_2$ treated recycled fine aggregate. *Cem. Concr. Res.* **2020**, *136*, 106062. [CrossRef]
89. Kapoor, K.; Singh, S.P.; Singh, B. Evaluating the durability properties of self compacting concrete made with coarse and fine recycled concrete aggregates. *Eur. J. Environ. Civ. Eng.* **2020**, *24*, 2383–2399. [CrossRef]
90. Siddique, R.; Klaus, J. Influence of metakaolin on the properties of mortar and concrete: A review. *Appl. Clay Sci.* **2008**, *43*, 392–400. [CrossRef]
91. Hossain, M.; Karim, M.; Hasan, M.; Hossain, M.K.; Zain, M.F.M. Durability of mortar and concrete made up of pozzolans as a partial replacement of cement: A review. *Constr. Build. Mater.* **2016**, *116*, 128–140. [CrossRef]

92. Givi, A.N.; Rashid, S.A.; Aziz FN, A.; Salleh MA, M. Contribution of rice husk ash to the properties of mortar and concrete: A review. *J. Am. Sci.* **2010**, *6*, 157–165.
93. Bie, R.-S.; Song, X.-F.; Liu, Q.-Q.; Ji, X.-Y.; Chen, P. Studies on effects of burning conditions and rice husk ash (RHA) blending amount on the mechanical behavior of cement. *Cem. Concr. Compos.* **2015**, *55*, 162–168. [CrossRef]
94. He, Z.-H.; Li, L.-Y.; Du, S.-G. Creep analysis of concrete containing rice husk ash. *Cem. Concr. Compos.* **2017**, *80*, 190–199. [CrossRef]
95. Msinjili, N.S.; Schmidt, W.; Rogge, A.; Kühne, H.C. Performance of rice husk ash blended cementitious systems with added superplasticizers. *Cem. Concr. Compos.* **2017**, *100*, 202–208. [CrossRef]
96. Rattanachu, P.; Toolkasikorn, P.; Tangchirapat, W.; Chindaprasirt, P.; Jaturapitakkul, C. Performance of recycled aggregate concrete with rice husk ash as cement binder. *Cem. Concr. Compos.* **2020**, *108*, 103533. [CrossRef]
97. Hwang, C.-L.; Vo, D.-H.; Tran, V.-A.; Yehualaw, M.-D. Effect of high MgO content on the performance of alkali-activated fine slag under water and air curing conditions. *Constr. Build. Mater.* **2018**, *186*, 503–513. [CrossRef]
98. Bravo, M.; De Brito, J.; Evangelista, L.; Pacheco, J. Durability and shrinkage of concrete with CDW as recycled aggregates: Benefits from superplasticizer's incorporation and influence of CDW composition. *Constr. Build. Mater.* **2018**, *168*, 818–830. [CrossRef]
99. Debnath, B.; Li, M.; Liu, S.; Pan, T.; Ma, C.; Qiu, D. Melatonin-mediate acid rain stress tolerance mechanism through alteration of transcriptional factors and secondary metabolites gene expression in tomato. *Ecotoxicol. Environ. Saf.* **2020**, *200*, 110720. [CrossRef]
100. Berredjem, L.; Arabi, N.; Molez, L. Mechanical and durability properties of concrete based on recycled coarse and fine aggregates produced from demolished concrete. *Constr. Build. Mater.* **2020**, *246*, 118421. [CrossRef]
101. Vigneshwari, M.; Arunachalam, K.; Angayarkanni, A. Replacement of silica fume with thermally treated rice husk ash in Reactive Powder Concrete. *J. Clean. Prod.* **2018**, *188*, 264–277. [CrossRef]
102. Kumar, B.V.; Ananthan, H.; Balaji, K. Experimental studies on utilization of recycled coarse and fine aggregates in high performance concrete mixes. *Alex. Eng. J.* **2018**, *57*, 1749–1759. [CrossRef]
103. Huang, J.; Zou, C.; Sun, D.; Yang, B.; Yan, J. Effect of recycled fine aggregates on alkali-activated slag concrete properties. *Structures* **2021**, *30*, 89–99. [CrossRef]
104. Omrane, M.; Kenai, S.; Kadri, E.-H.; Aït-Mokhtar, M. Performance and durability of self compacting concrete using recycled concrete aggregates and natural pozzolan. *J. Clean. Prod.* **2017**, *165*, 415–430. [CrossRef]
105. Richardson, A.; Coventry, K.; Bacon, J. Freeze/thaw durability of concrete with recycled demolition aggregate compared to virgin aggregate concrete. *J. Clean. Prod.* **2011**, *19*, 272–277. [CrossRef]
106. Farzampour, A. Compressive behavior of concrete under environmental effects. In *Compressive Strength of Concrete*; IntechOpen: London, UK, 2019.
107. Powers, T.C.; Willis, T. The air requirement of frost resistant concrete. *Highw. Res. Board Proc.* **1950**, *29*, 184–211.
108. Bager, D.H. Qualitative description of the micro ice body freeze–thaw damage mechanism in concrete. *Free. -Thaw Test. Concr.* **2010**, *1*, 9–36.
109. Bogas, J.A.; De Brito, J.; Ramos, D. Freeze–thaw resistance of concrete produced with fine recycled concrete aggregates. *J. Clean. Prod.* **2016**, *115*, 294–306. [CrossRef]
110. Yildirim, S.T.; Meyer, C.; Herfellner, S. Effects of internal curing on the strength, drying shrinkage and freeze–thaw resistance of concrete containing recycled concrete aggregates. *Constr. Build. Mater.* **2015**, *91*, 288–296. [CrossRef]
111. Bentz, D.P.; Weiss, W.J. *Internal Curing: A 2010 State-of-the-Art Review*; US Department of Commerce, National Institute of Standards and Technology: Gaithersburg, MD, USA, 2011.
112. El-Hawary, M.; Al-Sulily, A. Internal curing of recycled aggregates concrete. *J. Clean. Prod.* **2020**, *275*, 122911. [CrossRef]
113. Ghafari, S.; Moghadas Nejad, F.; Corbu, O. Durability and Mechanical Properties of Self-Compacting Concrete Incorporating Recovered Filler from Hot Mix Asphalt Plants and Recycled Fine Aggregate. *Key Eng. Mater.* **2021**, *894*, 95–101. [CrossRef]

*materials*

*Article*

# Uniaxial Compressive Stress–Strain Model of Jujube Nucleus Concrete following Exposure to Elevated Temperatures

**Jieqi Li [1], Mingming Jia [2], Shan Gao [2,*] and Jian Yuan [3]**

[1] School of Civil Engineering, Xijing University, Xi'an 710123, China
[2] Key Lab of Structures Dynamic Behavior and Control of the Ministry of Education, Harbin Institute of Technology, Harbin 150090, China
[3] Academy of Combat Support, Rocket Force University of Engineering, Xi'an 710025, China
* Correspondence: gaoshan@hit.edu.cn

**Abstract:** Aiming to provide a solution for natural resource consumption and agricultural waste pollution, jujube nucleus is utilized as a substitute for coarse aggregate in the preparation of lightweight aggregate concrete. The effect of the jujube nucleus (JN) replacement ratio and the elevated temperature on the uniaxial compressive stress–strain curves of jujube nucleus concrete (JNC) are experimentally studied. The results show that the failure of the JNC prisms became more serious with the increase in the JN replacement ratio. The linear proportion in ascending branch and the descending slope of the stress–strain curves for JNC increased gradually with the increase in the JN replacement ratio and elevated temperature, which is probably owing to the higher porosity and lower stiffness of the jujube nucleus, compared to natural aggregate. Moreover, as the JN replacement ratio and the elevated temperature increase, the peak stress and elastic modulus in the stress–strain curves of JNC decrease gradually, whilst an increase in the peak strain shows up, which is possibly due to the growth of hydrate calcium silicate and calcium hydroxide hampered by sucrose molecules. Based on the test results, a series of theoretical formulas are proposed to predict the compressive performance of JNC. A material constitutive model is developed for describing the stress–strain relationship of JNC by considering the JN replacement ratio and elevated temperature.

**Keywords:** agricultural waste; coarse aggregate; elevated temperature; modified concrete; stress–strain relationships

check for updates

**Citation:** Li, J.; Jia, M.; Gao, S.; Yuan, J. Uniaxial Compressive Stress–Strain Model of Jujube Nucleus Concrete following Exposure to Elevated Temperatures. *Materials* **2023**, *16*, 1037. https://doi.org/10.3390/ma16031037

Academic Editor: Carlos Leiva

Received: 15 December 2022
Revised: 11 January 2023
Accepted: 16 January 2023
Published: 24 January 2023



## 1. Introduction

Statistics show that, over the past few decades, the annual consumption of concrete for the construction industry has been up to 8 billion tons globally [1], of which China accounts for approximately 25% [2]. Accordingly, natural aggregates considered as a non-renewable resource tend to deplete as a result of over-exploitation of minerals [3]. It can be seen that gravel and sand is mined beyond 1.1 to 1.4 billion $m^3$ in China each year [4], and they make up 70% of concrete production [5]. In addition, large amounts of waste materials are dumped and burned as a result of rapid urbanization, industrialization, and population growth, which is a severe burden on waste management and environmental pollution [6]. As a result, the feasible study of using agricultural by-products as a substitute for coarse aggregate in concrete production has been proposed for alleviating the pressure from ever-increasing natural resource consumption and agricultural waste pollution [7].

In China, jujube nucleus, considered as one of the major agricultural by-products, would be discharged by more than 3 billion tons every year [8]. How to utilize these waste jujube nuclei has become a serious topic in the agricultural industry. Although Xie et al. [9] and Tabrez et al. [10] indicated that jujube nuclei waste can be carbonized to prepare the activated carbon adsorbent in pursuit of adsorbing pigments and their impurities in industrial wastewater, whilst the carbonization process for jujube nuclei has high energy consumption. In contrast, the jujube nucleus has the characteristic of low density and a reasonable gradation feature, which offers the possibility to replace natural coarse aggregates to prepare agricultural waste concrete. Jujube nucleus as coarse aggregate in structural concrete not only avoids the secondary treatment of the waste, but also reduces the self-weight of structural concrete [11].

The mechanical properties of agricultural waste concrete have been investigated for decades. However, study on the stress–strain relationship of agricultural waste concrete under uniaxial compressive loading is rarely involved. The effect of coconut shell (CS) as an alternative coarse aggregate for concrete materials was studied by Tomas and Ganiron [12]. The results showed that as the CS replacement ratio increases, the compressive strength of coconut shell concrete decreases gradually. Gunasekaran et al. [13] pointed out that the utilization of coconut shells as coarse aggregate in agricultural waste concrete reduces the bond between cement and coarse aggregate, which causes the compressive strength degradation of coconut shell concrete. Alengaram et al. [14] investigated the application of palm shell (OPS) as a substitute for coarse aggregate in concrete production. The results showed that palm shell concrete has comparable mechanical properties with ordinary concrete when the palm shell replacement ratio is low. However, the compressive strength and splitting tensile strength of palm shell concrete decrease rapidly as the palm shell replacement ratio increases. Khan et al. [15] found that palm shells are less bonded to cement compared with natural aggregate, which is the reason for the strength degradation of palm shell concrete. Muthusamy et al. [16] concluded that the mechanical properties of rubber concrete gradually decrease as the rubber replacement ratio increases, because the rubber and the cement paste does not cover the aggregates uniformly. Memon et al. [17] investigated the compressive performance of agricultural waste concrete using corncob ash as a substitute for fine aggregate. The results showed that the compressive strength degradation of the concrete may be related to the high initial free water content and the inherently weaker mechanical properties of corncob ash.

The fire resistance of concrete materials is one of the indices to evaluate the durability of concrete in engineering structures [18], which features structural design. Therefore, a number of studies on the effect of high temperature on the mechanical properties of agricultural waste concrete have been conducted [19]. However, the effect of high temperature on the stress–strain relationship of agricultural waste concrete is barely involved. Mo et al. [20] investigated the compressive strength of palm shell concrete after elevated temperatures. The results show that after exposure to elevated temperature within the range of 200 °C to 400 °C, the loss in compressive strength of palm shell concrete is about 30% to 55%, respectively. The compressive strength of coconut concrete after elevated temperatures was experimentally studied by Gunasekaran et al. [21]. The results show that, according to the exposure duration, the compressive strength attenuation of coconut concrete is approximately 25–50% and 55–75% at temperatures of 200 °C and 400 °C, respectively.

Based on the above-mentioned review, a lack of data on the stress–strain relationship of agricultural waste concrete under uniaxial compressive behavior is indicated. Therefore, the jujube nucleus is utilized herein as a coarse aggregate in the preparation of agricultural waste concrete, as shown in Figure 1. The failure phenomena, uniaxial compressive performance, and the stress–strain model of jujube nucleus concrete (JNC) under the effect of the jujube nucleus (JN) replacement ratio and the elevated temperature are investigated experimentally and discussed in detail. The prediction formulas for the stress–strain constitutive model of JNC are proposed by considering the JN replacement ratio and the elevated temperature.

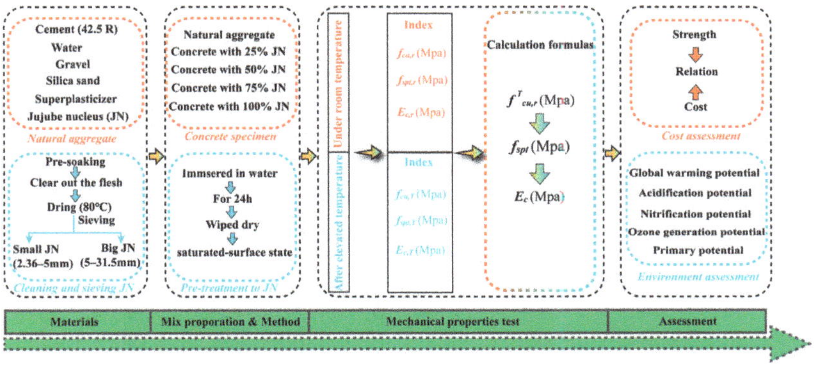


**Figure 1.** Research flow chart of the experiment.

## 2. Materials and Methods

### *2.1. Materials*

#### 2.1.1. Natural Aggregate, Cement, and Water

The concrete mixture was prepared by using natural silica sand and natural limestone gravel sourced from local suppliers, Xi'an, China. Following China's standard JGJ 52-2006 [22], the physical properties of the natural coarse aggregates were tested as shown in Table 1; 42.5R ordinary Portland cement and tap water were used to act as an adhesive for the concrete mixture, while a superplasticizer with 3% of the cement weight was used to reinforce the workability of the mixture.

**Table 1.** Physical properties of natural aggregates.

| | Gradation/mm | Apparent Density/(kg/m$^3$) | Water Absorption/% | Crushing Index/% |
|---|---|---|---|---|
| Natural gravel | 5–25 | 2660 | 1.4 | 16.9 |
| Silica sand | 0.25–2.5 | 2580 | 5.6 | 7.8 |

#### 2.1.2. Jujube Nucleus

The jujube nucleus used herein as a substitute for natural coarse aggregate was collected from a fruit processing facility located in Shaanxi province, China. The surface of the jujube nucleus was generally enveloped by large amounts of unripped flesh that was abundant in saccharides such as glucose, fructose, sucrose, oligosaccharide, arabinose, and galactose [10]. Saccharides that dissolved in the water during the concrete preparation process could interfere with the hydration of the cement, which has a noticeable influence on the mechanical properties of concrete [23,24]. Referring to previous studies [25], these jujube nuclei were first soaked in water for 24 h, and then rinsed by a high-pressure water gun in order to remove these fleshes. These jujube nuclei were then baked in a high-temperature oven for 2 h to make their surface completely dry (the elevated temperature was 80 °C). As shown in Figure 2a, the surface of these treated jujube nuclei is rough without the residual flesh.

Following the Chinese standard GB/T 17431.2-2010 [26], these jujube nuclei were sieved for the preparation of lightweight aggregate concrete. The particle sizes of the sieved jujube nuclei were in the range of 2.36–31.5 mm, of which 5–31.5 mm was large jujube nucleus, and 2.36–5 mm was small jujube nucleus, as seen in Figure 2b.

The physical properties of the jujube nucleus were tested, as shown in Table 2. Compared with the natural coarse aggregates, the jujube nucleus showed higher water absorption, but lower apparent density, implying porous structural characteristics. In order to remove the influence of water absorption on test results [27], these jujube nuclei were immersed in water for 24 h prior to specimen preparation to make them absorb water adequately, and they were then wiped to attain a saturated surface dry state for specimen preparation, as shown in Figure 2c.

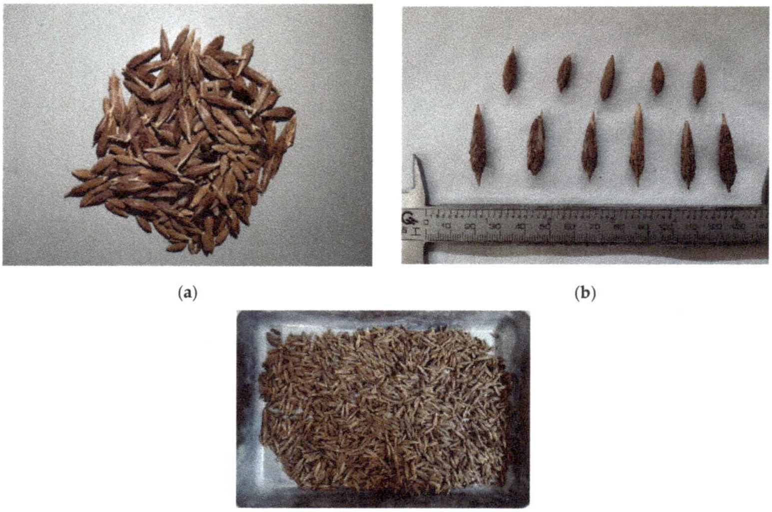

**Figure 2.** Jujube nuclei used in the test. (**a**) Jujube nuclei after pretreatment. (**b**) The particle size of jujube nuclei. (**c**) Jujube nuclei used in the specimen preparation.

**Table 2.** Physical properties of jujube nucleus.

| | Gradation (mm) | Apparent Density ($kg/m^3$) | Bulk Density ($kg/m^3$) | Water Absorption (by mass) | Tube Compression Strength (MPa) |
|---|---|---|---|---|---|
| Big jujube nuclei | 5–31.5 | 1048.56 | 447.50 | 32.50% | 0.52 |
| Small jujube nuclei | 2.36–5 | 1098.63 | 598.50 | 19.50% | 0.92 |

*2.2. Mixture Design and Specimen Preparation*

2.2.1. Mixture Proportions of Concrete

For the ordinary concrete without jujube nucleus, the mixture proportion was designed according to China's JGJ55-2011 standard [28]. Under the same mixture proportion design, concrete with jujube nucleus was prepared using the waste jujube nucleus as a substitute for natural coarse aggregate with the replacement ratio of 25%, 50%, 75%, and 100%, respectively (by volume) in the preparation of jujube nucleus concrete. The mixture proportions of ordinary concrete and jujube nucleus concrete are listed in Table 3.

**Table 3.** Concrete mixture proportions in 1 $m^3$.

| Mixture Mode *1 | r *2/% | Mixture Proportions /($kg/m^3$) | | | | | |
|---|---|---|---|---|---|---|---|
| | | Cement | Water | Sand | Gravel | Big Jujube Nucleus | Small Jujube Nucleus |
| C30-r0 | 0 | 380 | 185 | 648 | 1198 | 0 | 0 |
| C30-r25 | 25 | 380 | 185 | 648 | 898.50 | 18.15 | 102.85 |
| C30-r50 | 50 | 380 | 185 | 648 | 599 | 36.30 | 205.69 |
| C30-r75 | 75 | 380 | 185 | 648 | 299.50 | 54.45 | 308.55 |
| C30-r100 | 100 | 380 | 185 | 648 | 0 | 72.59 | 411.39 |

*1: For the nomenclature, taking 'C30-r0' as an example, C30 is ordinary concrete grade; r0 stands for 0% jujube nucleus replacement ratio. *2: r stands for jujube nucleus replacement ratio.

2.2.2. Preparation of Specimens

Prisms of 150 mm × 150 mm × 300 mm were prepared for testing the uniaxial compressive strength, elastic modulus, peak strain, and the whole stress–strain curves of JNC (three parallel specimens for each concrete mixture proportion). Specimen identification, as listed in Table 4, begins with specimen type followed by the target elevated temperature and the JN replacement rate.

**Table 4.** Specimen identification.

| *r* | Temperature | | | | |
|---|---|---|---|---|---|
| | **20 °C** | **100 °C** | **200 °C** | **400 °C** | **500 °C** |
| C30-r0 | P-20-r0 | P-100-r0 | P-200-r0 | P-400-r0 | P-500-r0 |
| C30-r25 | P-20-r25 | P-100-r25 | P-200-r25 | P-400-r25 | P-500-r25 |
| C30-r50 | P-20-r50 | P-100-r50 | P-200-r50 | P-400-r50 | P-500-r50 |
| C30-r75 | P-20-r75 | P-100-r75 | P-200-r75 | P-400-r75 | P-500-r75 |
| C30-r100 | P-20-r100 | P-100-r100 | P-200-r100 | P-400-r100 | P-500-r100 |

## 2.3. *Testing Procedure*

2.3.1. Mechanical Properties Test under Room Temperature

The uniaxial compression test of JNC was performed by a hydraulic compression machine with a maximum test load of 2000 kN. Specifically, a force control method was used to impose compression on the sample at a speed of 1 kN/s. Once the load reached 15 kN, the force-controlled method was used via the displacement-controlled method at a rate of 0.025 mm/s until the specimens failed. Two strain gauges were arranged on each side of the specimen to monitor the longitudinal strain, as shown in Figure 3.

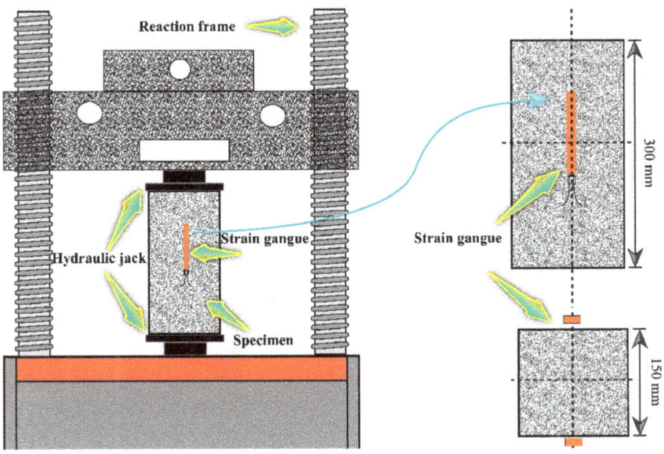


**Figure 3.** Test machine and arrangement method.

2.3.2. Heating Regime

The high temperature furnace in which the concrete specimens were preheated is illustrated in Figure 4a. The maximum output of the high temperature furnace was 18 kW. The dimensions of the chamber were 0.5 m × 0.5 m × 0.5 m.

Prior to the uniaxial compressive test, the concrete specimens were heated from room temperature to a specific target temperature (100 °C, 200 °C, 400 °C, and 500 °C, respectively) at a rate of 10 °C/min. Mohamedhai et al. [29] pointed out that the influence of high temperature on the mechanical properties of concrete materials occurs primarily in the first couple of hours after exposure to temperature. For this reason, all concrete specimens exposed to high temperature were held in the high temperature furnace for 2 h after reaching the target temperature. The concrete specimens were then cooled down to room temperature by nature-cooling in the furnace. The heating, holding, and cooling temperature regime in this test is illustrated in Figure 4b.

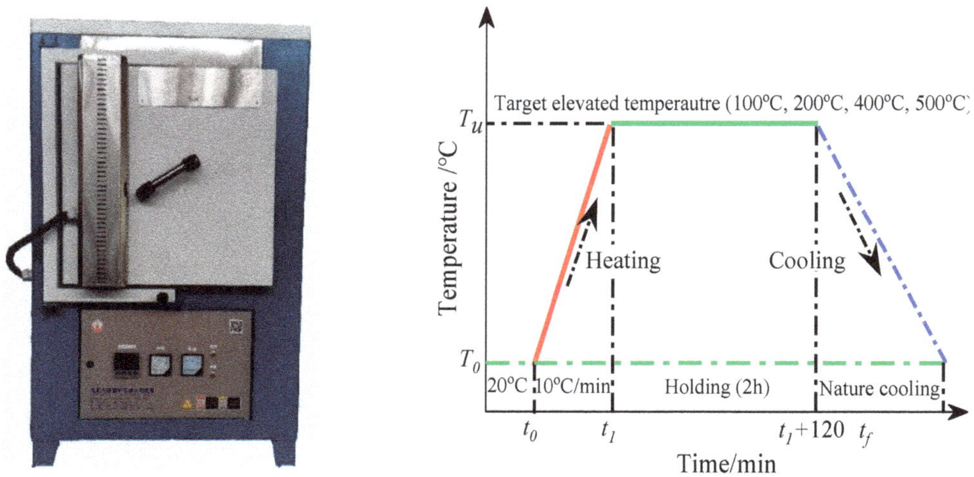


(a) (b)

**Figure 4.** High temperature loading procedure. (**a**) High temperature furnace. (**b**) Heating, holding, and cooling process.

## 3. Results and Discussions

### *3.1. Failure Pattern*

The failure phenomena of all the specimens tested under the uniaxial compressive loading for JNC are shown in Figure 5.

For ordinary concrete, namely without jujube nucleus, shear failure was observed in all the specimens, and a main diagonal crack occurred as a result of uniaxial compression. As the elevated temperature increased, more vertical cracks appeared in all the specimens besides the diagonal crack. Under heating, the evaporation of free water and the decomposition of hydrate calcium silicate (C-S-H) and calcium hydroxide resulted in the growing number of pores in the concrete. Thus, the crack propagation of the concrete after elevated temperature under the uniaxial compressive loading only needed to dissipate a little energy, compared to that of ordinary concrete without heating [30].

For jujube nucleus concrete (JNC), the number and width of diagonal and vertical cracks increase with the increase in the jujube nucleus replacement ratio and elevated temperature, which is mainly due to the low stiffness and high porosity of jujube nucleus, which may result in the crack propagation pattern being transferred from the damaged interfacial transition zone to aggregate penetrated under the uniaxial compressive loading, reflecting the weaker energy dissipation capacity of JNC [31]. However, the number of diagonal and vertical cracks for JNC decreased gradually with the increase in the elevated temperature, which is probably because the jujube nucleus was burned out under the effect of the elevated temperature, causing the generation of a large pore structure. It was noteworthy that when the jujube nucleus replacement ratio exceeded 25%, the JNC was prone to being squashed, which is due to the ineffectiveness of cementitious materials under high JN replacement ratio [32].

*3.2. Uniaxial Compressive Stress–strain Curves*

3.2.1. Effects of the Elevated Temperature on Stress Strain Curves for JNC

The whole stress–strain curves of JNC under the same JN replacement ratio after experiencing various elevated temperatures is presented in Figure 6.

It can be seen that as the elevated temperature increases, the peak stress and peak strain of JNC decrease gradually. Additionally, the slope of the stress–strain curves, for both the ascending and descending branches, decreases with the rise in the elevated temperature. It is noted that when the elevated temperature exceeds 400 °C, the peak stress and peak strain of ordinary concrete show a larger reduction than those of JNC. This may be because when the elevated temperature exceeds 400 °C, the Ca $(OH)_2$ crystal, having a larger volume in the concrete, begins dissociation, which causes cement paste dehydration [33], as shown in Figure 7. At the same time, the most noticeable thermal expansion promotes crack propagation. The difference is that during the heating procedure, the porosity and rough texture of jujube nucleus may result in less damage to the concrete.

3.2.2. Effects of the Replacement Ratio on Stress–Strain Curves for JNC

The whole stress–strain curves of JNC under the same elevated temperature with the various JN replacement ratio are presented in Figure 8.

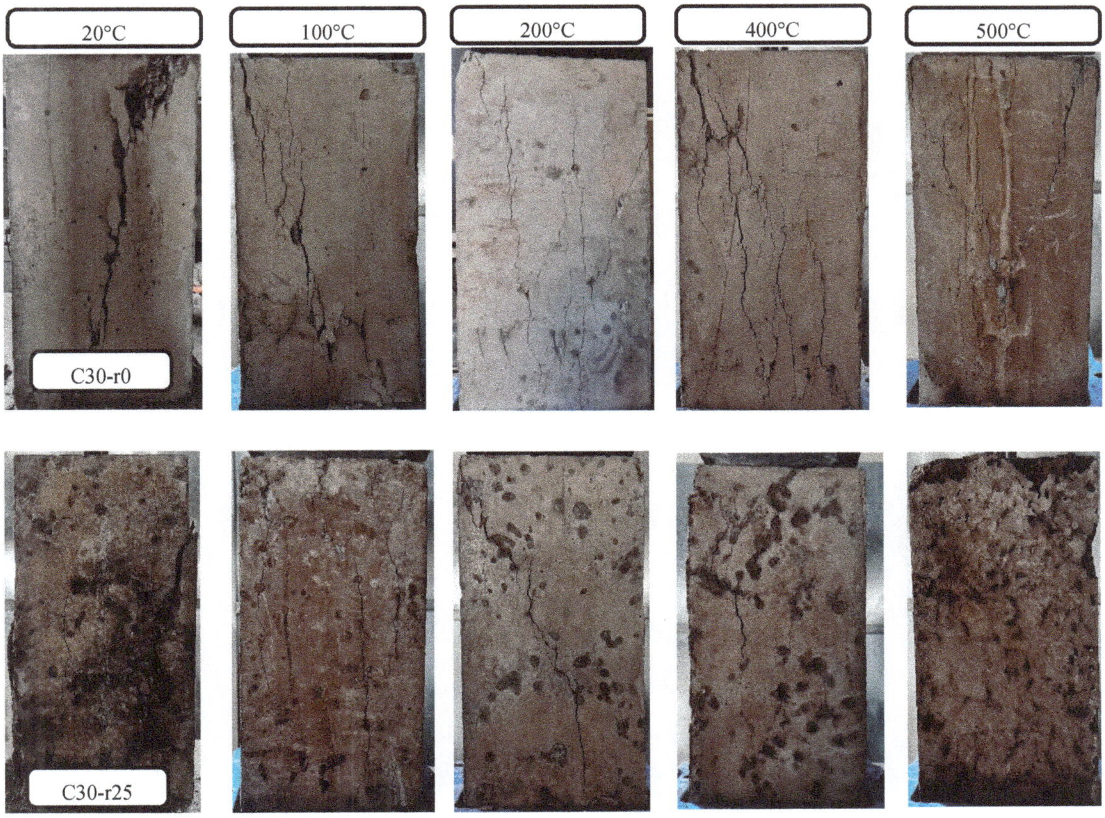


**Figure 5.** *Cont.*

**Figure 5.** Failure phenomenon of the uniaxial compressive strength test.

It can be seen from Figure 8 that as the JN replacement ratio increases, the peak stress and ascending branch slope of the mean curves gradually decrease whilst the peak strain and descending branch slope of the mean curves increase relatively under various elevated temperatures. Furthermore, the linear proportion of the curves in the ascending branch increases with the rise in the JN replacement ratio. Fan et al. [31] pointed out that on account of having lower elastic modulus compared to natural aggregate, the lightweight aggregate in concrete is inclined to fracture. For ordinary concrete (without jujube nucleus), the compressive damage is the destruction of the bonding surface of natural aggregate and cement mortar, namely the damaged interfacial transition zone, which proceeds in three stages: bond surface cracks, coalescence, and crack propagation. For lightweight aggregate concrete, the compressive damage is the penetrating of lightweight aggregates, which proceeds in three stages: bond surface cracks, penetrating of lightweight aggregate, and crack propagation, as shown in Figure 9.

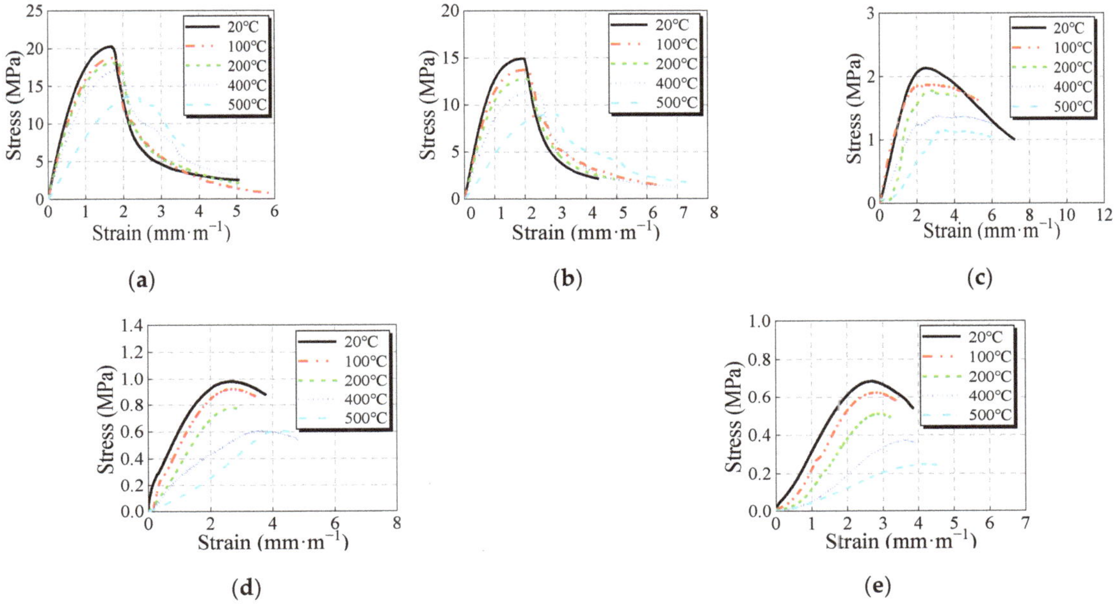


**Figure 6.** Stress–strain curves for JNC under the same replacement ratio after experiencing the various elevated temperatures. (**a**) $r$ = 0%. (**b**) $r$ = 25%. (**c**) $r$ = 50%. (**d**) $r$ = 75%. (**e**) $r$ = 100%.

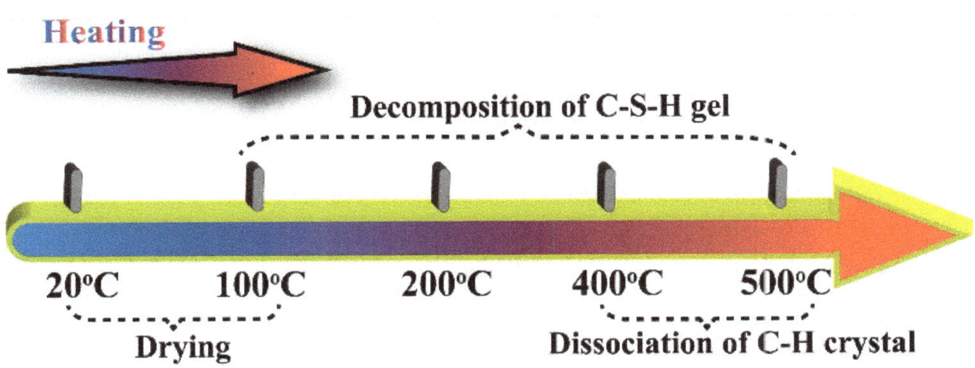


**Figure 7.** Effect of the elevated temperature on concrete.

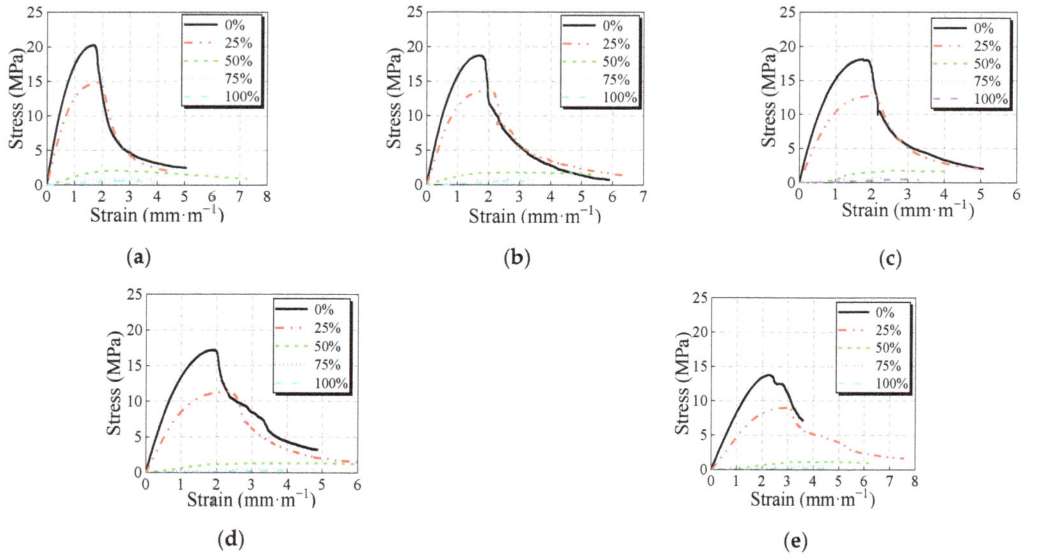


**Figure 8.** Stress–strain curves with the various jujube nucleus replacement ratio after elevated temperatures. (**a**) $T$ = 20 °C. (**b**) $T$ = 100 °C. (**c**) $T$ = 200 °C. (**d**) $T$ = 400 °C. (**e**) $T$ = 500 °C.

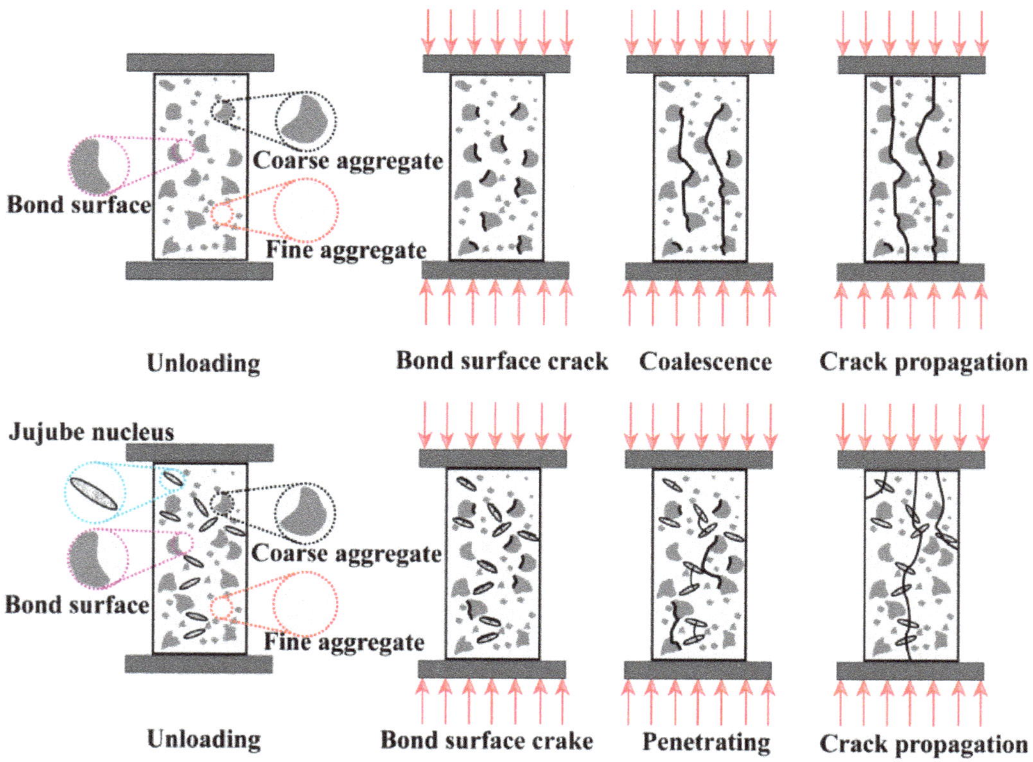


**Figure 9.** The fracture mechanism of jujube nucleus concrete.

### 3.3. Stress Strain Characteristic

After a 28-day curing process, the stress–strain characteristics with various jujube nucleus (JN) replacement ratios after experiencing various elevated temperatures were derived from the stress–strain curves of JNC, including the peak stress, peak strain, and elastic modulus, as tabulated in Table 5.

**Table 5.** The residual mechanical properties of JNC.

| Properties | Number | Ambient | 100 °C | 200 °C | 400 °C | 500 °C |
|---|---|---|---|---|---|---|
| $f_c$ (Mpa) | C30-r0 | 20.25 | 18.75 | 18.09 | 17.22 | 13.77 |
| | C30-r25 | 15.08 | 13.52 | 13.18 | 12.38 | 10.02 |
| | C30-r50 | 3.01 | 2.87 | 2.79 | 2.38 | 2.16 |
| | C30-r75 | 0.98 | 0.92 | 0.77 | 0.71 | 0.67 |
| | C30-r100 | 0.69 | 0.62 | 0.51 | 0.47 | 0.45 |
| $E_c$ (GPa) | C30-r0 | 19.43 | 19.11 | 18.96 | 17.13 | 15.61 |
| | C30-r25 | 17.24 | 16.52 | 14.83 | 14.30 | 13.70 |
| | C30-r50 | 10.13 | 9.72 | 9.05 | 9.44 | 8.79 |
| | C30-r75 | 8.24 | 8.23 | 7.40 | 7.92 | 7.13 |
| | C30-r100 | 7.87 | 7.42 | 6.69 | 6.64 | 6.39 |
| $\varepsilon_c$ (με) | C30-r0 | 1708.00 | 1720.00 | 1750.00 | 1920.00 | 2300.00 |
| | C30-r25 | 1980.00 | 2030.00 | 2090.00 | 2350.00 | 2370.00 |
| | C30-r50 | 2505.87 | 2542.07 | 2616.47 | 3154.00 | 3321.53 |
| | C30-r75 | 2698.07 | 2742.07 | 2776.73 | 3553.27 | 3672.00 |
| | C30-r100 | 2856.47 | 2952.67 | 2965.33 | 3627.40 | 3909.00 |

Note: $f_c$ is the peak stress of JNC; $E_c$ is the elastic modulus of JNC. $\varepsilon_c$ is peak strain of JNC.

#### 3.3.1. Peak Stress of JNC

As shown in Figure 9, each point datum presents the measured average value of the peak stress and its relative value ($f_{c,r}^T/f_{c,r}^{20}$ and $f_{c,r}^T/f_{c,0}^T$) for JNC under the effect of the JN replacement ratio and the elevated temperature. $f_{c,r}^T/f_{c,r}^{20}$ is the ratio of the peak stress after experiencing various elevated temperatures to the peak stress without heating under the same JN replacement ratio. $f_{c,r}^T/f_{c,0}^T$ is the ratio of the peak stress with various JN replacement ratios to the peak stress without replacement ratio under the same elevated temperature.

It can be seen from Figure 10a that as the elevated temperature increases, the peak stress of JNC degrades gradually under various JN replacement ratios. The difference in the peak stress between two adjacent temperatures becomes larger when the elevated temperature exceeds 400 °C, which is due to the dissociation of C-H crystal [30]. It can be seen from Figure 10b that the increase in the JN replacement ratio would increase the peak stress of JNC under the various elevated temperatures. The difference in the peak stress between the two adjacent JN replacement ratios becomes larger when the JN replacement ratio exceeds 25%, which is attributed to the rise in sucrose content [32]. Figure 10c shows the peak stress deterioration under the effect of the JN replacement ratio and the elevated temperature. It can be seen that the surface of the peak stress slips down precipitously at the bottom with the rise in the JN replacement ratio and the elevated temperature.

In order to explicitly express the relationship between the elevated temperature, the JN replacement ratio, and the peak stress, a calculation formula is proposed based on the test results through nonlinear fitting, as given in Equations (1)–(3). Figure 10d shows the comparison of the tested value and the predicted value by using Equations (1)–(3). It can be seen that the deviation of the proposed calculation formula is basically within ±10%.

$$f_{c,r}^T = \eta_c \lambda_c f_{c,0}^{20} \tag{1}$$

$$\eta_c = \begin{cases} 1 - 0.41(\frac{T-20}{500}) + 0.29(\frac{T-20}{500})^2 & 20°C \le T \le 400°C \\ 3.31 - 2.68(\frac{T-20}{500}) & 400°C < T \le 500°C \end{cases} \tag{2}$$

$$\lambda_c = \begin{cases} 1 - 1.08r & 0 \le r \le 25\% \\ 0.67 - 1.43r + 0.79r^2 & 25\% < r \le 100\% \end{cases} \tag{3}$$

where $\eta_c$ is the reduction factor of the relative value ($f_{c,r}^T / f_{c,r}^{20}$), $\lambda_c$ is the reduction factor of the relative value ($f_{c,r}^T / f_{c,0}^T$), $f_{c,r}^T$ is the peak stress of JNC under the effect of the JN replacement ratio and the elevated temperature, and $f_{c,0}^{20}$ is the peak stress of JNC without jujube nucleus and heating.

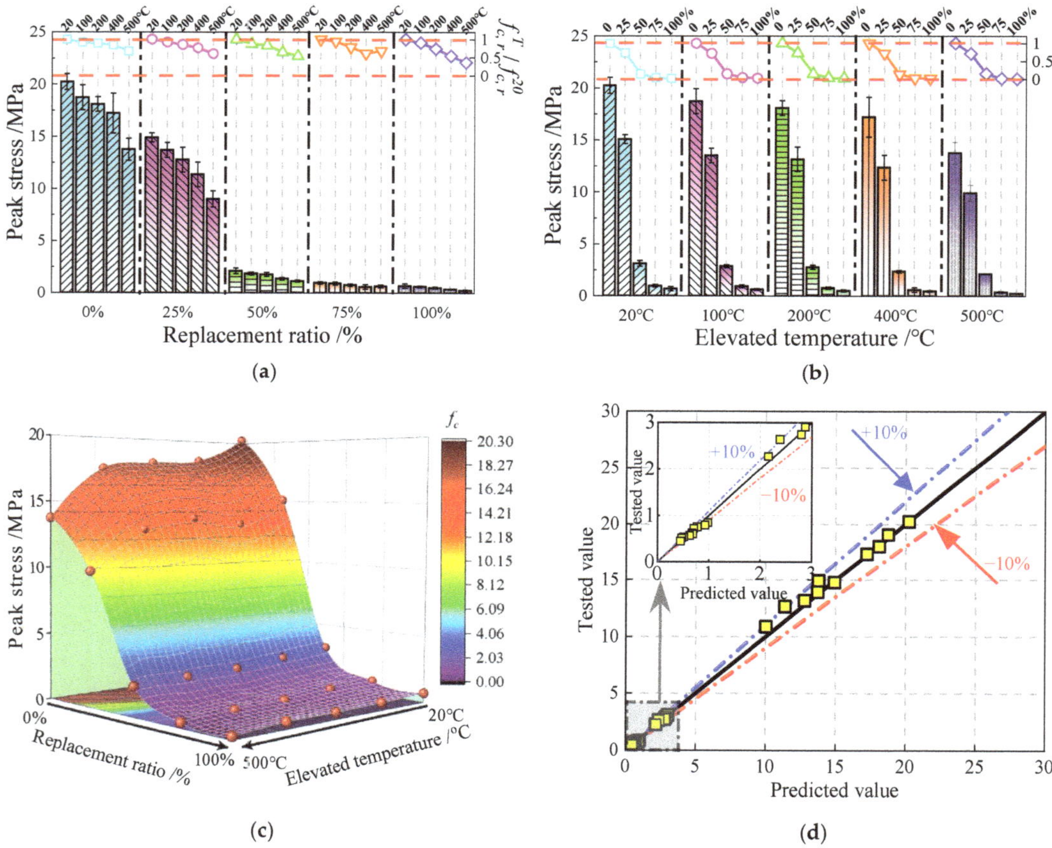


**Figure 10.** Peak stress of JNC. (**a**) Effect elevated temperature on $f_c$. (**b**) Effect of replacement ratio on $f_c$. (**c**) Surface of $f_c$ under the effect of replacement ratios and elevated temperatures. (**d**) Comparison of tested values and predicted values.

### 3.3.2. Elastic Modules of JNC

The elastic modulus of JNC is taken as the secant modulus from the origin to 40% of the peak stress. Figure 11 presents the measured average value of elastic modulus for JNC under the effect of the jujube nucleus (JN) replacement ratio and the elevated temperature.

Analogous to the peak stress of JNC, the elastic modulus of JNC decrease gradually with the rise in the JN replacement ratio and the elevated temperature, as shown in Figure 11a–c. To be specific, when the elevated temperature exceeds 400 °C, the elastic modulus of JNC within the various replacement ratios (0% (control group), 25%, 50%, 75%, and 100%) is reduced by about 20%, 29%, 55%, 63%, and 67%, respectively, whilst the elastic modulus of JNC after experiencing various elevated temperature (20 °C (control group), 100 °C, 200 °C, 400 °C, and 500 °C) is decreased by approximately 11%, 15%, 24%, 26%, and 29%, respectively, when the JN replacement ratio reaches 25%.

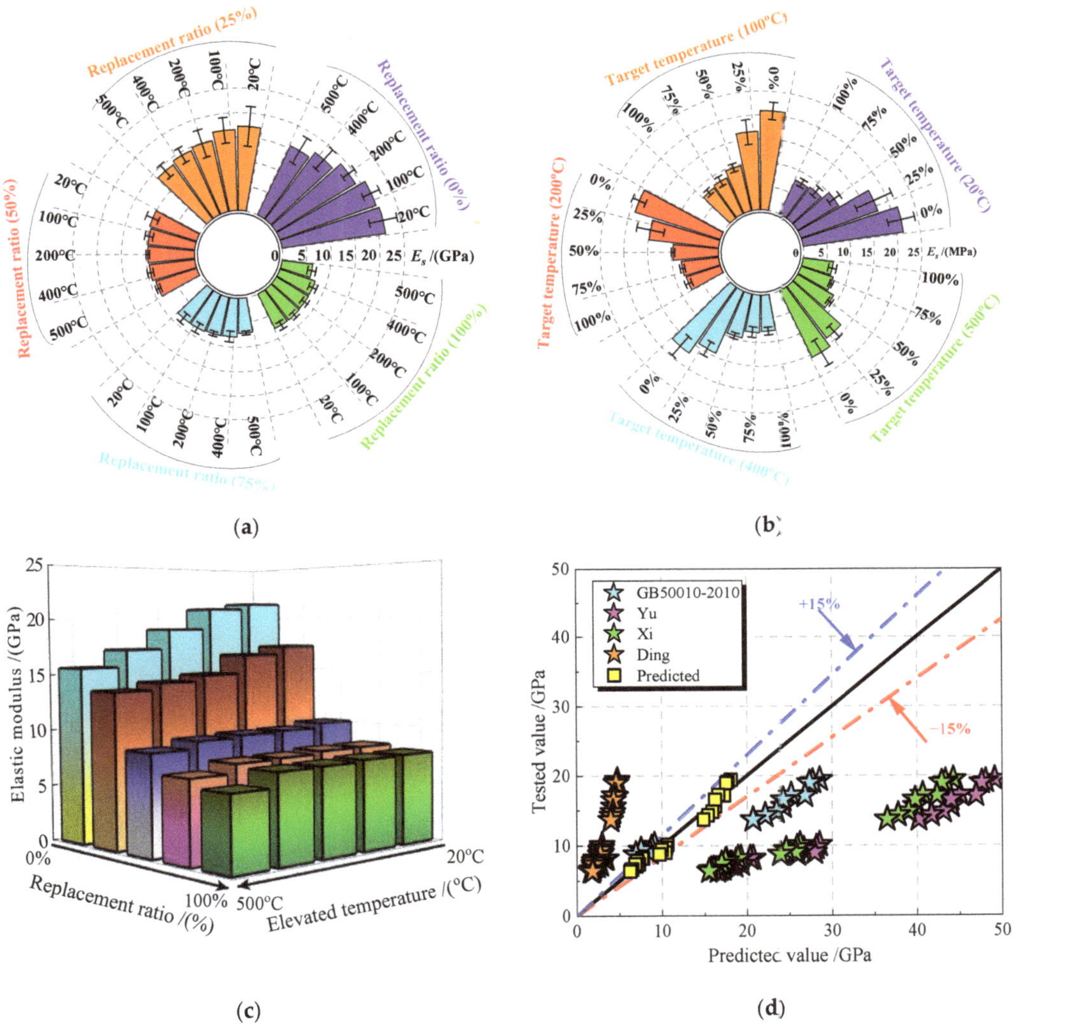


**Figure 11.** Elastic modulus test of JNC. (**a**) Effect of elevated temperature on $E_c$. (**b**) Effect of replacement ratio on $E_c$. (**c**) Variation of $E_c$ under the effect of replacement ratios and elevated temperatures. (**d**) Comparison of the tested values and the predicted values [34–37].

To predict the elastic modulus of concrete materials, Equations (4) and (5) are recommended by GB/T 50010-2010 [34] and Yu [35] (a = 20805, b = 0.29), respectively, to predict the elastic modulus of ordinary concrete, whilst Equation (6) is put forward by Xi [36] (a = 1.60, b = 1.347, and c = 0.275) and Ding [37] (a = 2.09, b = 1, c = 0.25) to calculate the elastic modulus of lightweight aggregate concrete. Figure 11d shows the comparison of the tested values and the predicted values. It can be seen that Ding is prone to overestimate the elastic modulus of JNC, while the formulas from Yu, Xi, and GB 50010-2010 are inclined to underestimate the elastic modulus of JNC. Therefore, Equation (7) is proposed based on the test results to predict the elastic modulus of JNC, namely, a = 2.38, b = 1.16, and c = 0.28. The discrepancy of the proposed constants is basically within ± 15%, as shown in Figure 11d.

$$E_c = \frac{10^5}{2.2 + \frac{26.37}{f_c}} \tag{4}$$

$$E_c = a\, f_c^b \tag{5}$$

$$E_c = a\, \rho^b f_c^c \tag{6}$$

$$E_c = 2.38\, \rho^{1.16} f_c^{0.28} \tag{7}$$

where, a and b are constants, and $\rho$ is the density of the jujube nucleus.

### 3.3.3. Peak Strain of JNC

The peak strain is the strain at maximum uniaxial compressive strength. Figure 12 shows the measured average value of JNC under the effect of the JN replacement ratio and the elevated temperature.

It can be seen from Figure 12a that the increase in the elevated temperature would increase the peak strain of JNC under various JN replacement ratios (0% (control group), 25%, 50%, 75%, and 100%). The difference in peak strain between the two adjacent temperatures becomes larger when the elevated temperature exceeds 400 °C, increased by about 35%, 39%, 94%, 115%, and 129%, respectively. As shown in Figure 12b, the rise in the JN replacement ratio increase similarly to the peak strain of JNC under the various elevated temperature (20 °C (control group), 100 °C, 200 °C, 400 °C, and 500 °C). The difference in the peak strain between two adjacent replacement ratios becomes larger when the JN replacement ratio reaches 25%, increased by approximately 15%, 19%, 22%, 38%, and 39%, respectively. Figure 12c shows the peak strain of JNC under the effect of the JN replacement ratio and the elevated temperature. It can be seen that the rise in the JN replacement ratio and the elevated temperature increases the peak strain of JNC. It is worth noting that the JN replacement ratio also has a notable effect on the peak strain of JNC, which is analogous to the reason for the peak stress deterioration for JNC.

To predict the peak strain of concrete materials, Equation (8) is recommended by Yu [35] (a = 520, b = 0.33) and Ding [37] (a = 760, b = 0.33), while Equation (9) is used by GB50010-2010 [34] (a = 172, b = 0.5, and c = 700) and Wang [38] (a = 17.48, b = 1, and c = 1788). Figure 12d shows the comparison of the tested values and the predicted values. It can be seen that all the formulas proposed tend to overestimate the peak strain of JNC. Based on the test results, Equation (10) is proposed to predict the peak strain of JNC, namely, a = −1510.82, b = 0.2, and c = 4611.82. The error of the proposed constants is basically within ±15%, as shown in Figure 12d.

$$\varepsilon_c = a\, f_c^b \tag{8}$$

$$\varepsilon_c = a\, f_c^b + c \tag{9}$$

$$\varepsilon_c = (4611.82 - 1510.82 f_c^{0.2}) \times 10^{-6} \tag{10}$$

where a and b are constants, and $\rho$ is the density of the jujube nucleus.

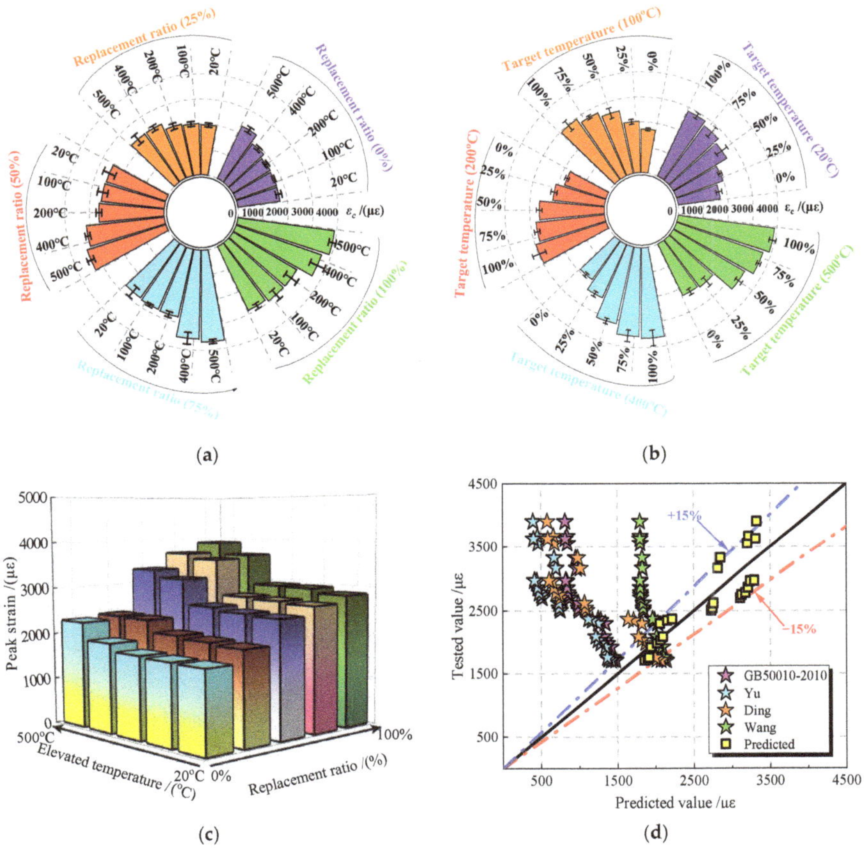


**Figure 12.** Peak strain of JNC. (**a**) Effect of elevated temperature on $\varepsilon_c$. (**b**) Effect of replacement ratio on $\varepsilon_c$. (**c**) Variation of $\varepsilon_c$ under the effect of replacement ratios and elevated temperatures. (**d**) Comparison of the tested values and the predicted values [34,35,37,38].

## 4. Constitutive Model of JNC

### *4.1. Suggested Stress–Strain Model for JNC*

In this paper, the shape of the stress–strain curves for jujube nucleus concrete (JNC) is changed by the addition of jujube nucleus (JN). To tackle the phenomenon, a constitutive model of JNC corresponding to the requirement of structure design is needed, which is important for both practical engineering design and numerical calculations.

As shown in Figures 6 and 8, the stress–strain curves of JNC present a longer linear stage than ordinary concrete with the rise in the JN replacement ratio and the elevated temperature and show more obvious elastic characteristics, which is typical for lightweight aggregate concrete. At present, large amounts of constitutive models concerning lightweight aggregate concrete have been researched. Therefore, based on the experimental test presented herein, a two-stage model adapted from Ding's model [37] is used to describe the stress–strain relationship of JNC by considering various JN replacement ratios and elevated temperatures. A two-stage model (see Equation (11)) is defined by five parameters: the peak stress of concrete, $f_c$, the peak strain of concrete, $\varepsilon_c$, the parameter of ascending branch, $A$, $B$, and the parameter of descending branch, $\alpha_c$. The peak stress of concrete, $f_c$, can be calculated by Equations (1)–(3). The peak strain of concrete, $\varepsilon_c$, can be expressed by Equation (10). Other parameters (ascending branch parameters, $A$ and $B$, descending branch parameter, $\alpha$) need modification by considering the effect of replacement ratio and elevated temperature.

$$\sigma = \begin{cases} f_c \frac{A(\varepsilon/\varepsilon_c)+(B-1)(\varepsilon/\varepsilon_c)}{1+(A-2)(\varepsilon/\varepsilon_c)+B(\varepsilon/\varepsilon_c)^2} & 0 \le \frac{\varepsilon}{\varepsilon_c} \le 1 \\ f_c \frac{\varepsilon/\varepsilon_c}{\varepsilon/\varepsilon_c+\alpha_c((\varepsilon/\varepsilon_c)-1)^2} & \frac{\varepsilon}{\varepsilon_c} > 1 \end{cases} \quad (11)$$

$$A = 1.64 \times 10^{-3}\rho f_c^{-1/6} \quad (12)$$

$$B = 1.68(A-1)^2 \quad (13)$$

$$\alpha_c = 3.67 \times 10^{-5} f_c^3 \quad (14)$$

## 4.2. Analysis of Model Parameters

The values of parameters $A$, $B$, and $\alpha_c$ can be determined through a nonlinear curve fitting for the stress–strain curves of jujube nucleus concrete (JNC) under the effect of the jujube nucleus (JN) replacement ratio and elevated temperature. Specific data are tabulated in Table 6.

**Table 6.** Parameters of constitutive equation.

| Sort | Parameters | 20 °C | 100 °C | 200 °C | 400 °C | 500 °C |
|---|---|---|---|---|---|---|
| C30-r0 | $A$ | 2.31 | 2.26 | 2.21 | 2.18 | 2.05 |
| | $B$ | 2.19 | 2.11 | 2.01 | 1.97 | 1.59 |
| | $\alpha_c$ | 8.16 | 8.17 | 7.44 | 7.35 | 6.62 |
| C30-r25 | $A$ | 2.51 | 2.47 | 2.33 | 2.27 | 2.08 |
| | $B$ | 2.08 | 2.01 | 1.96 | 1.88 | 1.61 |
| | $\alpha_c$ | 8.39 | 8.41 | 7.51 | 7.42 | 6.76 |
| C30-r50 | $A$ | 1.29 | 1.21 | 1.11 | 1.07 | 1.01 |
| | $B$ | 1.05 | 0.98 | 0.94 | 0.83 | 0.79 |
| | $\alpha_c$ | 1.92 | 1.75 | 1.8 | 1.56 | 1.32 |
| C30-r75 | $A$ | 1.05 | 1.02 | 0.97 | 0.85 | 0.85 |
| | $B$ | 1.01 | 0.93 | 0.89 | 0.81 | 0.69 |
| | $\alpha_c$ | 0.82 | 0.78 | 0.68 | 0.61 | 0.52 |
| C30-r100 | $A$ | 0.91 | 0.88 | 0.84 | 0.81 | 0.75 |
| | $B$ | 0.94 | 0.92 | 0.86 | 0.81 | 0.62 |
| | $\alpha_c$ | 0.62 | 0.6 | 0.41 | 0.40 | 0.20 |

### 4.2.1. Ascending Branch Parameter A

The definition of parameter $A$ is the ratio of elastic modulus to peak secant modulus for concrete. Ding et al. [37] shows that the larger the value of parameter $A$ is, the longer the linear stage in the ascending branch of stress–strain curve is, and the more the accumulated elastic strain energy of the lightweight aggregate is. Figure 11 shows the values of ascending branch parameter $A$ under the effect of the JN replacement ratio and the elevated temperature. It can be seen from Figure 13a that the rise in the JN replacement ratio and the elevated temperature would make the values of the parameter $A$ climb to peak when the JN replacement ratio is no more than 25%. It can be concluded that the jujube nucleus has more elastic strain energy under the uniaxial compressive loading than natural aggregate. However, the further rise in the JN replacement ratio and the elevated temperature increases the sucrose molecule content in the JNC, and the values of the parameter $A$ decrease rapidly.

Ding et al. [37] shows that parameter $A$ can be expressed by the peak stress. However, the predicted values of parameter $A$ are discrete when using Equation (12), as shown in Figure 13b. Therefore, Equation (15) is proposed based on the test results for predicting the values of parameter $A$. Figure 13b shows the comparison of the tested values and the predicted values by using Equation (15). It can be seen that the discrepancy of the formula calculation is in the range of ±15%.

$$A = 8.95 \times 10^{-4}\rho f_c^{0.31} \quad (15)$$

where $\rho$ stands for the density of the jujube nucleus.

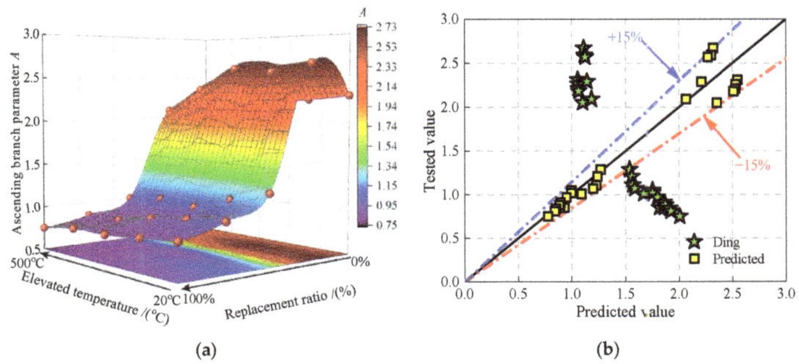


**Figure 13.** The values of ascending branch parameter $A$ for JNC. (**a**) Surface of $A$ under the effect of replacement ratios and elevated temperatures. (**b**) Comparison of the tested values and the predicted values [37].

### 4.2.2. Ascending Branch Parameter B

Parameter $B$ reflects the degradation degree of the elastic modulus for concrete [37]. Figure 14a presents the values of ascending branch parameter $B$ under the effect of JN replacement and elevated temperature. It can be seen that the values of parameter $B$ slip down gradually to the bottom with the rise in the JN replacement ratio and the elevated temperature. The difference is that the values of parameter $B$ decrease rapidly when the JN replacement ratio exceeds 25%, for the same reason for the decline of the values of parameter $A$.

Ding et al. [37] shows that parameter $B$ can be expressed by parameter $A$. However, the predicted values of parameter $B$ are overestimated by using Equation (13), as shown in Figure 14b. Therefore, Equation (16) is proposed based on test results to predict the values of parameter $B$. Figure 14b shows the comparison of the predicted values and the test values by using Equation (16). It can be seen that the discrepancy of the formula calculation is in the range of $\pm15\%$.

$$B = 0.15(A + 1.51)^{1.90} \tag{16}$$

where $A$ is obtained from Equation (15).

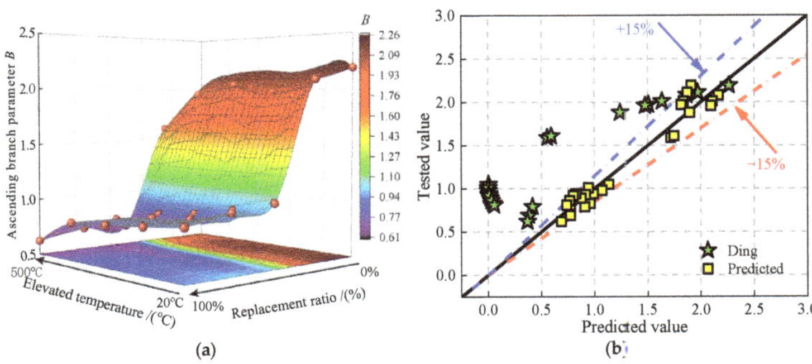


**Figure 14.** The values of ascending branch parameter $B$ for JNC. (**a**) Surface of $B$ under the effect of replacement ratios and the elevated temperature. (**b**) Comparison of the tested values and the predicted values [37].

### 4.2.3. Descending Branch Parameters $\alpha_c$

Parameter $\alpha_c$ reflects the slope of descending branch curve for concrete [37]. This means that the larger the value of parameter $\alpha_c$ is, the more obvious the brittle characteristic is. Figure 15a presents the values of descending branch parameter $\alpha_c$ under the effect of JN replacement ratio and elevated temperature. It can be seen that analogous to parameter $A$, the rise in the JN replacement ratio and elevated temperature would make the values of parameter $\alpha_c$ climb to the peak when the JN replacement ratio is no more than 25%, which may be because the jujube nucleus has lower stiffness than natural aggregate [31]. When the JN replacement ratio exceeds 25%, the values of parameter $\alpha_c$ slips down to the bottom rapidly, which is attributed to the same reason for the parameter $A$ deterioration.

Similar to parameter $A$, descending branch parameter $\alpha_c$ can also be expressed by the cubic compressive strength [37]. However, the values of parameter $\alpha_c$ are overestimated by using Equation (14), as shown in Figure 15b. Therefore, Equation (17) is proposed based on test results to predict the values of parameter $B$. Figure 15b shows the comparison of the predicted values and the test values by using Equation (17). It can be seen that the discrepancy of the formula calculation is in the range of ±15%.

$$\alpha_c = \begin{cases} 2.13\, f_c^{0.51} - 1.19 & r \leq 25\% \\ 0.72 f_c^{0.83} & r > 25\% \end{cases} \tag{17}$$

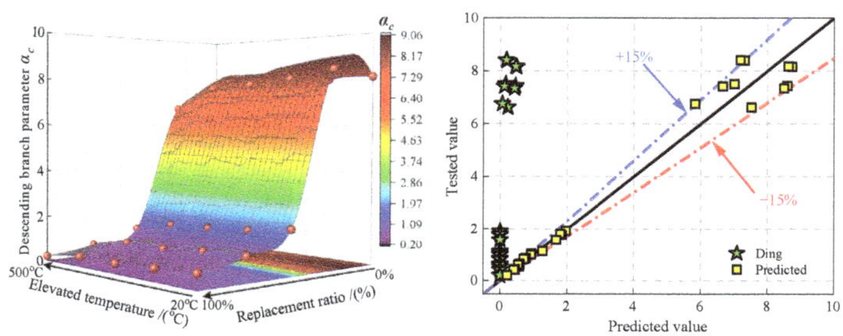


(a) (b)

**Figure 15.** The values of descending branch parameter $\alpha_c$ for JNC. (**a**) Surface of $\alpha_c$ under the effect of replacement ratios and elevated temperatures. (**b**) Comparison of the tested values and the predicted values [37].

## *4.3. Model Validation*

As shown in Figure 16, the predicted curves of JNC by using the proposed two-stage model (see Equation (11)) match well with the tested curves of JNC under the effect of the JN replacement ratio and the elevated temperature. Consequently, the proposed two-stage constitutive model herein can be adopted to perform accurate numerical calculation in the future study.

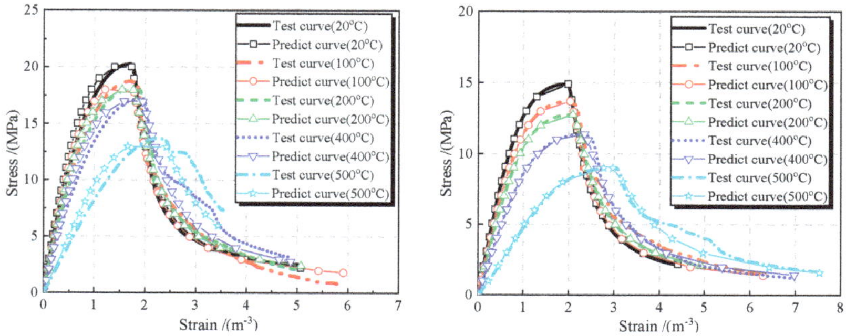


(a) (b)

**Figure 16.** *Cont.*

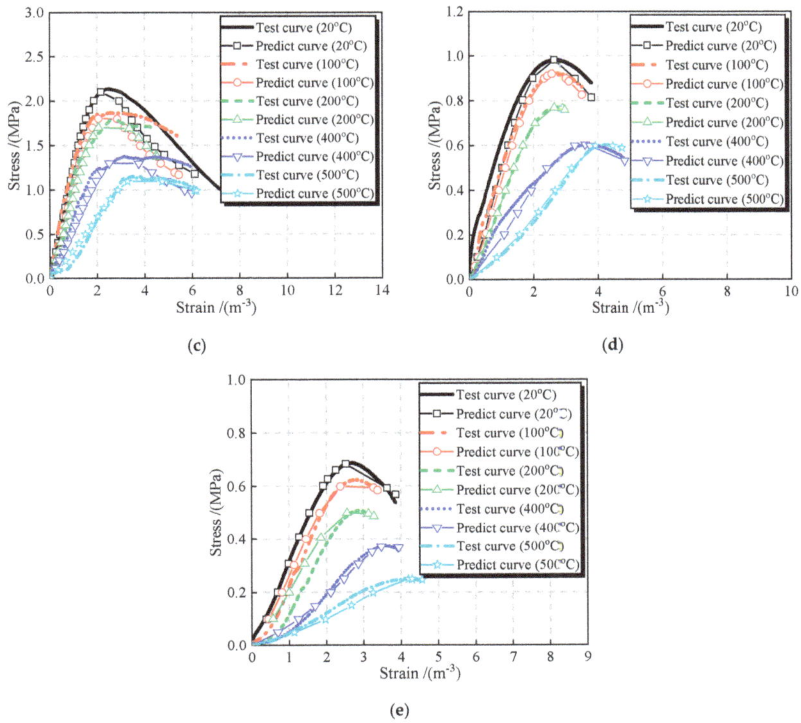


**Figure 16.** Model validation. (**a**) $r$ = 0%. (**b**) $r$ = 25%. (**c**) $r$ = 50%. (**d**) $r$ = 75%. (**e**) $r$ = 100%.

## 5. Conclusions and Next Steps in the Research

This paper investigates the feasibility of using jujube nucleus as a substitute for coarse aggregate in green lightweight aggregate concrete, aiming to provide a solution for natural resources consumption and agricultural waste pollution. The uniaxial compressive behavior of jujube nucleus concrete (JNC) was tested and discussed in detail by considering the effect of the jujube nucleus (JN) replacement ratio and the elevated temperature.

(1) Different from ordinary concrete, the increase in the JN replacement ratio and the elevated temperature would result in the growing number and width of the vertical and diagonal cracks for JNC whilst the linear proportion in the ascending branch and the descending slope increased gradually for the stress–strain curves of JNC, probably due to the low stiffness and high porosity of the jujube nucleus.

(2) The rise in the JN replacement ratio and elevated temperature would decrease the peak stress and elastic modulus but would cause a rapid increase in the peak strain in the stress–strain curves of JNC, which is ascribed possibly to the fact that sucrose molecules can hinder the growth of calcium silicate hydrate and calcium hydroxide to a larger content.

(3) When the JN replacement ratio reaches 25%, the reduction in the peak stress and elastic modulus is 26.36% and 11.27%, respectively, whilst the reduction in the peak stress and elastic modulus of JNC under 100% replacement ratio is 96.60% and 59.48%, respectively. The peak strain of JNC under 25% and 100% replacement ratio increases by 15.93% and 67.24%, respectively.

(4) When the elevated temperature is up to 100 °C, the peak stress and elastic modulus for JNC under 25% replacement ratio decrease by approximately 8.00% and 4.16%, respectively, whilst the reduction increases to 33.00% and 20.52%, respectively, under

400 °C heating. The peak strain of JNC with 25% replacement ratio increases from 2.53% under 100 °C to 19.70% under 400 °C.

(5) A material constitutive model is developed based on the test results for describing the stress–strain relationship of JNC under the effect of the JN replacement ratio and the elevated temperature. The predicted stress–strain curves using the model can match well with the tested stress–strain curves under the different JN replacement ratios and elevated temperatures.

It must be mentioned that the mechanical properties of JNC are acceptable with a 25% substitution ratio, indicating the potential utilization of JNC in pervious, sound-proof, and insolation members towards eco-friendly design. However, the mechanical properties of JNC decrease rather too significantly under a substitution ratio larger than 50% to be of practical interest. Thus, it is necessary to continue the experimental research to refine the formulas for the compressive performance and the stress–strain model of JNC, under substitution ratios between 0% and 35% at small intervals to have more data in this range of ratios. Because normally the durability of lightweight aggregate concrete is not as good as ordinary concrete [39], more studies regarding the durability of JNC are needed in the future. In addition, some reinforcement methods such as hybrid fiber [40] are also worth investigating to improve the mechanical properties of JNC under large substitution ratios.

**Author Contributions:** Resources and Supervision, J.Y.; Writing—original draft and investigation J.L.; Writing—review & editing and Conception S.G.; Investigation and data curation M.J. All authors have read and agreed to the published version of the manuscript.

**Funding:** This research was funded by National innovation and entrepreneurship training program for College Students (S202012715006) and the Youth Innovation Team of Shaanxi Universities (21JP138 and 22JP098).

**Institutional Review Board Statement:** Not applicable.

**Informed Consent Statement:** Not applicable.

**Data Availability Statement:** The raw/processed data required to reproduce these findings cannot be shared at this time as the data also forms part of an ongoing study.

**Acknowledgments:** The project is supported by National innovation and entrepreneurship training program for College Students (S202012715006) and the Youth Innovation Team of Shaanxi Universities (21JP138 and 22JP098) which are gratefully acknowledged.

**Conflicts of Interest:** We declare that we do not have any commercial or associative interest that represents a conflict of interest in connection with the work submitted.

## References


1. Tam, V.W.Y.; Soomro, M.; Evangelista, A.C.J. A review of recycled aggregate in concrete applications (2000–2017). *Constr. Build. Mater.* **2018**, *172*, 272–292. [CrossRef]
2. Xiao, J.; Li, W.; Fan, Y.; Huang, X. An overview of study on recycled aggregate concrete in China (1996–2011). *Constr. Build. Mater.* **2012**, *31*, 364–383. [CrossRef]
3. Gao, S.; Zhao, G.; Guo, L.; Zhou, L.; Yuan, K. Utilization of coal gangue as coarse aggregates in structural concrete. *Constr. Build. Mater.* **2021**, *268*, 121212. [CrossRef]
4. Gao, S.; Zhao, G.; Guo, L.; Zhou, L.; Cui, X.; Yang, H. Mechanical properties of circular thin-tubed molybdenum tailing concrete stubs. *Constr. Build. Mater.* **2021**, *268*, 121215. [CrossRef]
5. Tam, V.W.Y.; Soomro, M.; Evangelista, A.C.J. Quality improvement of recycled concrete aggregate by removal of residual mortar: A comprehensive review of approaches adopted. *Constr. Build. Mater.* **2021**, *288*, 123066. [CrossRef]
6. Gao, S.; Cui, X.; Kang, S.; Ding, Y. Sustainable applications for utilizing molybdenum tailings in concrete. *J. Clean. Prod.* **2020**, *266*, 122020. [CrossRef]
7. Wang, W.; Wei, W.; Gao, S.; Chen, G.; Yuan, J.; Li, Y. Agricultural and Aquaculture Wastes as Concrete Components: A Review. *Front. Mater.* **2021**, *8*, 762568. [CrossRef]
8. Liu, C.; Hai, N.; Zhang, Y. The content variation of phytochemicals in different parts of Chinese jujube from different Cultivars. *Mod. Food Sci. Technol.* **2014**, *30*, 258–261+205. (In Chinese)
9. Xie, Y.; Chu, H.H.; Li, G.Y.; Tao, P.; Lin, X. Preparation and uranium adsorption property of activated carbon from jujube nucleus. *J. Southwest Univ. Sci. Technol.* **2021**, *36*, 1–8. (In Chinese)

10. Alam Khan, T.; Nouman, M.; Dua, D.; Khan, S.A.; Alharthi, S.S. Adsorptive scavenging of cationic dyes from aquatic phase by $H_3PO_4$ activated indian jujube (*ziziphus mauritiana*) seeds based activated carbon: Isotherm, kinetics, and thermodynamic study. *J. Saudi Chem. Soc.* **2022**, *26*, 101417. [CrossRef]
11. Lu, W.; Cai, H.; Yu, W.; Xu, H.; Han, X. Mechanical properties of jujube pit/linear low density polyethylene composites. *Acta Mater. Compos. Sin.* **2021**, *38*, 1775–1783. (In Chinese)
12. Ganiron, T.U., Jr. Sustainable management of waste coconut shells as aggregates in concrete mixture. *J. Eng. Sci. Technol. Rev.* **2013**, *6*, 7–14. [CrossRef]
13. Gunaskearan, K.; Kumar, P.S.; Lakshmipathy, M. Mechanical and bond properties of coconut shell concrete. *Constr. Build. Mater.* **2011**, *25*, 92–98. [CrossRef]
14. Alengaram, U.J.; Munit, B.A.A.; Jumaat, M.Z.B. Utilization of oil palm kernel shell as lightweight aggregate in concrete—A review. *Constr. Build. Mater.* **2013**, *38*, 161–172. [CrossRef]
15. Khan, M.M.H.; Guong, W.L.; Deepak, T.J.; Nair, S. Use of oil palm shell as replacement of coarse aggregate for investigating properties of concrete. *Int. J. Appl. Eng. Res.* **2016**, *11*, 2379–2383.
16. Muthusamy, K.; Nordin, N.; Vesuvapateran, G.; Ali, M.; Mohd Annual, N.A.; Harun, H.; Ullap, H. Exploratory study of rubber seed shell as partial coarse aggregate replacement in concrete. *Res. J. Appl. Sci. Eng. Technol.* **2014**, *7*, 1199–1202. [CrossRef]
17. Memon, S.A.; Javed, U.; Khushnood, R.A. Eco-friendly utilization of corncob ash as partial replacement of sand in concrete. *Constr. Build. Mater.* **2019**, *195*, 165–177. [CrossRef]
18. Netinger, I.; Kesegic, I.; Guljas, I. The effect of high temperatures on the mechanical properties of concrete made with different types of aggregates. *Fire Saf. J.* **2011**, *46*, 425–430. [CrossRef]
19. Abdulhameed, H.; Mansi, A.; Mohammed, A.; Abdulhameed, A.; Hanoon, A. Study the use of Nano-limestone and Egg-shell Ash in Eco-friendly SCC: An Experimental and Statistical Evaluation Based on Computer Programming. In Proceedings of the 2021 14th International Conference on Developments in eSystems Engineering (DeSE), Sharjah, United Arab Emirates, 7–10 December 2021; pp. 509–514.
20. Mo, K.H.; Alengaram, U.J.; Jumaat, M.Z.; Liu, M.J.Y.; Lim, J. Assessing some durability properties of sustainable lightweight oil palm shell concrete incorporating slag and manufactured sand. *J. Clean. Prod.* **2016**, *112*, 763–770. [CrossRef]
21. Gunasekaran, K.; Annadurai, R.; Kumar, P.S. A study on some durability properties of coconut shell aggregate concrete. *Mater. Struct.* **2015**, *48*, 1253–1264. [CrossRef]
22. *JGJ 52-2006*; Standard for Technical Requirements and Test Method of Sand and Crushed Stone (or Gravel) for Ordinary Concrete. Ministry of Construction of the People's Republic of China: Beijing, China, 2007.
23. Zhang, Q.; Wang, Z.; Shan, G.; Zhu, E.; Ly, F.; Zhou, G. Research progress in chemical constituents of *Zizyphus jujuba*. *Acta Agric. Jiangxi* **2013**, *25*, 25–29. (In Chinese)
24. Hanehara, S.; Yamada, K. Interaction between cement and chemical admixture from the point of cement hydration, absorption behaviour of admixture, and paste rheology. *Cem. Concr. Res.* **1999**, *29*, 1159–1165. [CrossRef]
25. Long, W.; Wang, Y. Effect of pine needle fiber reinforcement on the mechanical properties of concrete. *Constr. Build. Mater.* **2021**, *278*, 122333. [CrossRef]
26. *GB/T 17431.2-2010*; Lightweight Aggregates and Its Test Methods—Part 2: Test Methods for Lightweight Aggregates. General Administration of Quality Supervision, Inspection and Quarantine of the People's Republic of China: Beijing, China, 2011.
27. Cui, Z.; Li, J. Effect of coarse aggregate with different water absorption on strength and drying shrinkage of concrete. *Bull. Chin. Ceram. Soc.* **2016**, *36*, 2396–2399. (In Chinese)
28. *JGJ 55-2011*; Specification for Mix Proportion Design of Ordinary Concrete. Ministry of Housing and Urban-Rural Development of the People's Republic of China: Beijing, China, 2011.
29. Mohamedbhai, G.T.G. Effect of exposure time and rates of heating and cooling on residual strength of heated concrete. *Mag. Concr. Res.* **1986**, *38*, 151–158+66. [CrossRef]
30. Liu, J.C.; Tan, K.H.; Yao, Y. A new perspective on nature of fire-induced spalling in concrete. *Constr. Build. Mater.* **2018**, *184*, 581–590. [CrossRef]
31. Fan, L.D.; Li, P.T.; Yu, Y.Q.; Zhang, Z.J. Research progress on property, constitutive relation and failure mechanism of lightweight aggregate concrete. *Concrete* **2017**, *11*, 44–49. (In Chinese)
32. Shen, W.; Zhou, M. Influence Mechanism of Sucrose on Hydration Process of Portland Cement. *J. Build. Mater.* **2007**, *10*, 566–572. (In Chinese)
33. Tanyildizi, H.; Coskun, A. The effect of high temperature on compressive strength and splitting tensile strength of structural lightweight concrete containing fly ash. *Constr. Build. Mater.* **2008**, *22*, 2269–2275. [CrossRef]
34. *GB/T 50010-2010*; Code for Design of Concrete Structures. Ministry of Housing and Urban-Rural Development of the People's Republic of China: Beijing, China, 2010.
35. Yu, Z.W.; Ding, F.X. Unified calculation method of compressive mechanical properties of concrete. *J. Build. Struct.* **2003**, *24*, 41–46. (In Chinese)
36. Xi, Y.L. Material and structural properties of coal gangue and coal gangue concrete. *Ind. Constr.* **1983**, *3*, 30–40. (In Chinese)
37. Ding, F.X.; Ying, X.W.; Yu, Z.W. Unified calculation method of uniaxial mechanical properties of lightweight aggregate concrete. *J. Cent. South Univ.* **2010**, *41*, 1973–1979. (In Chinese)

38. Wang, Z.Y.; Ding, J.T.; Guo, Y.S. Stress-strain curves of structural lightweight aggregate concrete. *Concrete* **2005**, *3*, 39–41+66. (In Chinese)
39. Al-Kharabsheh, B.N.; Arbili, M.M.; Majdi, A.; Majdi, A.; Alogla, S.M.; Hakamy, A.; Ahmad, J.; Deifalla, A.F. Basalt Fiber Reinforced Concrete: A Compressive Review on Durability Aspects. *Materials* **2023**, *16*, 429. [CrossRef] [PubMed]
40. Abdulhameed, A.A.; Al-Zuhairi, A.H.; Al Zaidee, S.R.; Hanoon, A.N.; Al Zand, A.W.; Hason, M.M.; Abdulhameed, H.A. The Behavior of Hybrid Fiber-Reinforced Concrete Elements: A New Stress-Strain Model Using an Evolutionary Approach. *Appl. Sci.* **2022**, *12*, 2245. [CrossRef]

*materials* MDPI

*Article*

# Quantification of Ceramsite Granules in Lightweight Concrete Panels through an Image Analysis Technique

**Changming Bu [1,2], Haiyan Yang [1,2], Lei Liu [1,2], Dongxu Zhu [1,2], Yi Sun [1,2,*], Linwen Yu [3], Yuhui Ouyang [1,2], Xuemei Cao [1,2] and Qike Wei [4]**

[1] School of Civil Engineering and Architecture, Chongqing University of Science & Technology, Chongqing 401331, China; buchangming@cqust.edu.cn (C.B.); 2019206041@cqust.edu.cn (H.Y.); 2020206111@cqust.edu.cn (L.L.); 2020206029@cqust.edu.cn (D.Z.); 2020206038@cqust.edu.cn (Y.O.); 2020206051@cqust.edu.cn (X.C.)
[2] Chongqing Key Laboratory of Energy Engineering Mechanics & Disaster Prevention and Mitigation, Chongqing 401331, China
[3] College of Materials Science and Engineering, Chongqing University, Chongqing 400044, China; linwen.yu@cqu.edu.cn
[4] China Metallurgical Construction Engineering Group Construction, Ltd., Chongqing 400084, China; weiqike@cmccltd.com
* Correspondence: sunyi@cqust.edu.cn; Tel.: +86-135-9416-9610

**Abstract:** Ceramsite particles are an important component of lightweight ceramsite concrete wall panels, and the density of the aggregate is much lower than the density of the slurry. It is generally accepted that there are inhomogeneities in the distribution of ceramsite particles in wall panels. Ceramsite concrete wallboard material is a research hotspot in the field of fabricated building materials at home and abroad; however, there is no effective way to quantify their inhomogeneity. Based on the application of image recognition technology in concrete homogeneity, a method to quantitatively evaluate the distribution of light aggregates in wall panels was developed. Three commercial lightweight vitrified concrete wall panels were cut into 324 cubes. The four cut surfaces of each specimen were photographed to analyze the proportion of ceramsite particle area, while the density, ultrasonic pulse velocity, and compressive strength of the specimens were tested. The results demonstrated that the image analysis method could effectively describe the homogeneity of the panels. The proportion of particle area of aggregate in the section of the cube had a strong correlation with the compressive strength, ultrasonic pulse velocity, and density, and there was an obvious linear relationship with the height of the plate where the cube was located. Based on this, the correlation equations of the proportion of particle area of aggregate, density, ultrasonic pulse velocity, compressive strength, and the height where the specimen was located were proposed. The quantitative parameters of the relevant properties of the wall panels were also obtained: the maximum difference between the proportion of particle area of the aggregate was 24%, the maximum difference between the density at the top and bottom of the wall panels was 115 $kg/m^3$, and the maximum difference in the strength reached 5 MPa.

**Keywords:** ceramsite concrete; lightweight concrete wall panels; density; compressive strength; ultrasonic pulse velocity; quantitative study

check for updates

**Citation:** Bu, C.; Yang, H.; Liu, L.; Zhu, D.; Sun, Y.; Yu, L.; Ouyang, Y.; Cao, X.; Wei, Q. Quantification of Ceramsite Granules in Lightweight Concrete Panels through an Image Analysis Technique. *Materials* **2022**, *15*, 1063. https://doi.org/10.3390/ma15031063

Academic Editor: Alessandro P. Fantilli

Received: 22 December 2021
Accepted: 27 January 2022
Published: 29 January 2022



## 1. Introduction

Lightweight concrete wall panels are mostly used as non-load-bearing structural elements in assembled buildings, usually as partition walls [1] or as composite panels with other materials for exterior walls [2]. As an assembled building, being lightweight is the primary requirement [3], and a lighter wall panel weight can help reduce the overall weight of the assembled building while reducing the permanent load of the load-bearing panels in the structure. Ceramsite concrete wall panels have become mainstream in the fabricated building wallboard market due to their low heat conductivity [4], low density [5], good permeability resistance [6], low early age autogenous shrinkage [7], and early and faster gains in strength [8].

However, as a lightweight concrete wallboard, its strength index also needs to be fully considered. Some studies have summarized the relevant laws to draw general conclusions that the strength of lightweight concrete materials increases with increasing density [9], with a different emphasis for different structural parts [10], such as the compressive strength of compression members and the flexural strength of tension bending members. Although normal-weight aggregate concrete performed better, lightweight aggregate concrete weighed less and had a positive effect on the environment [11], lower early age autogenous shrinkage [12], relatively good thermal insulation performance [13], less detrimental impacts of solid waste on the environment, and also addressed the shortage of natural resources [14]. Some scholars have discussed the influence of substitution rate [8], shape [15], materials [16], and type [17] of the ceramsite, silica fume [18], and CNTs [19] on the compressive strength of lightweight aggregate concrete.

Some scholars have shown that there is a significant density difference between ceramsite and cement paste [20], the antisegregation of lightweight aggregate is weak, and the segregation phenomenon of ceramsite concrete is more significant [21], so the aggregate cannot be evenly distributed in the concrete, thus affecting the strength of the specimens [22]. Compared with ordinary concrete, one of the biggest disadvantages of lightweight concrete is the unevenness of aggregate in the mixture. The uneven distribution of aggregate will seriously affect the working performance of concrete materials and have a negative impact. In order to reduce the segregation phenomenon [23,24], most engineers adopt a way of optimizing the means of vibrating and reducing the time of vibrating [25,26], which requires the construction personnel to finish a large degree of slurry vibration in a short time [21]. In addition, the goal of reducing the segregation phenomenon can also be achieved by modifying the ceramsite granules. Most often, prewetted ceramsite granules are used to reduce the difference in density [27], adjust the ceramsite granule gradation [28], and reduce the aggregate particle size [29]. Increasing the surface roughness of the aggregate can also help to reduce the segregation phenomenon [30]. In recent years, image recognition technology has become an effective technique for identifying the surface and internal structure of structures and is widely used for rapid crack detection, damage monitoring [31], and safety assessment of bridge structures, highways [32], and subway tunnels [33]. Image recognition techniques can also be used for the detection of homogeneity in vitrified concrete wall panels and can help to determine the distribution of vitrified particles in strong panel material [34]. Although there are many methods to study the unevenness of concrete, applying these methods to the quantification of commercial ceramsite concrete wallboard is still lacking.

Based on the actual engineering needs, this paper quantifies the uneven distribution of ceramsite in wallboard by using the uniformity evaluation method. The wallboard was cut into cubes and its physical and mechanical properties were tested. In addition, the distribution of ceramsite was studied by image analysis technology. The main purpose was to reveal and quantify the uneven distribution of ceramsite concrete wallboard and provide reference data for its application in engineering.

## 2. Experimental Details

### 2.1. Geometry of the Wall Panels

The size of the wall panels was 3000 $\times$ 600 (width $\times$ height, in mm), and their thicknesses were 90 mm, 100 mm, 120 mm, and 150 mm. The thickness of the wall panels selected for this study was 100 mm. The schematic diagram of the wall panel cutting is shown in Figure 1.

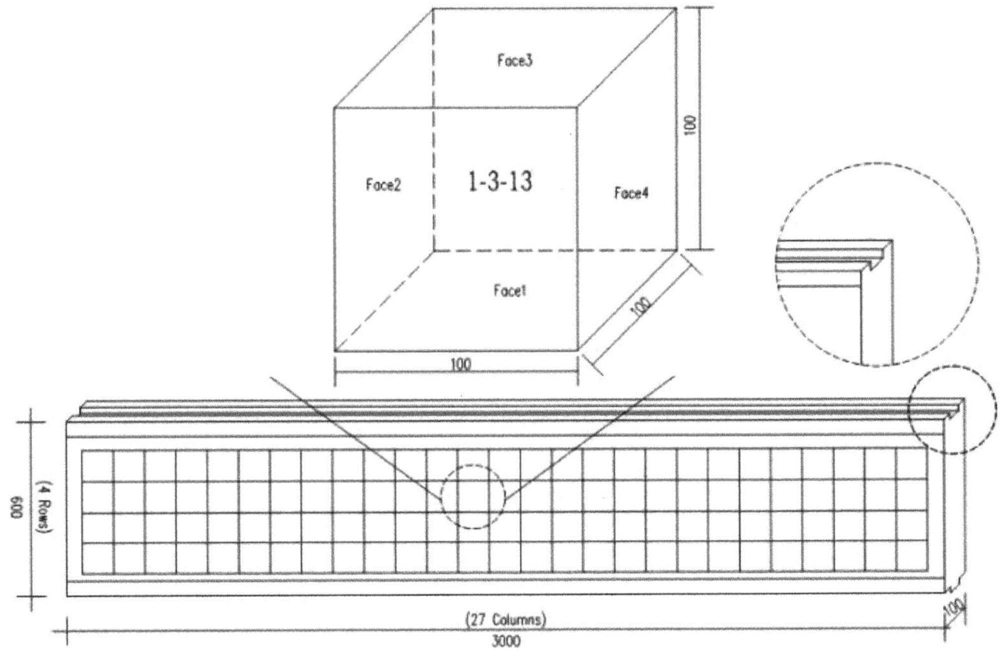


**Figure 1.** Schematic diagram of wall panel cuttings.

### 2.2. Material Composition of Wall Panels

The wall panels used in the present study were developed by Chongqing Chengwei Lightweight Wallboard Co., Ltd. The mix ratio shown in Table 1 was also provided by the company, and was obtained many times by the factory for research, development, and improvement. The water-cement ratio was 0.4, the aggregate of the wallboard used in this paper was clay ceramsite, and the relevant parameters were as follows: the particle size range was 0–10 mm, the density grade was 500, the bulk density was 476 kg/m$^3$, and the moisture content was 16.63%. The chemical composition of the clay used to make ceramsite included quartz, montmorillonite, chlorite, etc. The materials and mix ratios used in the slab are shown in Table 1.The slump value of the freshly poured concrete complied with Chinese standard GB/T50080-2016. Fresh concrete was mixed well and poured into each formwork below from above, and due to the presence of light aggregates, a small action of plugging and pounding was carried out to avoid serious delamination and segregation.

**Table 1.** The proportions of the ceramsite concrete mixture (kg/m$^3$).

| Component | Proportion |
|---|---|
| Cement | 750 |
| Ceramsite | 400 |
| Filament | 2 |
| Foaming agent | 0.5 |
| Additive | 0.7 |

### 2.3. Test Procedures

The test procedures followed in this paper are shown in Figure 2.

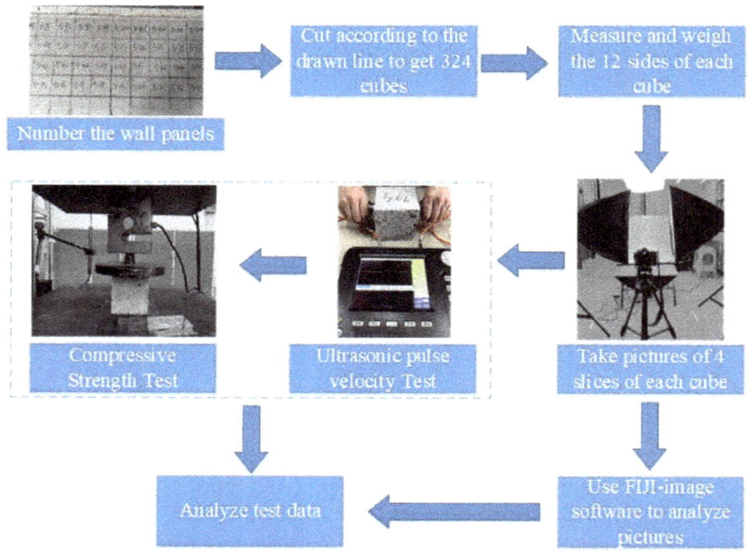


**Figure 2.** Test procedures.

### 2.3.1. Preparation of Specimens

To investigate the inhomogeneous distribution of ceramsite particles in commercial ceramsite concrete wall panels, three 100 mm thick commercial ceramsite concrete wall panels were cut into 100 mm × 100 mm × 100 mm cubic specimens, and all three wall panels were hardened under natural curing conditions. The upper part of the wall panels had a groove for connection, and the lower part had a projection for connection. The upper and lower parts of the wall panels had 100 mm in an inclined state, therefore, the above factors were excluded when cutting the wallboards. At the same time, during the cutting process, the thickness of the cutting blade caused some wear and tear. Finally, each wallboard was cut into 4 rows and 27 columns of samples and marked as A-B-CC, with A indicating the number of the wallboard (1 digit), B indicating the row number (1 digit), and CC indicating the column number (2 digits), as shown in Figure 1.

The method of cutting the wall panels was important for the image identification results; to reduce the effect of specimen cutting on the image identification results, no water was poured during the cutting. Because of the certain strength, this test used a cutting table for the preparation of specimens. In specimen preparation, due to the manual cutting of workers and the strength of the specimen, some specimens had certain cutting scratches on the surface, which was unavoidable.

### 2.3.2. Photography and Analysis

In each cubic specimen, there were four faces that showed the distribution of ceramsite particles in the concrete. The photos of each face were taken from the direction perpendicular to the face. In order to reduce the error in the image recognition analysis, the position of the camera (SONY ILCE-7RM3 includes FE 24-70mm F2.8 GM, Japan, Tokyo) and the placement table was fixed in the photograph of the specimen, and the photograph of each face was taken from the direction perpendicular to the face. Therefore, the distance between the photographic surface and the camera was the same for all specimens, and the actual length of the specimen matched the number of pixels in the graph. Fiji-ImageJ (ImageJ-win64)software was used to complete the recognition of image particles; the main steps of the image recognition are shown below and in Figure 3.

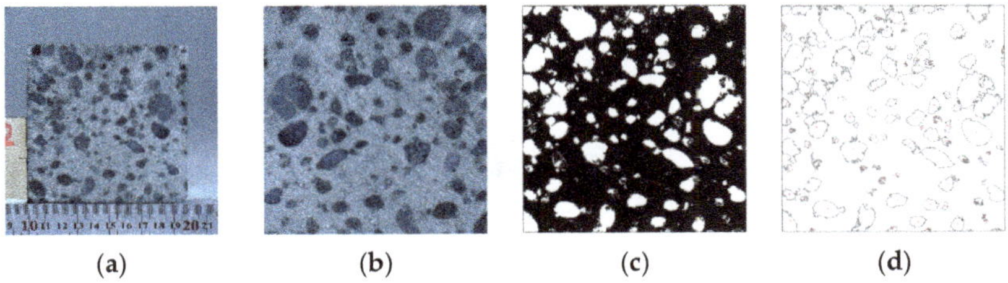


**Figure 3.** Image analysis process: (**a**) original photo; (**b**) cropped photo; (**c**) after binarization; (**d**) outlines of ceramsite beads.

1. The standard scale of the image was used to obtain the relationship between the pixels and the actual size of the image.
2. The rectangular area at the edge was cropped off by Photoshop (PS, Adobe Photoshop CC 2019) to make sure that only the part of vitrified concrete was left in the picture.
3. To distinguish the cement paste from the ceramsite particles, the colored pictures were changed to gray, different algorithms were analyzed and compared, and the automatic global threshold processing method was considered most appropriate.
4. Using the image analysis software program ImageJ, the area of surface ceramsite particles was calculated.

2.3.3. Density, Ultrasonic Pulse Velocity and Compressive Strength Test

The density of ceramsite concrete was obtained by measuring the actual length of 12 sides of each specimen with vernier calipers, accurate to 1 mm, and weighing all cubic specimens in their natural environment.

A nonmetallic ultrasonic detector (ZBL-U520, produced by Beijing Zhibolian Technology Co., Ltd., Beijing, China) was used to test the ultrasonic pulse velocity. Two transducers were aligned to both sides of the specimen, respectively, the ultrasonic pulse wave was recorded through the specimen for a time, and the ultrasonic pulse velocity was calculated according to Equation (1).

$$V = \frac{L}{T} \tag{1}$$

where $V$ is the ultrasonic pulse velocity of the specimens (km/s), $L$ is the distance of the ultrasonic wave passing through, and $T$ is the time of ultrasonic wave passing (s).

Every specimen was put into a testing machine with a 300 kN capacity flatwise, keeping the two facings at the bottom and top to obtain the compressive strength of the core, and the compressive strength was obtained by a dynamic high-speed data acquisition system (DH8303), and the loading velocity was 0.2 MPa/s.

## 3. Test Results and Discussion

### *3.1. Distribution of Ceramsite Granules*

Although the total amount of ceramsite particles in the concrete was determined by the mix ratio of the wall panels, the percentage of particles in each specimen was inconsistent. The three-dimensional distribution information of particles within the specimen was analyzed by two-dimensional images of the surface of the cut specimen, and when hundreds of images were used to analyze and compare this information, a quantitative distribution of particles could be achieved.

As shown in Figure 1, each cube had six faces, and the distribution of ceramsite particles can be seen on all four faces, except for the top and bottom faces of the specimen. The four faces of ceramsite concrete were photographed and analyzed using ImageJ software. The distribution of the percentage of ceramsite particles is discussed below.

3.1.1. Determination of Available Image Areas

A standard ruler was placed on the photo storage table to obtain the relationship between the actual length of the photo and the pixel points of the picture, and then the information of the shape of ceramsite particles was analyzed. Due to the use of manual cutting, the boundary of the specimens could not be idealized into a straight line, and some particles at the edge had a certain impact, so these areas were excluded from the analysis.

The effects of the areas of the selected regions on the results are discussed. A picture of the cut face was selected at random, and 11 rectangular regions of different edge lengths, from 1258 × 1265 to 1258 × 1275 pixels with increments of 10 pixels, were selected for analysis. The results show that with the enlargement of the area of the selected region, the percentage of the area of the ceramsite beads to the area of the selected region was almost the same: the mean value was 37.94%, the standard deviation was 0.01%, and the coefficient of variation was only 0.03.

By selecting rectangular areas with different edge lengths, we found that with a slight change of section area, the percentage of ceramsite particles in the area remained almost unchanged. Therefore, it was reasonable to manually select the image area.

3.1.2. Image Analysis Results

In this paper, the $X$-axis in the cloud chart represents the abscissa of the sample, and the $Y$-axis represents the ordinate of the sample.

Combined with the location information of the specimens, the proportional area of ceramsite particles on each wall panel could be expressed on four sides, as shown in Figure 4. The results show that the distribution of particles on the four sides of each wall panel strongly correlated. The proportion of particles at the lower part of each side of each wall panel was lower than that at the upper part. With increased height, the proportion of ceramsite particles gradually increased, with only a small aggregate of ceramsite particles at the top of the wallboard. The minimum particle areas of the three wallboards were 24.68%, 18.9%, and 21.3%, respectively, the maximum particle areas were 37.62%, 42.5%, and 40.35%, and the differences between the maximum and minimum values were 13%, 23.6%, and 19.05%. The floating of the aggregate is caused by the density difference between the mortar and the aggregate. An aggregate with a larger particle size can obtain greater buoyancy, thus causing unevenness of the aggregate. Therefore, the image analysis method can reflect the unevenness in the concrete to a certain extent [35].

Faces 1 and 3 and faces 2 and 4 of the sample are opposite faces. When the thickness of the cut sample is small enough (up to dx), the particle area ratios of faces 1 and 3 and faces 2 and 4 should be consistent. In this paper, since the thickness of the specimen was 100 mm, there were some differences in the opposite sides, but the general trend remained the same. For example, in Figure 3a, when the abscissa of the sample was 10 and the ordinate was about 3, although the particle area ratios of each surface were quite different, they all reached the larger value of the surface where they were located. It can be seen that the proportions of prominent larger and smaller particle areas were consistent in faces 1 and 3 as well as faces 2 and 4 in the tangential grain area distribution of each wallboard.

At the same time, there are some abnormal situations in the figures. For example, in the part of the middle height of wallboard 2, the proportion of particles is greater than that of the surrounding particle area. This was found to be caused by the insertion and tamping process after pouring, and the insertion and tamping process of wallboard pouring were completed by a hooked reinforcement. When the wallboard is rammed with steel bars, the ceramsite in the lower part lifts, causing the ceramsite to gather in the upper part of the middle of the wallboard.

In order to discuss the effect of height on the proportion of particles, the height variation trend was studied. We used an analysis of variance (ANOVA) to verify the differences among the sample test groups. After the analysis of variance, the results showed that F = 14.24, $p = 7 \times 10^{-6} < 0.05$ for wall panel 1, F = 4.85, $p = 0.04 < 0.05$ for wall panel 2, F = 6.15, $p = 6.8 \times 10^{-4} < 0.05$ for wall panel 3; with the change in the height of the sample from the ground, the difference in the proportion of particle area was statistically significant. Since the distribution trend of the particle area ratio of the four faces of each wall panel was similar, the average value of the particle area ratio of the four faces was obtained, and then the average value of the particle area ratio of the 27 specimens in each row of the wall panel was calculated. As shown in Figure 5, the area ratio of particles increases with the increase in height. After linear fitting, the relationship between the area ratio of particles and the height was obtained, as shown in Equation (2):

$$0.14A = 4.15 + hi \tag{2}$$

where $h_i$ is the distance from the top of the specimens to the ground and $A$ is the particle area proportion of the specimens at $h_i$.

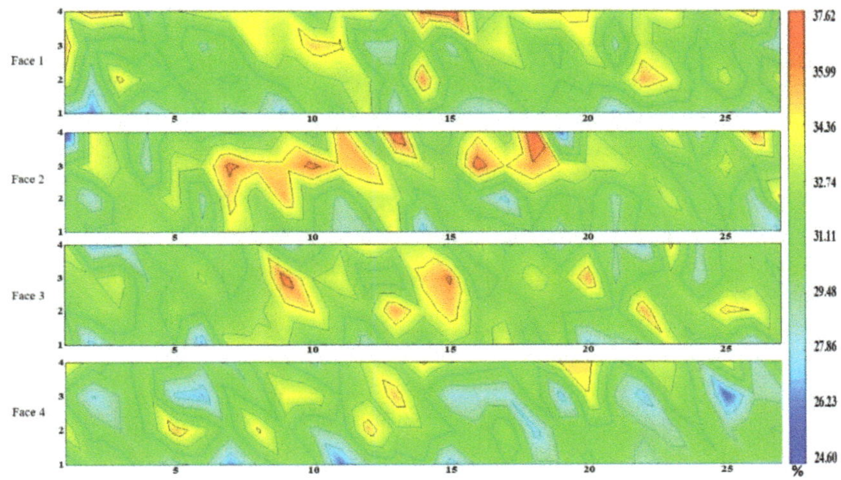


(**a**)

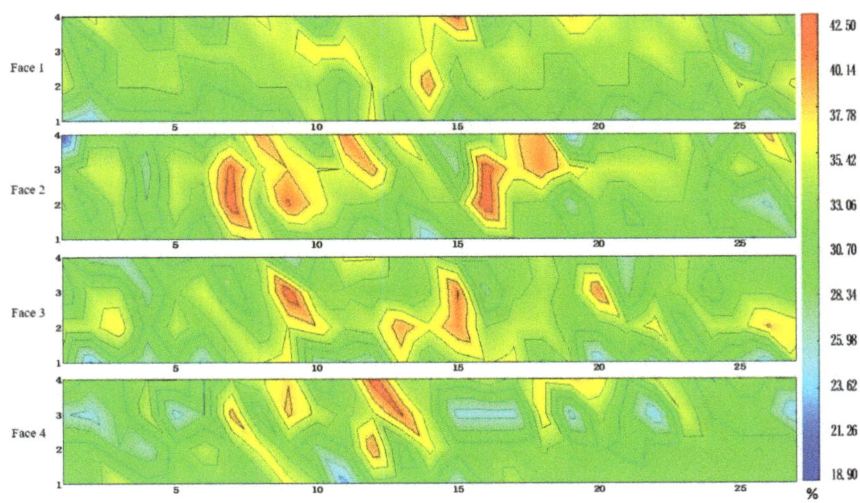


(**b**)

**Figure 4.** *Cont.*

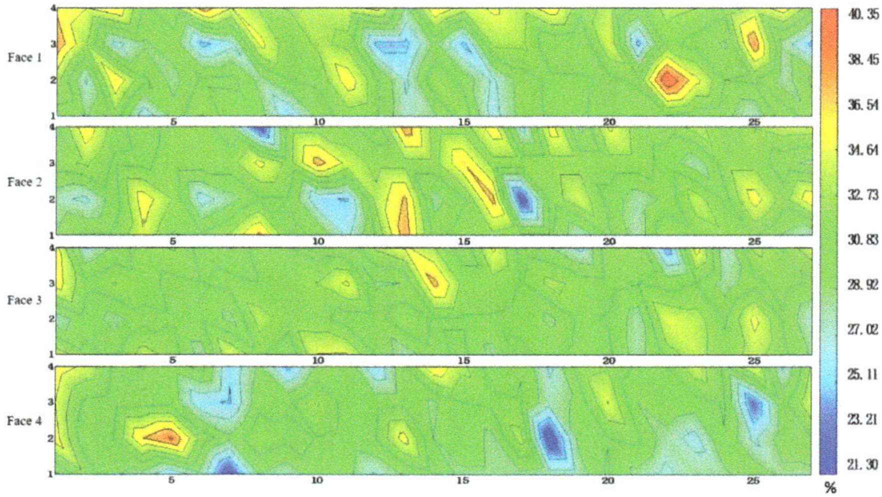


(**c**)

**Figure 4.** Contours of the area ratios of the ceramsite beads in the cut faces: (**a**) panel 1, (**b**) panel 2, (**c**) panel 3.

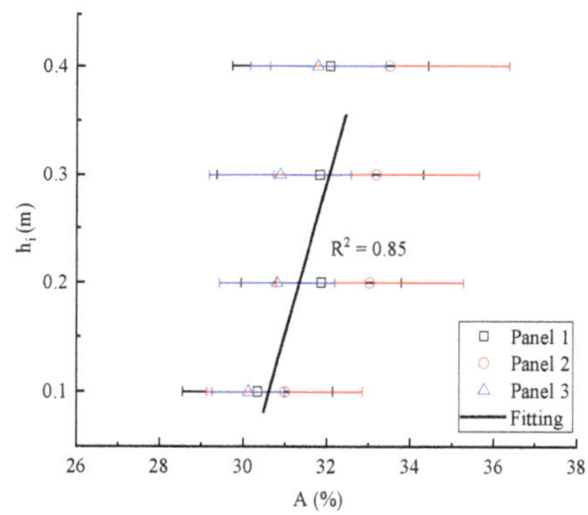


**Figure 5.** Relationship between proportion of ceramsite area and height.

## 3.2. Ultrasonic Pulse Velocity

Nonmetallic, ultrasonic, nondestructive testing technology can be used to analyze the filling capacity and homogeneity of the concrete. The purpose of this experimental research was to analyze the correlation between the height and the ultrasonic pulse velocity of the specimen. The ultrasonic velocity of the specimen was calculated and analyzed by testing the duration of passage of the ultrasonic pulse through the specimen.

The ultrasonic pulse velocity distribution nephogram of the three wallboards is shown in Figure 6. Each wallboard was cut into 108 specimens, named 1–108, respectively. With the increase in height of the specimens from the ground, the ultrasonic pulse velocity of the specimens decreased gradually. Because of the increase in ceramsite particles in the specimens, the time required for the ultrasonic wave to pass through the specimens increases and the ultrasonic pulse velocity decreases. This phenomenon is consistent with the conclusion of the previous section, because there are more particles at the top, the ultrasonic velocity at the top is smaller.

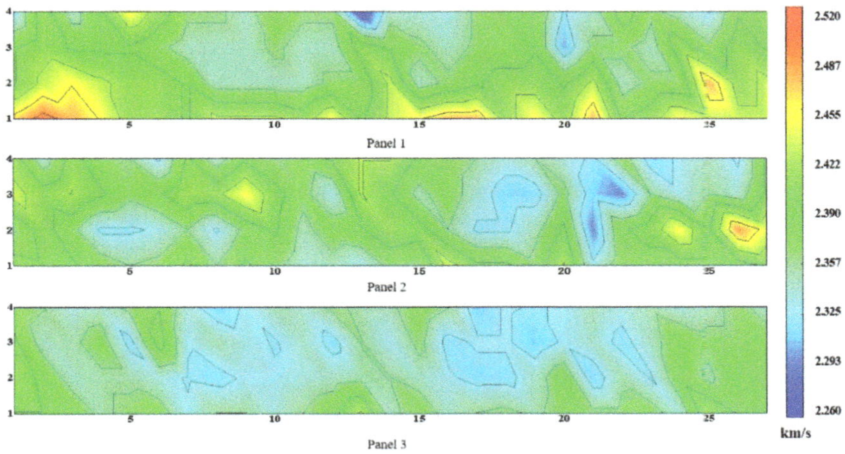


**Figure 6.** Ultrasonic pulse velocity distribution nephogram.

It can be seen from Figure 6 that the ultrasonic pulse velocity at the bottom is basically the highest for the three wallboards. On the one hand, there is a large amount of cement paste at the bottom and the cement paste is the component that really changes the velocity [36]. The energy consumed by ultrasonic waves passing through cement paste is lower than that passing through the lightweight aggregate; therefore, the time for the ultrasonic pulse wave to pass through the former is lower than that passing through the latter. On the other hand, the agglomeration of the upper ceramsite results in a lack of tight bonding with the cement paste, resulting in excessive pores, and the wave propagation speed in the air will be much slower than that in the solid.

In order to discuss the effect of height on ultrasonic pulse velocity, the trend of its variation with height was analyzed, and its calculation method was consistent with the calculation of the correlation between the percentage of particle area and height. Through the analysis of variance, the $p$ values of the sample data were all less than 0.05, and the coefficients of variation were 0.015, 0.013, and 0.01, respectively. As shown in Figure 7, with the increase in height, the ultrasonic pulse velocity of the specimens gradually decreased. The relationship between ultrasonic pulse velocity and height was obtained, as shown in Equation (3):

$$7.46V = 18.0 - hi \tag{3}$$

where $h_i$ is the distance from the top of the specimens to the ground, and $V$ is the ultrasonic pulse velocity of the specimens at $h_i$.

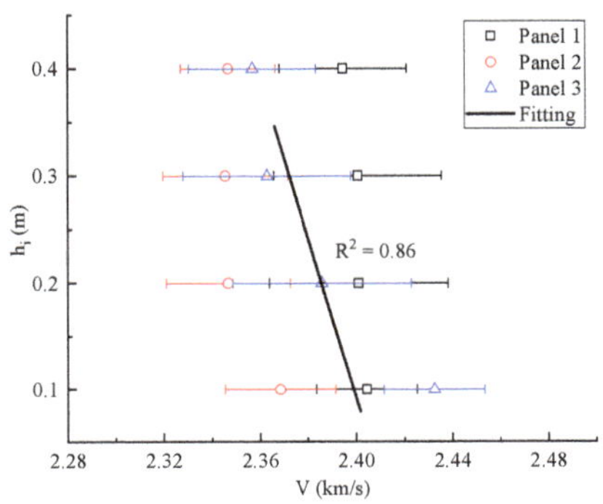


**Figure 7.** Variation trend of ultrasonic pulse velocity with height.

*3.3. Distribution of Density and Compressive Strength*

Figures 8 and 9 show the density distribution cloud map and compressive strength distribution cloud map of the wallboard, respectively. It can be clearly seen that the density and compressive strength of the bottom of the wall panel are larger, and the opposite is true at the top. From the bottom to the top, the density decreased from 1010 kg/m$^3$ to 896 kg/m$^3$, and the compressive strength decreased from 10 MPa to 5 MPa. Due to the nonuniformity of the particles, the density of the middle part of plates 2 and 3 was relatively discrete, the compressive strength and density of the specimens were inversely proportional to the proportion of particle area, the proportion of particle area increased, and the compressive strength and density of the specimens decreased.

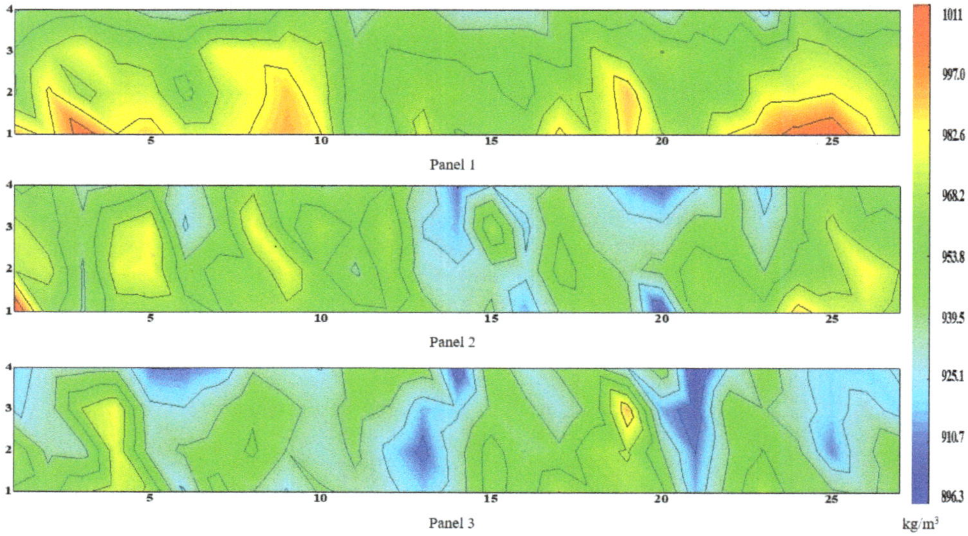


**Figure 8.** Density distribution nephogram.

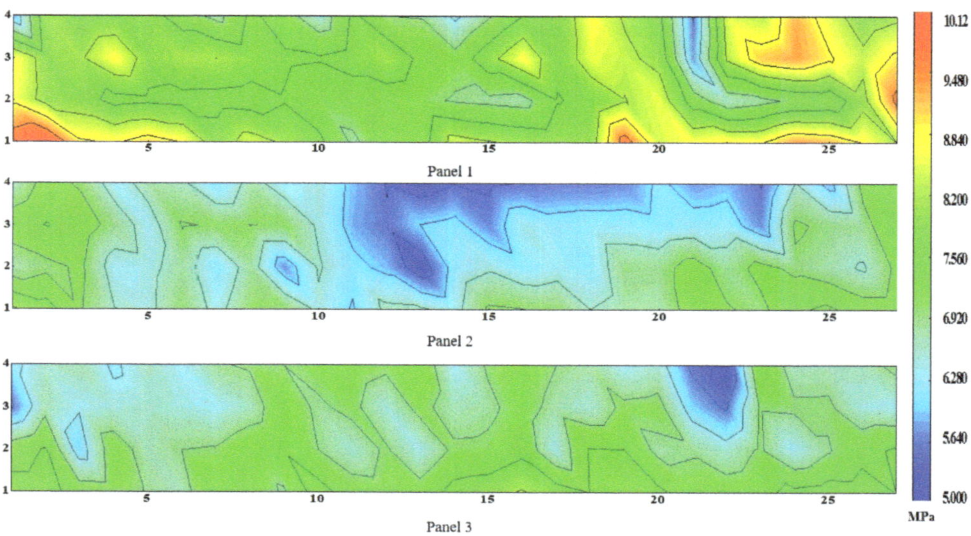


**Figure 9.** Compressive strength distribution nephogram.

As shown in Figure 10, a specimen's density and compressive strength tended to decrease with increasing height. The *Y*-axis represents the height, which is the top of the specimen from the ground, and the *X*-axis represents density and compressive strength in Figure 10a,b, respectively. By taking the average of 27 specimens in each row, the density and compressive strength of the specimens were obtained in relation to the height. As shown by the fitting curve in the figure, as the height decreased, the density and compressive strength of the specimen gradually increased.

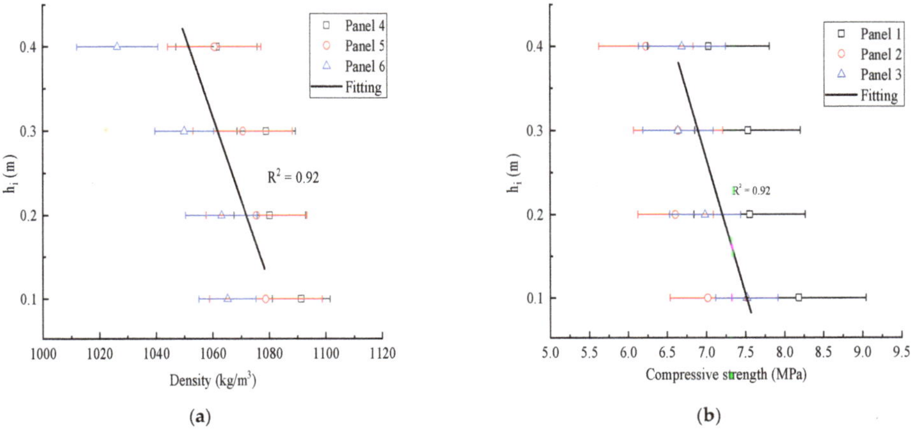


**Figure 10.** Graphs of parameter variation with height: (**a**) variation trend of density with height; (**b**) variation trend of compressive strength with height.

By varying six different factors on density and compressive strength, Lu et al. [37] found a positive correlation between the changes in density and compressive strength. Meanwhile, Zhang et al. [38] suggested that the strength increases with increasing density in a basically linear relationship and Ahmad et al. [39] considered that density has a direct relationship with the strength (the higher the density, the higher the strength). Previous studies [37–39] have shown that the compressive strength increases as the density of the specimen increases, which is consistent with the phenomena shown in this paper (Figure 11). The decrease in specimen density with height is because there are more aggregates at the top, and at the same time the thickness of the gel between the aggregates becomes smaller, resulting in a decrease in the compression resistance of the specimen.

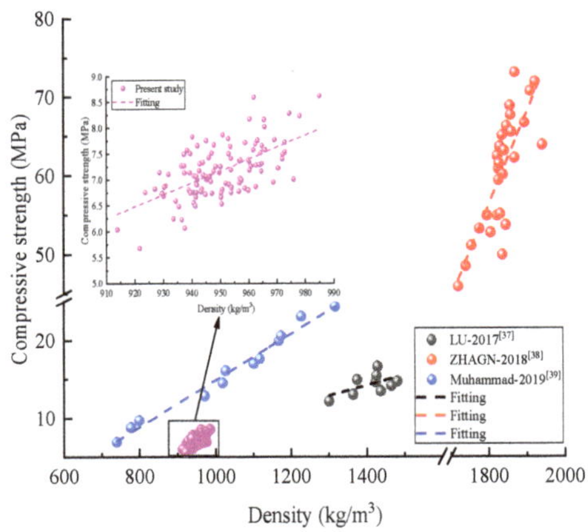


**Figure 11.** Comparison with previous studies [37–39].

The average of 27 specimens in each row of each wall panel can be used to obtain the average density and height distribution of compressive strength in the corresponding row. It can be seen that these values seem to have a linear relationship with the height, and the changes in density and compressive strength with the increase in height comply with Equations (4) and (5):

$$0.01\rho = 12.27 - hi \tag{4}$$

$$0.32fc = 2.53 - hi \tag{5}$$

where $h_i$ is the distance from the top of the specimens to the ground, $\rho$ is the density of the specimens at $h_i$, and $f_c$ is the compressive strength of the specimens at $h_i$.

Integrating the above equations, the relationship between the density, the compressive strength, the ultrasonic pulse velocity and the percentage of the particle area of the specimen can be obtained as shown in Equations (6)–(8) below. There is a certain correlation between the density, compressive strength, and ultrasonic pulse velocity of the specimens and the proportion of the specimen's particle area. With an increase in the proportion of particle area, the density, compressive strength, and ultrasonic pulse velocity of the specimens gradually decreased, as shown in Figure 12.

$$0.01\rho = 16.42 - 0.14A \tag{6}$$

$$7.46V = 13.85 - 0.14A \tag{7}$$

$$0.32fc = 6.68 - 0.14A \tag{8}$$

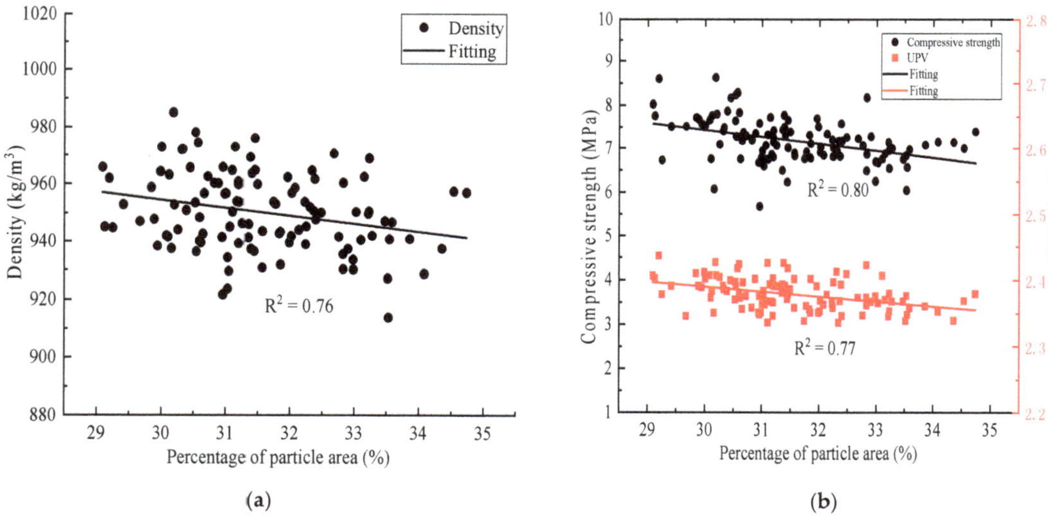


**Figure 12.** Parameter Correlation: (**a**) relationship between percentage of particle area and density; (**b**) percentage of particle area in relation to compressive strength and UPV.

Here, $h_i$ is the distance from the top of the specimens to the ground, $\rho$ is the density of the specimens at $h_i$, $f_c$ is the compressive strength of the specimens at $h_i$, and $V$ is the ultrasonic pulse velocity of the specimens at *hi*.

The proportion of particle area, density, and strength distribution of commercial ceramsite concrete wallboards were quantified, and the corresponding correlation formula was established, which has great reference value for the performance optimization, production, and installation of commercial ceramsite concrete wallboards. The proportion difference of ceramsite particles is due to the uneven distribution of wallboard particles, resulting in uneven distribution of particle area, density, ultrasonic velocity and compressive strength of wallboard.

## 4. Conclusions

This paper quantified the inhomogeneity of commercial ceramsite concrete wall panels through test methods such as image recognition, compressive strength, and ultrasonic velocity. We have provided relevant reference data for the production and use of wallboards that can help factories to optimize the production of wallboards. The following conclusions can be drawn based on the experimental results and discussion.

1. The results demonstrate that the image analysis method can effectively describe the homogeneity of panels. From the image analysis results of the cut specimens of the wallboards, the commercial ceramsite concrete wallboard showed nonuniformity.
2. There was a serious nonuniform distribution of the aggregate of the commercial ceramsite concrete wallboard; the highest particle area increased from 18.9% at the bottom to 42.5% at the top, with a difference of approximately 24%.
3. From the bottom to the top of the wallboard, the density of ceramsite concrete specimens decreased from 1010 $kg/m^3$ to 895 $kg/m^3$, the corresponding compressive strength decreased from 10 MPa to 5 MPa, and the ultrasonic pulse velocity decreased from 2.5 km/s to 2.3 km/s.
4. According to experiments and regressions in this paper, the distribution of compressive strength, density, and ultrasonic pulse velocity had a linear relationship with the height of the specimens; they decreased with the increase in height from the specimens to the ground. They were negatively correlated with the proportion of ceramsite particle area. This result is dependent on the manufacturing process; different manufacturing processes may cause somewhat different results. In future research, more tests and analyses are needed to verify the same regular pattern for different manufacturing processes.

**Author Contributions:** Conceptualization, validation, and project administration, C.B.; formal analysis, data curation, writing—original draft preparation, and visualization, H.Y. and L.L.; writing—review and editing and investigation, D.Z. and Y.O.; methodology, software, and funding acquisition, Y.S.; supervision, methodology, L.Y.; Writing—review and editing, Investigation, X.C. resources, funding acquisition, Q.W. All authors have read and agreed to the published version of the manuscript.

**Funding:** This research was funded by the Project of Natural Science Foundation of Chongqing Municipality (grant numbers: cstc2021jcyj-msxmX0444); the Project of Chongqing Construction Science and Technology Plan (grant numbers: 2021 No. 1–6); the Project of Chongqing Bureau of Human Resources and Social Security (grant numbers: cx2020008), the Open Fund of Chongqing Key Laboratory of Energy Engineering Mechanics & Disaster Prevention and Reduction (grant numbers: EEMDPM2021103); the Cooperation Project of Ministry of Education "Chunhui Planning" (grant numbers: Z2015147); the Fund of State Key Laboratory of Bridge Engineering Structural Dynamics, Key Laboratory of Bridge Earthquake Resistance Technology, Ministry of Communications, PRC; the third author acknowledges financial support from the Postgraduate Science and Technology Innovation Project in Chongqing University of Science and Technology (No. YKJCX2120601).

**Institutional Review Board Statement:** Not applicable.

**Informed Consent Statement:** Not applicable.

**Acknowledgments:** Thanks to L.Y. for the guidance during the experiment; to Q.W. for the experimental materials and guidance; to all the teachers and classmates who worked hard during the experiment.

**Conflicts of Interest:** The authors declare that they have no known competing financial interests or personal relationships that could have appeared to influence the work reported in this paper.

## References


1. Evola, G.; Marletta, L. The effectiveness of PCM wallboards for the energy refurbishment of lightweight buildings. *Energy Procedia* **2014**, *62*, 13–21. [CrossRef]
2. Liu, P.; Gong, Y.F.; Tian, G.H.; Miao, Z.K. Preparation and experimental study on the thermal characteristics of lightweight prefabricated nano-silica aerogel foam concrete wallboards. *Constr. Build. Mater.* **2021**, *272*, 121895. [CrossRef]
3. Kuznik, F.; Virgone, J.; Noel, J. Optimization of a phase change material wallboard for building use. *Appl. Therm. Eng.* **2008**, *28*, 1291–1298. [CrossRef]
4. Han, R.; Qing, Y. Study on the material performance of ceramsite concrete roof brick. *Procedia Eng.* **2017**, *205*, 642–649. [CrossRef]
5. Zhang, Y.G.; Shi, Y.X.; Shi, J.B.; Wang, Q.X.; Ni, K.; Zhang, F.S. An experimental research on basic properties of ceramsite cellular concrete. *Adv. Mater. Res.* **2017**, *1142*, 329–333. [CrossRef]
6. Fan, L.; Zhang, Z.; Yu, Y.; Li, P.; Cosgrove, T. Effect of elevated curing temperature on ceramsite concrete performance. *Constr. Build. Mater.* **2017**, *153*, 423–429. [CrossRef]
7. Ji, T.; Zheng, D.D.; Chen, X.F.; Lin, X.J.; Wu, H.C. Effect of prewetting degree of ceramsite on the early-age autogenous shrinkage of lightweight aggregate concrete. *Constr. Build. Mater.* **2015**, *98*, 102–111. [CrossRef]
8. Chen, Y.; Hui, Q.; Zhang, H.; Zhu, Z.; Wang, C.; Zhao, J. Experiment and application of ceramsite concrete used to maintain roadway in coal mine. *Meas. Control* **2020**, *53*, 1832–1840. [CrossRef]
9. Yew, M.K.; Yew, M.C.; Beh, J.H.; Saw, L.H.; Lim, S.K. Effects of pre-treated on dura shell tenera shell for high strength lightweight concrete. *J. Build. Eng.* **2021**, *42*, 102493. [CrossRef]
10. Dong, J.; Zhang, T.; Fu, Y.; Zheng, B.; Chai, Y. Study on connection and properties of green assembled building steel structure. *Results Phys.* **2019**, *12*, 1126–1132. [CrossRef]
11. Shen, Y.; Liu, B.; Lv, J.; Shen, M. Mechanical properties and resistance to acid corrosion of polymer concrete incorporating ceramsite, fly ash and glass fibers. *Materials* **2019**, *12*, 2441. [CrossRef] [PubMed]
12. Zhao, H.; Liu, H.; Wan, Y.; Ghantous, R.M.; Li, J.; Liu, Y.; Ni, Y.; Guan, J. Mechanical properties and autogenous deformation behavior of early-age concrete containing pre-wetted ceramsite and CaO-based expansive agent. *Constr. Build. Mater.* **2021**, *267*, 120992. [CrossRef]
13. Adhikary, S.K.; Ashish, D.K.; Rudžionis, Ž. Aerogel based thermal insulating cementitious composites: A review. *Energy Build.* **2021**, *245*, 111058. [CrossRef]
14. Xie, J.; Liu, J.; Liu, F.; Wang, J.; Huang, P. Investigation of a new lightweight green concrete containing sludge ceramsite and recycled fine aggregates. *J. Clean. Prod.* **2019**, *235*, 1240–1254. [CrossRef]
15. Yang, Y.; Wu, W.; Fu, S.; Zhang, H. Study of a novel ceramsite-based shape-stabilized composite phase change material (PCM) for energy conservation in buildings. *Constr. Build. Mater.* **2020**, *246*, 118479. [CrossRef]
16. Sun, Y.; Li, J.; Chen, Z.; Xue, Q.; Sun, Q.; Zhou, Y.; Chen, X.; Liu, L.; Poon, C.S. Production of lightweight aggregate ceramsite from red mud and municipal solid waste incineration bottom ash: Mechanism and optimization. *Constr. Build. Mater.* **2021**, *287*, 122993. [CrossRef]
17. Zhuang, Y.Z.; Chen, C.Y.; Ji, T. Effect of shale ceramsite type on the tensile creep of lightweight aggregate concrete. *Constr. Build. Mater.* **2013**, *46*, 13–18. [CrossRef]
18. Ahmad, M.R.; Chen, B. Experimental research on the performance of lightweight concrete containing foam and expanded clay aggregate. *Compos. Part B Eng.* **2019**, *171*, 46–60. [CrossRef]
19. Adhikary, S.K.; Rudžionis, Ž.; Tučkutė, S.; Ashish, D.K. Effects of carbon nanotubes on expanded glass and silica aerogel based lightweight concrete. *Sci. Rep.* **2021**, *11*, 2104. [CrossRef]
20. Mi, H.; Yi, L.; Wu, Q.; Xia, J.; Zhang, B. Preparation of high-strength ceramsite from red mud, fly ash, and bentonite. *Ceram. Int.* **2021**, *47*, 18218–18229. [CrossRef]
21. Li, Y.; Ding, J. Segregation capability of fresh High Strength Lightweight Aggregate Concrete (HSLWAC). *Concrete* **2005**, *3*, 42–45.
22. Tenza-Abril, A.J.; Villacampa, Y.; Solak, A.M.; Baeza-Brotons, F. Prediction and sensitivity analysis of compressive strength in segregated lightweight concrete based on artificial neural network using ultrasonic pulse velocity. *Constr. Build. Mater.* **2018**, *189*, 1173–1183. [CrossRef]
23. Jiang, L.; Lai, X.; Jiao, H. Concrete relative velocity prediction to prevent mortar segregation for safe gravity transportation. *Alex. Eng. J.* **2020**, *59*, 5155–5164. [CrossRef]
24. Cano-Pleite, E.; Hernández-Jiménez, F.; Acosta-Iborra, A.; Tsuji, T.; Müller, C. Segregation of equal-sized particles of different densities in a vertically vibrated fluidized bed. *Powder Technol.* **2017**, *316*, 101–110. [CrossRef]
25. Yan, W.; Cui, W.; Qi, L. Effect of aggregate gradation and mortar rheology on static segregation of self-compacting concrete. *Constr. Build. Mater.* **2020**, *259*, 119816. [CrossRef]
26. González-Taboada, I.; González-Fonteboa, B.; Pérez-Ordóñez, J.L.; Eiras-López, J. Prediction of self-compacting recycled concrete mechanical properties using vibrated recycled concrete experience. *Constr. Build. Mater.* **2017**, *131*, 641–654. [CrossRef]
27. Yu, L.L.; Chen, J.D.; Wu, Y.W. The performance of the ceramsite concrete block and its application research. *Appl. Mech. Mater.* **2015**, *744*, 1543–1546. [CrossRef]
28. Wang, S.; Ming, Y.; Han, Z.; Dong, W.; Fei, L. Experimental Research on Crushed Ceramsite Concrete. *Ind. Constr.* **2013**, *43*, 111–114.

29. Navarrete, I.; Lopez, M. Understanding the relationship between the segregation of concrete and coarse aggregate density and size. *Constr. Build. Mater.* **2017**, *149*, 741–748. [CrossRef]
30. Huang, H.; Yuan, Y.; Zhang, W.; Liu, B.; Viani, A.; Macova, P. Microstructure investigation of the interface between lightweight concrete and normal-weight concrete. *Mater. Today Commun.* **2019**, *21*, 100640. [CrossRef]
31. German, S.; Brilakis, I.; DesRoches, R. Rapid entropy-based detection and properties measurement of concrete spalling with machine vision for post-earthquake safety assessments. *Adv. Eng. Inform.* **2012**, *26*, 846–858. [CrossRef]
32. Zhou, K.; Lei, D.; He, J.; Zhang, P.; Bai, P.; Zhu, F. Single micro-damage identification and evaluation in concrete using digital image correlation technology and wavelet analysis. *Constr. Build. Mater.* **2021**, *267*, 120951. [CrossRef]
33. Lei, M.; Liu, L.; Shi, C.; Tan, Y.; Lin, Y.; Wang, W. A novel tunnel-lining crack recognition system based on digital image technology. *Tunn. Undergr. Space Technol.* **2021**, *108*, 103724. [CrossRef]
34. Sun, Y.; You, J.; Zhou, J.; Liu, X.; Yu, L.; Bu, C.; Yan, Z.; Chen, X. Quantified research on the non-uniform distribution of expanded polystyrene beads in sandwich panels. *Constr. Build. Mater.* **2020**, *263*, 120672. [CrossRef]
35. Wang, G.; Kong, Y.; Sun, T.; Shui, Z. Effect of water–binder ratio and fly ash on the homogeneity of concrete. *Constr. Build. Mater.* **2013**, *38*, 1129–1134. [CrossRef]
36. Chen, T.T.; Wang, W.C.; Wang, H.Y. Mechanical properties and ultrasonic velocity of lightweight aggregate concrete containing mineral powder materials. *Constr. Build. Mater.* **2020**, *258*, 119550. [CrossRef]
37. Lu, K.; Xia, D.; Yang, S. Experimental study on material m ix of new ceramic concrete. *Concrete* **2017**, *4*, 147–150.
38. Zhang, Y.; Liu, B.; Wu, T.; Chen, X. Mechanical properties and microstructure of high-strength lightweight aggregate concrete with shale ceramist. *J. Xi'an Univ. Archit. Technol. Nat. Sci. Ed.* **2018**, *50*, 225–232.
39. Ahmad, M.R.; Chen, B.; Shah, S.F.A. Investigate the influence of expanded clay aggregate and silica fume on the properties of lightweight concrete. *Constr. Build. Mater.* **2019**, *220*, 253–266. [CrossRef]

*materials*

MDPI

*Article*

# Effect of Microstructure on the Mechanical Properties of Steel Fiber-Reinforced Recycled Concretes

**Hanquan Yuan [1], Lihua Zhu [1,2,*], Xiaopeng Wang [1] and Hongtao Yang [3]**

[1] School of Civil Engineering, Xi'an University of Architecture & Technology, 13 Yanta Road, Xi'an 710055, China; yuanhanquan@126.com (H.Y.); wangxiaopengxa@126.com (X.W.)
[2] State Key Laboratory of Green Building in Western China, Xi'an University of Architecture & Technology, 13 Yanta Road, Xi'an 710055, China
[3] Shaanxi Hong Wei Ecological Environmental Protection Technology Co., Xi'an 710100, China; yanghongtaoxa@126.com
* Correspondence: zhulihuaxa@163.com; Tel.: +86-186-9182-8626

**Abstract:** A steel fiber-reinforced recycled concrete (SFRRC) is a porous material, and its macromechanical properties are affected by its microstructure. To elucidate the change rules and internal mechanisms of the mechanical properties of SFRRCs, the mechanical properties and failure modes of SFRRCs were studied at different water–cement ratio, replacement rate of recycled concrete aggregate (RCA), and steel fiber content. Moreover, the microstructures of the interface transition zones (ITZ) of the SFRRC specimens were tested by scanning electron microscopy and mercury intrusion, and the effect of the microscopic pore structure on the macromechanical properties of SFRRC was analyzed. The research results showed that an appropriate amount of steel fibers could reduce the size and number of cracks in the ITZ and improve the pore structure of an SFRRC. Based on the fractal dimension, porosity and other factors, the quantitative relationship between the macromechanical properties and microscopic pore structure parameters of SFRRCs was established.

**Keywords:** steel fiber-reinforced recycled concrete; macromechanical properties; microstructure; interface transition zone



**Citation:** Yuan, H.; Zhu, L.; Wang, X.; Yang, H. Effect of Microstructure on the Mechanical Properties of Steel Fiber-Reinforced Recycled Concretes. *Materials* **2022**, *15*, 4018. https://doi.org/10.3390/ma15114018

Academic Editor: Jean-Marc Tulliani

Received: 25 April 2022
Accepted: 29 May 2022
Published: 6 June 2022



## 1. Introduction

Since the beginning of the 21st century, concrete has been the most important material basis for urban construction, and its demand has also been increasing. Currently, the annual output of concrete globally is approximately 4 billion $m^3$, and the output of China accounts for approximately 70% of the global total. In urban construction, the demolition of old buildings produces a large amount of construction and demolition waste, which causes environmental pollution [1,2]. Concurrently, the demand for aggregates for concrete production is increasing. Recycled concrete aggregates (RCAs) produced from waste concrete by crushing, screening, and cleaning are used to replace natural concrete aggregates (NCAs), which can effectively solve the dual problems of environment and resources, with very significant economic, social, and environmental benefits [3].

Research results show that RCAs are damaged during the production process, which produces numerous microcracks inside them [4]. Simultaneously, RCAs feature more edges and corners than NCAs, and hardened cement mortar can become attached to their surfaces, as shown in Figure 1. Because of these factors, RCAs have higher porosity and water absorption than NCAs, and their physical properties are poor [5]. To improve the mechanical properties of recycled aggregate concrete (RAC), scholars usually use concrete-modification technology. These modification techniques mainly include removing the cement mortar on the surface of the RCA (physical method) [6–8], filling the pores and cracks inside the RCA with chemical solutions (chemical method) [9–11], and doping fibers to improve the pore structure of the interface transition zone (ITZ) [12–14]. Compared with the complicated operation process of physical and chemical methods, the method of doping fibers is more convenient.

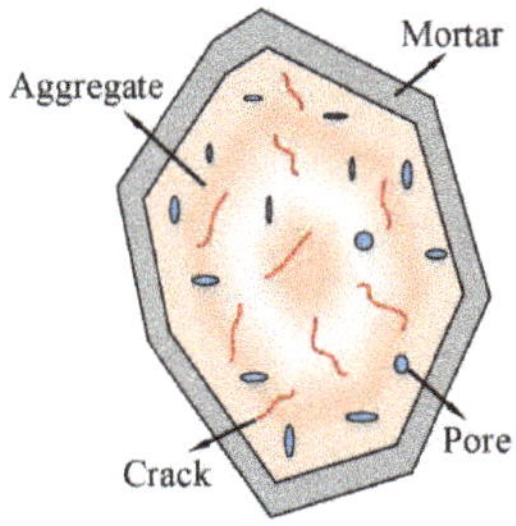


**Figure 1.** Recycled concrete aggregates.

In engineering practice, commonly used fibers are steel fiber (SF) [15], polypropylene fiber [16], copper fiber [17] and basalt fiber [18]. Because SFs can improve and inhibit the development of concrete cracks, improve the quality of RAC, and at the same time, steel fibers can be mass-produced and cheap, the research and application of steel fiber-reinforced concrete has developed rapidly in recent years [19]. According to Bentur, the mechanical properties of steel fiber-reinforced concrete is primarily determined by the bonding properties between the SFs and cement matrix [20]. The ITZ is a weak zone between the SF surfaces and the concrete matrix, as shown in Figure 2. The porosity of the ITZ is relatively high, and its width is approximately 40 μm. The pullout process of the SFs in the matrix is the relative slipping of the SFs caused by the destruction of the structure of the ITZ. Hannawi believes that the SFs have a more hydrophilic surface, and the surrounding cement paste will hydrate and fill the space between the SF and the cement matrix, making its interface transition zone more compact [21]. However, the research on SFRRCs mainly focuses on the macromechanical properties, which are mainly qualitative descriptions. SFRRC is a porous material whose pore structure determines the macromechanical properties. The relationship between the macromechanical properties of SFRRC and the microscopic pore structure has not been discussed in detail, which requires further quantitative research.

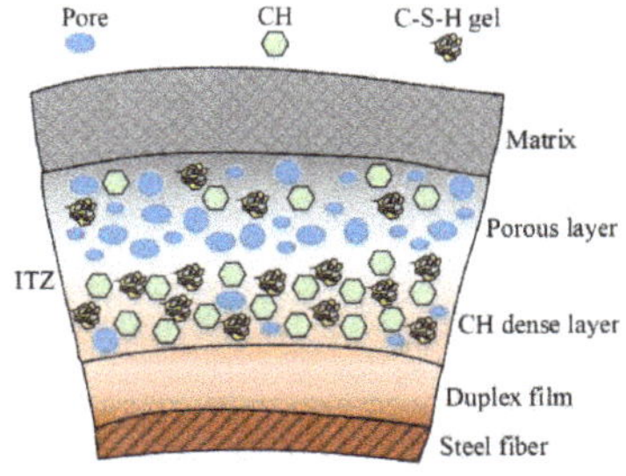


**Figure 2.** Interface transition zone of SFRRC.

Therefore, the main purpose of this study is to explore the relationship between the macromechanical properties of SFRRC and the microscopic pore structure. This work tested the macromechanical properties and microstructure of nine sets of SFRRC specimens with different water–cement ratios, RCAs replacement rates, and SFs content. Finally, a modified formula for the calculation of the strength of SFRRCs based on natural concrete is proposed, and the reasons for the variation of macromechanical properties of SFRRCs from the perspective of microstructure were revealed, and a quantitative relationship between the strength of SFRRCs and microscopic pore structure parameters was proposed.

## 2. Materials and Methods

### 2.1. Materials

The coarse aggregates (RCAs and NCAs) used in this study were purchased from Shaanxi Hong Wei Ecological Environmental Protection Technology Co., Ltd. (Xi'an, China), and the fine aggregates were natural river sand from Xi'an, Shaanxi Province. Among them, RCAs mainly come from the waste concrete blocks generated from the demolition of old buildings (20–30 years old) in Xi'an city, the strength of these concrete blocks is about C20 to C40, and the adhered mortar content of RCAs is in the range of 40–45%. The size ranges of the coarse and fine aggregates were 5–20 mm and 0–5 mm, respectively, and the particle dimension distributions of the aggregates are shown in Figure 3. It can be seen that the aggregates satisfied the ASTM-C33 [22] limit requirements. The primary physical properties of the employed RCAs, NCAs, and sand were tested in accordance with the Chinese code GB/T14685-2011 [23]; these are listed in Table 1. The results show that the density of the RCAs is lower than those of the NCAs and that the crushing index and water absorption are higher than those of the NCAs; this is related to the adhesion of the old bonding mortar to the surface of the RCAs and the internal microcracks.

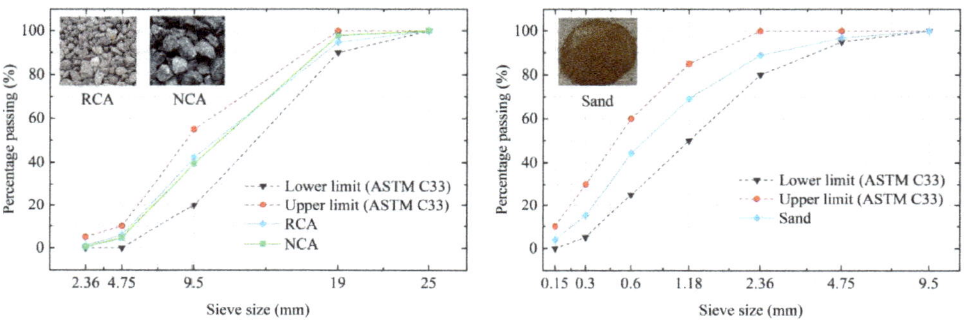


(a) (b)

**Figure 3.** Particle dimension distributions of aggregates: (**a**) Coarse aggregates; (**b**) Fine aggregates.

**Table 1.** Physical properties of RCAs, NCAs, and sand.

| Properties | RCA | | NCA | | Sand |
|---|---|---|---|---|---|
| Particle size (mm) | 5–10 | 10–20 | 5–10 | 10–20 | 0–5 |
| Apparent density (kg/m$^3$) | 2640 | 2650 | 2810 | 2820 | 2620 |
| Bulk Density (kg/m$^3$) | 1290 | 1330 | 1510 | 1550 | 1520 |
| Water absorption (%) | 7.3 | 5.6 | 0.8 | 0.3 | 0.7 |
| Crushing index (%) | — | 13.5 | — | 8.1 | — |
| Fineness modulus | — | — | — | — | 2.76 |

The shear-cut type of the SFs used in this study, which are wavy overall, is shown in Figure 4. The length ($l_f$) of the SFs is 38.0 mm, the equivalent diameter ($d_f$) is 0.6 mm, the aspect ratio ($l_f/d_f$) is 63.3, and the tensile strength is 1000 MPa. The cementitious materials used in all the mixtures were cement (P.O42.5), whose physical and chemical properties are listed in Table 2. The water-reducing agent used in this test is polycarboxylic acid high-performance water-reducing agent, the water reduction rate was approximately 25%, and its indexes are shown in Table 3.

**Figure 4.** Steel fibers.

**Table 2.** Physical and Chemical properties of cement.

| Setting Time (min) | | Chemical Composition (%) | | | | | | | Strength (MPa) | | | |
|---|---|---|---|---|---|---|---|---|---|---|---|---|
| | | | | | | | | | Compressive Strength | | Flexural Strength | |
| Initial Set | Final Set | CaO | $SiO_2$ | $Al_2O_3$ | $Fe_2O_3$ | $SO_3$ | MgO | $K_2O$ | 3d | 28d | 3d | 28d |
| 180 | 280 | 60.3 | 22.5 | 7.2 | 3.6 | 2.1 | 2.3 | 0.8 | 39.1 | 50.9 | 6.5 | 8.1 |

**Table 3.** Physical and chemical compositions of water-reducing agent.

| Density ($g/cm^3$) | Solid Content (%) | Chloride Ion Content (%) | Alkali Content (%) | pH Value | Water Reduction Rate (%) |
|---|---|---|---|---|---|
| 1.062 | 23.7 | 0 | 1.13 | 6.5 | 25 |

## 2.2. Mixture Proportions and Specimen Preparation

In this research, the effects of the water–cement ratio ($w/c$), replacement rate of the RCAs ($r_a$), and SF volume content ($v_{sf}$) on the mechanical properties of SFRRCs were studied. To effectively analyze the influence of various factors on the mechanical properties of SFRRCs, orthogonal experiments were conducted, as summarized in Table 4. Based on previous research [24], in these tests, the unit water consumption and the sand rate were set as 160 $kg/m^3$ and 0.44, respectively. The mixture proportions for the SFRRC specimens are listed in Table 5.

**Table 4.** Orthogonal test factors and levels of SFRC.

| Levels | Factors | | |
|---|---|---|---|
| | $w/c$ (A) | $r_a$ (B) | $v_{sf}$ (C) |
| 1 | 0.4 | 30% | 0% |
| 2 | 0.47 | 50% | 1% |
| 3 | 0.54 | 100% | 2% |

Note: $w/c$: water–cement ratio; $r_a$: replacement ratio of RCAs; $v_{sf}$: SF volume content.

**Table 5.** Mixture proportions of specimens (kg/m$^3$).

| Groups | Orthogonal Design | Water | Cement | Sand | NCA | RCA | Steel Fiber | Extra Water | Water-Reducing Agent |
|---|---|---|---|---|---|---|---|---|---|
| SFRRC-1 | A1B1C1D1 | 160 | 400 | 788 | 702 | 301 | 0 | 18 | 4 |
| SFRRC-2 | A1B2C2D2 | 160 | 400 | 753 | 479 | 479 | 78 | 29 | 4 |
| SFRRC-3 | A1B3C3D3 | 160 | 400 | 719 | 0 | 915 | 156 | 56 | 4 |
| SFRRC-4 | A2B1C2D3 | 160 | 340 | 779 | 694 | 298 | 78 | 18 | 3.4 |
| SFRRC-5 | A2B2C3D1 | 160 | 340 | 745 | 474 | 474 | 156 | 29 | 3.4 |
| SFRRC-6 | A2B3C1D2 | 160 | 340 | 814 | 0 | 1036 | 0 | 63 | 3.4 |
| SFRRC-7 | A3B1C3D2 | 160 | 296 | 765 | 681 | 292 | 156 | 18 | 2.96 |
| SFRRC-8 | A3B2C1D3 | 160 | 296 | 833 | 530 | 530 | 0 | 32 | 2.96 |
| SFRRC-9 | A3B3C2D1 | 160 | 296 | 799 | 0 | 1017 | 78 | 62 | 2.96 |

Note: The extra water is calculated from the water absorption of the RCA and is absorbed by the RCA and does not participate in the hydration reaction of the cement.

The mixing process to form SFRRC specimens with improved properties is shown in Figure 5, which can be summarized as the following phases:

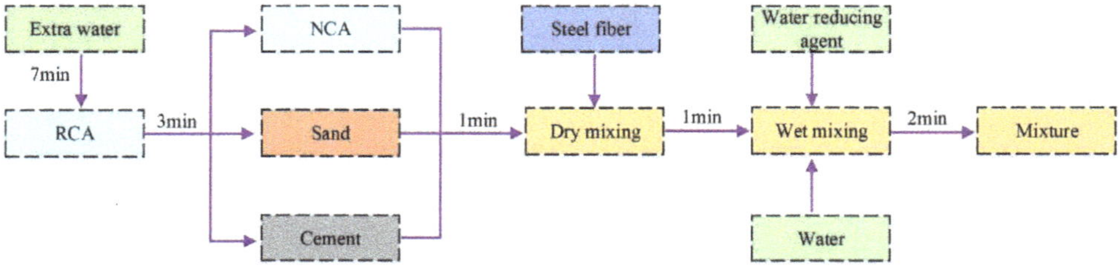


**Figure 5.** SFRC mixing process.

Phase I: First, the RCAs and extra water are mixed for 7 min to ensure the RCAs reach a saturated dry surface. Subsequently, they are mixed for 3 min.

Phase II: The NCAs, sand and cement are mixed for 2 min, and the SFs are added in the process of mixing in a manner that they are evenly dispersed in the dry mixture.

Phase III: Finally, water and a water-reducing agent are added to the mixer and stirred for 2 min, and subsequently the mixture is placed into molds.

For each mixture, six cubic and prismatic specimens were prepared for testing the cubic compressive strength ($f_{cu}$), splitting tensile strength ($f_{ts}$), axial compressive strength ($f_c$), and elastic modulus ($E_c$) of the SFRRCs. Their dimensions were 100 mm × 100 mm × 100 mm and 150 mm × 150 mm × 300 mm, respectively.

### 2.3. *Test Procedures*

#### 2.3.1. Mechanical Properties Test

A TYE-2000E machine was used for all tests, which is shown in Figure 6. The test method was in accordance with the "Standard for test method of mechanical properties on ordinary concrete" (GB/T50081-2002) [25]. The loading rate of the testing machine was set as 0.5 MPa/s until failure of the cubic and prism specimens occurred to obtain $f_{cu}$ and $f_c$. When testing $f_{ts}$, the loading rate was 0.05 MPa/s, and $f_{ts}$ is calculated using Equation (1).

$$f_{ts} = \frac{2F}{\pi A} = 0.637\frac{F}{A} \quad (1)$$

where $F$ is the ultimate load and $A$ is the splitting surface area. Because the cube test block was a nonstandard test block, $f_{ts}$ was multiplied by a conversion factor of 0.85 [25].

**Figure 6.** Pressure-testing machine.

Figure 7 depicts the mechanical property test setup. In the splitting tensile test, a circular arc steel shim and a wooden shim were placed between the pressure plate and the specimen. In the elastic modulus test, the test distance was 150 mm in the mid of the prism sample, and two dial gauges were placed on both sides of the test piece to measure the displacement.

(a) (b)

**Figure 7.** Mechanical property test setup: (**a**) Splitting tensile test; (**b**) Elastic modulus test.

The specimen needs to be preloaded to ensure axial loading. First, loading from 0 to 0.5 MPa ($F_0$) is applied, which is followed by a constant load for 60 s. Subsequently, a load equal to the stress of $f_c/3$ load ($F_a$) is applied, followed again by a constant load for 60 s to ensure that the deformation difference between the two sides is less than 20%. This loop is repeated thrice, following which the preloading ends. After unloading the load, the load is increased from 0 to $F_0$, the deformation value, $L_0$, is recorded; subsequently a load from $F_0$ to $F_a$ is added, and the deformation value, $L_a$, is recorded. Finally, the deformation measuring instrument is removed, and the specimen is loaded to failure. The difference between the failure load and the previously measured axial compressive strength should not exceed 20%. The elastic modulus is calculated using Equation (2).

$$E_c = \frac{F_a - F_0}{A} \times \frac{L}{L_a - L_0} \tag{2}$$

where $E_c$ is the elastic modulus, $F_a$ is taken as $f_c/3$, $F_0$ is taken as 0.5 MPa, $A$ is the pressure-bearing area of the specimen, $L$ is the test distance of the specimen, and $L_a$ and $L_0$ are the average deformation values on the two sides when the loads are $F_a$ and $F_0$, respectively.

#### 2.3.2. Microstructure Test

Concrete is a type of porous material, and it has various pore shapes. Mercury intrusion porosimetry (MIP) is frequently used to analyze the microscopic pore structure of materials. A scanning electron microscope has a high resolution, magnification, and large depth of field, and it can analyze the characteristics of a material at the nanometer level. Thus, the microscopic morphologies of the SFRRC specimens were analyzed using scanning electron microscopy (SEM), and their pore structures were analyzed via MIP. In this research, the SEM experiments employed a Gemini SEM300 scanning electron microscope (Oberkochen, Germany) with a resolution of 0.8 nm and an acceleration voltage of 30 KV. The MIP experiments were conducted using an American AutoPore IV 9500 (Norcross, GA, USA), and the pressure range was 0.003–300 MPa. The test instruments are shown in Figure 8.

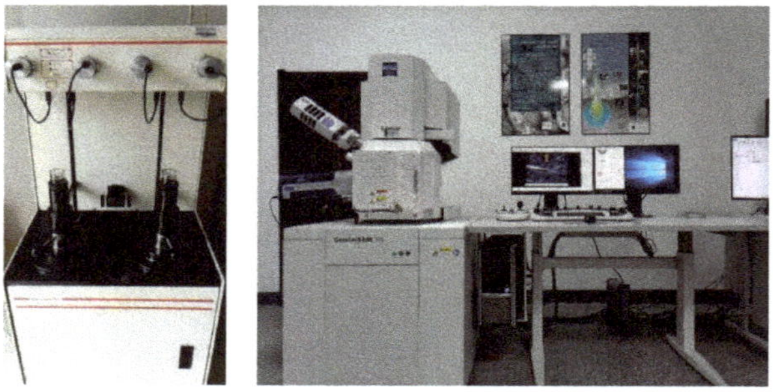

(a) (b)

**Figure 8.** Test instruments: (**a**) Mercury porosimeter; (**b**) Scanning electron microscope.

## 3. Results and Discussion

### *3.1. Failure Mode of SFRRC*

Before the tests, the specimens are placed in a curing chamber for 28 days. The temperature and relative humidity of the curing chamber are, respectively, 20 ± 2 °C and not less than 95%.

In the cubic compressive tests, a specimen without SFs undergoes compression deformation and transverse elongation deformation during compression. At the initial stages of loading, microcracks appear on the surface of the concrete and gradually develop into edges and corners. With an increase in the load, the rapid development of microcracks occurs, expanding both the concrete inside and the surface of the specimen, with the latter severely falling off. Finally, the concrete is crushed, and the failure mode exhibits a wedge shape on the surface of the specimen, which is understandably brittle failure. During the compressive failure process, cracks appear later in the specimens mixed with SFs than in the specimens without SFs. The cracks are significantly small in the early stages, and the microcracks gradually develop and become larger in the later stages, with reduced peeling of the surface of the test specimen. Until the end of the loading of a specimen with SFs, the specimen does not undergo wedge failure and maintains good integrity, showing remarkable plastic failure. A large amount of SFs implies fewer cracks, less shedding, and good integrity. It is found that the addition of SFs improves the failure mode of the concrete and transforms it from brittle failure to plastic failure. The failure modes in the axial compressive strength tests are basically the same as those in the cubic compressive strength tests. The typical failure modes in the cubic and prismatic axial compressive tests of the SFRRC specimens are shown in Figures 9 and 10, respectively. The failure modes of the SFRRCs with different replacement ratios for the RCAs and water-to-binder ratios are similar.

(a) (b) (c)

**Figure 9.** Failure modes in cubic compressive tests: (**a**) $v_{sf} = 0\%$; (**b**) $v_{sf} = 1\%$; (**c**) $v_{sf} = 2\%$.

(a) (b) (c)

**Figure 10.** Failure modes in axial compression tests: (**a**) $v_{sf} = 0\%$; (**b**) $v_{sf} = 1\%$; (**c**) $v_{sf} = 2\%$.

In the splitting tensile tests, a specimen without SFs starts to show microcracks along the shims as the load is first increased. When the ultimate splitting strength is reached, the cracks rapidly penetrate the bottom and top surfaces of the specimen, at which time the load drops sharply, and the specimen is split into two parts. The damage is heard with a notable ringing sound, which belongs to brittle failure. The failure surface of the specimen shows that the fracture occurs mainly at the intersection of the aggregates and the cement mortar, and the aggregates themselves are not pulled off. The failure mode of a specimen changes significantly after the addition of SFs. As the load is increased, the cracks on the surface of the concrete gradually increase, a part of the concrete begins to fall off, and cracks develop through the bottom and top surfaces. However, the specimen is not split into two parts owing to the connection of the SFs between the cracks. The SFRRC has good integrity, and its failure mode belongs to plastic failure. The results show that the admixture of SFs effectively inhibits the development of cracks. The SFs consumes a large part of the energy, and cracks are produced mainly on the surface of the specimen during the failure process and the concrete is not severely peeled off and broken, and these phenomena become more significant with the increase in the SF admixture. The typical failure modes of the SFRRC specimens in the splitting tensile tests are shown in Figure 11.

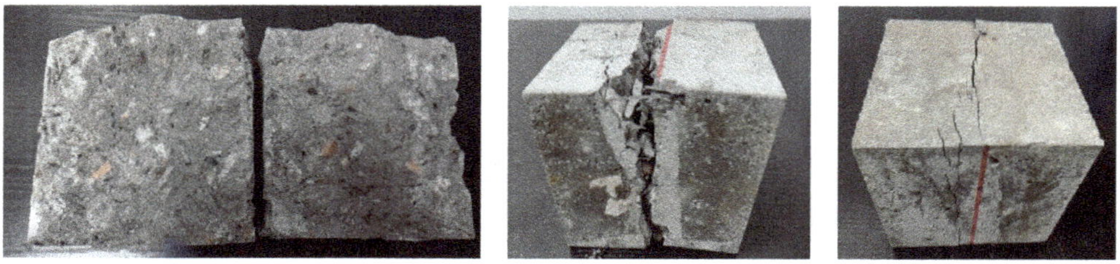

(a) (b) (c)

**Figure 11.** Failure modes in splitting tensile tests: (**a**) $v_{sf}$ = 0%; (**b**) $v_{sf}$ = 1%; (**c**) $v_{sf}$ = 2%.

*3.2. Mechanical Properties of SFRRC*

Loads are applied to the specimens by a pressure-testing machine, and the $f_{cu}$, $f_{ts}$, $f_c$, and $E_c$ of each group of specimens are obtained, as summarized in Table 6. The values in Table 6 are the averages of the test results of three specimens in each group.

**Table 6.** Test results of mechanical properties.

| Groups | Orthogonal Design | $f_{cu}$ (MPa) | $f_{ts}$ (MPa) | $f_{ts}/f_{cu}$ | $f_c$ (MPa) | $f_c/f_{cu}$ | $E_c$ (MPa) |
|---|---|---|---|---|---|---|---|
| SFRRC-1 | A1B1C1D1 | 48.4 | 3.74 | 0.077 | 34.4 | 0.71 | 32,900 |
| SFRRC-2 | A1B2C2D2 | 51.9 | 4.46 | 0.086 | 37.4 | 0.72 | 32,700 |
| SFRRC-3 | A1B3C3D3 | 47.1 | 5.10 | 0.108 | 35.6 | 0.79 | 26,400 |
| SFRRC-4 | A2B1C2D3 | 39.2 | 4.10 | 0.104 | 28.2 | 0.72 | 32,000 |
| SFRRC-5 | A2B2C3D1 | 37.0 | 4.29 | 0.116 | 25.2 | 0.71 | 29,200 |
| SFRRC-6 | A2B3C1D2 | 35.1 | 2.87 | 0.082 | 26.8 | 0.73 | 25,400 |
| SFRRC-7 | A3B1C3D2 | 36.0 | 4.39 | 0.122 | 24.8 | 0.72 | 30,500 |
| SFRRC-8 | A3B2C1D3 | 33.8 | 2.61 | 0.077 | 24.0 | 0.71 | 27,400 |
| SFRRC-9 | A3B3C2D1 | 32.7 | 3.29 | 0.100 | 23.9 | 0.74 | 24,100 |

It can be found that there is a significant linear relationship between $f_{cu}$ and $f_c$, with the ratio fluctuating at approximately 0.73, which is similar to that of natural concrete. The ratio of $f_{ts}$ to $f_c$ significantly fluctuates, which is mainly due to the large influence of the SFs on the splitting tensile behavior of the concrete. $f_{cu}$ and $f_{ts}$ are the main mechanical properties of concrete. The results in Table 6 are processed according to the orthogonal test analysis method, and the mean response analysis of the mechanical properties is shown in Figure 12.

It was found that $f_{cu}$ decreases with the increase in $w/c$ and $r_a$, which is consistent with the research results of [26]. The $f_{cu}$ and $f_{ts}$ of the SFRRCs only decreased by 6.38% and 1.06% when $r_a$ increased from 50% to 100%, so it is recommended that $r_a$ be increased to achieve more consumption of recycled aggregates in actual engineering. The SFs had a particularly significant influence on $f_{ts}$, when the SFs volume content is 1% and 2%, the $f_{ts}$ increases by 28.66% and 49.51%. At the same time, the $f_{cu}$ increased by 5.55% and 2.45%, indicating that too much SFs dosing will instead reduce the $f_{cu}$ of SFRRC, after considering $f_{cu}$ simultaneously, it is recommended that the amount of SFs should not be extremely large. Since SFs have a significant impact on the mechanical properties of recycled concrete, the microstructure of SFRRC will be analyzed in detail in the follow-up, and the reasons for the change in macromechanical properties will be explained through the microstructure.

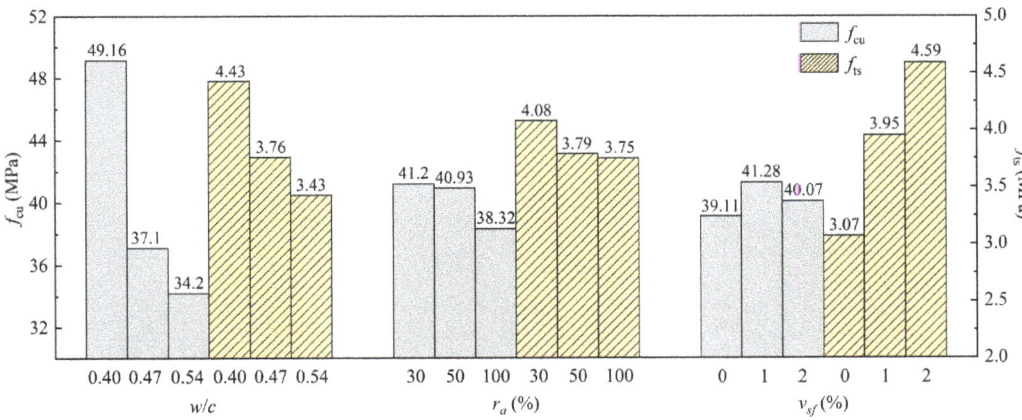


**Figure 12.** Relationships between various factors and mechanical properties.

*3.3. Strength Calculation Model*

Tensile strength is an important property for reflecting the cracking resistance of concrete, and most scholars advocate the use of splitting tensile tests of cylinders or cubes to indirectly determine the tensile strength of concrete [27]. Elastic modulus ($E_c$) reflects the relationship between the stress and deformation of a material in elastic state, and it is a necessary parameter for calculating deformation in concrete structure design [28]. The results of numerous studies have shown that there is a relationship between the $f_{ts}$, $E_c$, and $f_{cu}$ of concrete [29,30]. In this study, $f_{ts}$ and $E_c$ are evaluated according to Chinese code GB50010-2010 [31] as follows:

$$f_{ts} = 0.19 f_{cu}^{0.75} \tag{3}$$

$$E_c = \frac{10^5}{2.2 + 34.7/f_{cu}} \tag{4}$$

Similar formulas are available in the code, ACI318-14 [32], as expressed in Equations (5) and (6).

$$f_{ts} = 0.55 f_{cu}^{0.5} \tag{5}$$

$$E_c = 4700\sqrt{f_{cu}} \tag{6}$$

The calculation results of the above formula are compared with the test results of this study, and the connections between $f_{ts}$, $E_c$, and $f_{cu}$ are drawn as scatter points in Figure 13. It can be seen that the test values of $f_{ts}$ are larger than the standard calculated values, mainly because of the significant enhancement in $f_{ts}$ caused by the SFs. In general, SFs enhance the $E_c$ of concrete, whereas the RAC shows a suppressive effect, and a combined effect of the two factors in a test value of $E_c$ is a large deviation from the standard calculated values. Therefore, the calculation formula for natural concrete is unsuitable for the SFRRCs.

According to the formula in the standard, $f_{ts}$ and $E_c$ are still calculated from $f_{cu}$. Considering the influence of different factors in the orthogonal tests on the $f_{cu}$ of the SFRRCs, multiple parameters need to be defined to predict $f_{cu}$. Assuming that there are linear relationships between the $f_{cu}$ of SFRRCs and the water–cement ratio, replacement rate of the RCAs, and SF volume content, a calculation model for SFRRCs based on the following formula is proposed:

$$f_{cu} = b_1 + b_2(w/c) + b_3 r_a + b_4 \lambda_m \tag{7}$$

$$f_{ts} = c_1 f_{cu}^{c_2}(1 + c_3\lambda_m)(1 + c_4 r_a) \tag{8}$$

$$E_c = \frac{10^5}{2.2 + 34.7/f_{cu}}(1 + d_1\lambda_m)(1 + d_2 r_a) \tag{9}$$

where $b_i$ (I = 1–4) are the regression coefficients of the compressive strength, $w/c$ is the water–cement ratio, $r_a$ is the replacement rate of the RCAs, $\lambda_m$ is the product of the SF volume content and the aspect ratio of the SFs, $c_i$ (I = 1–4) are the regression coefficients of the splitting tensile strength, and $d_i$ (I = 1, 2) are the regression coefficients of the elastic modulus.

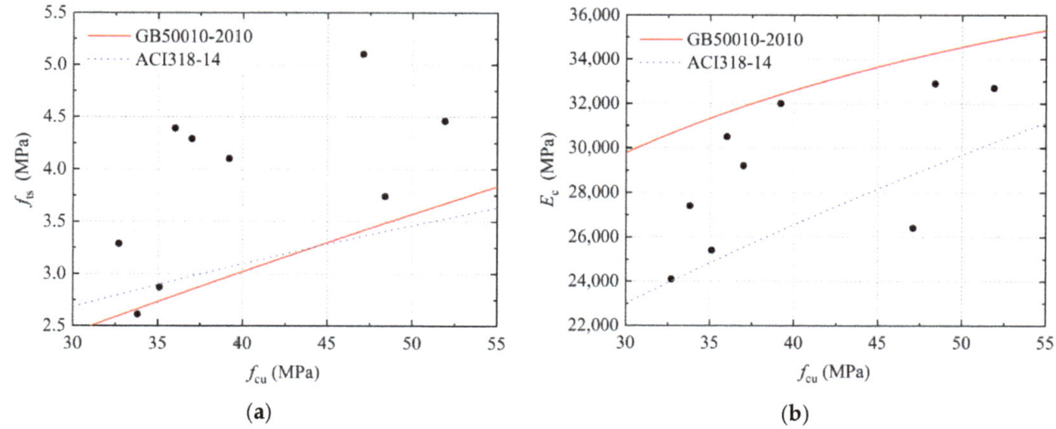


**Figure 13.** Comparisons between tests and standards: (**a**) Relationship between $f_{cu}$ and $f_{ts}$; (**b**) Relationship between $f_{cu}$ and $E_c$.

The formulas for the $f_{cu}$, $f_{ts}$, and $E_c$ of SFRRCs were established by regression analysis. As expected, Figure 14 shows that the fitting coefficients are sufficient for satisfying the requirement. To verify the reliability of Equations (7)–(9), the data of Gao [33,34], Carneiro [35], Yazc [36], Ahmadi [37], and Kachouh [38] were used to compare the calculation results of the formula with the test results. The results are listed in Table 7. It can be found that the predicted values are very close to the actual values. The maximum error of $f_{cu}$ and $f_{ts}$ is approximately 20%, the average error is approximately 5%, the maximum error of $E_c$ is approximately 10%, and the average error is approximately 2%. Consequently, the modified formulas can be applied to predict the $f_{cu}$, $f_{ts}$, and $E_c$ of SFRRCs with satisfactory accuracy.

### 3.4. Morphology of ITZ

Concrete is a composite material consisting of three parts: aggregates, mortar, and aggregate–mortar ITZ, where the ITZ is the weakest region in concrete and has a significant impact on the strength of concrete. Figure 15 shows the morphology of the ITZs of the SFRRCs with different SF volume contents.

From Figure 15a, it can be found that there is a microcrack between the SFs and the cement matrix. The SFs are pulled out from the cement matrix mainly because the ITZ structure is damaged, which eventually leads to the relative sliding of the SFs and the matrix. From Figure 15b,c, it can be found that the holes and cracks near the ITZ are significantly reduced at 1% SF doping compared to that in the specimens without SFs. However, the number and size of the cracks near the ITZ increases significantly when the SF dosing is 2%.

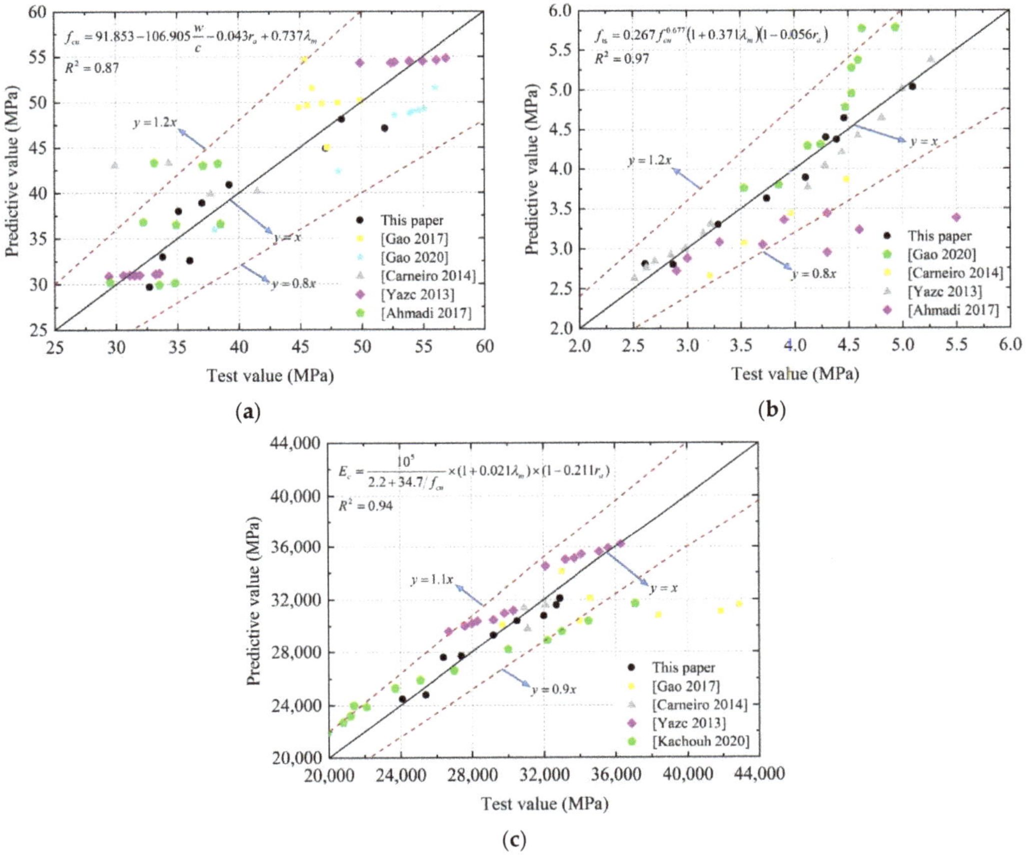


**Figure 14.** Accuracy of prediction model for mechanical properties of SFRRC: (**a**) Compressive strength; (**b**) Splitting tensile strength; (**c**) Elastic modulus.

### *3.5. Microstructural Analysis of the SFRRC*

For the aggregate ITZ, when cement is dissolved in water, the dissolved ions diffuse and enter the water film. The active order of ions is $Na^+ > K^+ > SO_4{}^{2-} > Al^{3+} > Ca^{2+} > Si^{4+}$, and the first products to be formed in the water film are hexagonal-shaped calcium hydroxide (CH) and needle-shaped ettringite (AFt). Compared to the cement matrix, the water–cement ratio in the ITZ is higher, and its ion saturation is lower. Thus, CH and AFt can grow freely and oriented on the aggregate surface without restriction. This prevents the calcium silicate hydrate (CSH) gel from contacting the aggregates and increases the porosity close to the ITZ. As the ion concentration in the ITZ decreases, the amount of the CSH gel is reduced, which eventually leads to the formation of a loose network structure in the ITZ. Therefore, the weakest position in the ITZ of the aggregates is that close to the surface of the aggregates. Similarly, the mechanical properties of an SFRRC also strongly depend on the microstructure of the ITZ, i.e., the area where the SFs are in contact with the cement matrix. Results of the SEM with energy-dispersive X-ray analysis (SEM-EDXA) for an SF-cement matrix-based ITZ are shown in Figure 16. Figure 16b,c show the effect of area A (25–35 μm from the SF surface) and area B (5–15 μm from the SF surface) magnified by 5000 times, respectively.

**Table 7.** Experimental data and information about prediction.

| Source | No. | Compressive Strength (MPa) $f_{cu}$=91.853+106.905($w/c$)+0.043$r_a$+0.737$\lambda_m$ ($R^2$ = 0.87) | | | Splitting Tensile Strength (MPa) $f_{ts}$=0.267$f_{cu}^{0.677}$(1+0.371$\lambda_m$)(1−0.056$r_a$) ($R^2$ = 0.97) | | | Elastic Modulus (MPa) $E_c=\frac{10^5}{2.2+34.7/f_{cu}}$(1+0.021$\lambda_m$)(1−0.211$r_a$) ($R^2$ = 0.94) | | |
|---|---|---|---|---|---|---|---|---|---|---|
| | | Calculated | Tested | Calculated/ Tested | Calculated | Tested | Calculated/ Tested | Calculated | Tested | Calculated/ Tested |
| Article | SFRRC-1 | 43.1 | 48.4 | 0.994 | 3.63 | 3.74 | 0.971 | 32,100 | 32,900 | 0.976 |
| | SFRRC-2 | 47.1 | 51.9 | 0.908 | 4.64 | 4.46 | 1.040 | 31,600 | 32,700 | 0.966 |
| | SFRRC-3 | 44.9 | 47.1 | 0.953 | 5.03 | 5.10 | 0.986 | 27,600 | 26,400 | 1.045 |
| | SFRRC-4 | 40.9 | 39.2 | 1.043 | 3.89 | 4.10 | 0.949 | 30,800 | 32,000 | 0.963 |
| | SFRRC-5 | 38.9 | 37.0 | 1.051 | 4.40 | 4.29 | 1.026 | 29,300 | 29,200 | 1.003 |
| | SFRRC-6 | 38.0 | 35.1 | 1.083 | 2.80 | 2.87 | 0.976 | 24,800 | 25,400 | 0.976 |
| | SFRRC-7 | 32.6 | 36.0 | 0.906 | 4.37 | 4.39 | 0.995 | 30,400 | 30,500 | 0.997 |
| | SFRRC-8 | 33.0 | 33.8 | 0.976 | 2.81 | 2.61 | 1.077 | 27,700 | 27,400 | 1.011 |
| | SFRRC-9 | 29.7 | 32.7 | 0.908 | 3.30 | 3.29 | 1.003 | 24,500 | 24,100 | 1.017 |
| Literature [33] | CR0F1 | 54.6 | 45.4 | 1.203 | — | — | — | 34,100 | 33,000 | 1.033 |
| | CR30F1 | 51.5 | 46.0 | 1.120 | — | — | — | 32,100 | 34,600 | 0.928 |
| | CR50F1 | 49.8 | 46.8 | 1.064 | — | — | — | 30,800 | 38,400 | 0.802 |
| | CR100F1 | 45.0 | 47.2 | 0.953 | — | — | — | 27,200 | 37,800 | 0.720 |
| | CR50F0 | 49.4 | 44.9 | 1.100 | — | — | — | 30,100 | 29,700 | 1.010 |
| | CR50F0.5 | 49.6 | 45.6 | 1.088 | — | — | — | 30,400 | 34,000 | 0.894 |
| | CR50F1.5 | 49.9 | 48.1 | 1.037 | — | — | — | 31,100 | 41,900 | 0.742 |
| | CR50F2 | 50.1 | 49.9 | 1.004 | — | — | — | 31,600 | 42,900 | 0.737 |
| Literature [34] | N-C30R50F1 | 36.0 | 38.0 | 0.947 | 3.76 | 3.53 | 1.065 | — | — | — |
| | N-C45R50F1 | 49.0 | 54.1 | 0.906 | 4.78 | 4.47 | 1.069 | — | — | — |
| | N-C60R50F1 | 57.6 | 71.5 | 0.806 | 5.78 | 4.94 | 1.170 | — | — | — |
| | N-C45R0F1 | 55.5 | 61.5 | 0.902 | 5.37 | 4.59 | 1.170 | — | — | — |
| | N-C45R30F1 | 51.60 | 56.0 | 0.921 | 4.95 | 4.53 | 1.093 | — | — | — |
| | N-C45R100F1 | 42.4 | 48.1 | 0.881 | 4.29 | 4.12 | 1.041 | — | — | — |
| | N-C45R50F0 | 48.6 | 52.7 | 0.922 | 3.80 | 3.85 | 0.987 | — | — | — |
| | N-C45R50F0.5 | 48.8 | 53.9 | 0.905 | 4.31 | 4.24 | 1.017 | — | — | — |
| | N-C45R50F1.5 | 49.1 | 54.6 | 0.899 | 5.27 | 4.53 | 1.163 | — | — | — |
| | N-C45R50F2.0 | 49.3 | 55.1 | 0.895 | 5.77 | 4.63 | 1.246 | — | — | — |
| Literature [35] | REF | 43.0 | 29.9 | 1.438 | 2.66 | 3.21 | 0.829 | 29,800 | 31,100 | 0.958 |
| | RFCA | 39.9 | 37.7 | 1.058 | 3.07 | 3.53 | 0.870 | 30,400 | 29,200 | 1.041 |
| | SF-REF | 43.3 | 34.3 | 1.262 | 3.44 | 3.96 | 0.869 | 31,400 | 30,900 | 1.016 |
| | SF-RFCA | 40.2 | 41.5 | 0.969 | 3.86 | 4.48 | 0.862 | 31,600 | 32,100 | 0.984 |
| Literature [36] | C20 | 30.9 | 29.4 | 1.051 | 2.63 | 2.51 | 1.048 | 29,600 | 26,700 | 1.109 |
| | C20-10K | 31 | 30.6 | 1.013 | 2.76 | 2.62 | 1.053 | 30,000 | 27,600 | 1.087 |
| | C20-20K | 31 | 31.1 | 0.997 | 2.84 | 2.7 | 1.052 | 30,200 | 28,000 | 1.079 |
| | C20-30K | 31 | 31.5 | 0.984 | 2.92 | 2.85 | 1.025 | 30,400 | 28,300 | 1.074 |
| | C20-40K | 31 | 31.9 | 0.972 | 3.00 | 2.98 | 1.007 | 30,500 | 29,200 | 1.045 |
| | C20-60K | 31.1 | 33.2 | 0.937 | 3.19 | 3.15 | 1.013 | 31,000 | 29,800 | 1.040 |
| | C20-80K | 31.2 | 33.5 | 0.931 | 3.31 | 3.22 | 1.028 | 31,200 | 30,300 | 1.030 |
| | C40 | 54.3 | 49.9 | 1.088 | 3.77 | 4.12 | 0.915 | 34,500 | 32,100 | 1.075 |
| | C40-10U | 54.3 | 52.4 | 1.036 | 4.04 | 4.28 | 0.944 | 35,000 | 33,200 | 1.054 |
| | C40-20U | 54.4 | 52.7 | 1.032 | 4.21 | 4.44 | 0.948 | 35,100 | 33,700 | 1.042 |
| | C40-30U | 54.5 | 53.9 | 1.011 | 4.42 | 4.59 | 0.963 | 35,400 | 34,100 | 1.038 |
| | C40-40U | 54.5 | 55 | 0.991 | 4.64 | 4.81 | 0.965 | 35,600 | 35,100 | 1.014 |
| | C40-60U | 54.6 | 56.1 | 0.973 | 5.01 | 5 | 1.002 | 35,900 | 35,600 | 1.008 |
| | C40-80U | 54.8 | 56.9 | 0.963 | 5.37 | 5.27 | 1.019 | 36,200 | 36,300 | 0.997 |
| Literature [37] | R0F0 | 43 | 37.1 | 1.159 | 3.08 | 3.30 | 0.933 | — | — | — |
| | R50F0 | 36.5 | 34.9 | 1.046 | 2.88 | 3.00 | 0.960 | — | — | — |
| | R100F0 | 29.9 | 33.5 | 0.893 | 2.72 | 2.90 | 0.938 | — | — | — |
| | R0F0/5 | 43.2 | 38.3 | 1.128 | 3.44 | 4.30 | 0.800 | — | — | — |
| | R50F0/5 | 36.50 | 38.5 | 0.951 | 3.36 | 3.90 | 0.861 | — | — | — |
| | R100F0/5 | 30.1 | 34.8 | 0.865 | 3.05 | 3.70 | 0.824 | — | — | — |
| | R0F1 | 43.3 | 33.1 | 1.308 | 3.38 | 5.50 | 0.615 | — | — | — |
| | R50F1 | 36.8 | 32.2 | 1.143 | 3.23 | 4.60 | 0.702 | — | — | — |
| | R100F1 | 30.2 | 29.5 | 1.024 | 2.95 | 4.30 | 0.686 | — | — | — |
| Literature [38] | R0SF0 | — | — | — | — | — | — | 31,700 | 37,100 | 0.854 |
| | R30SF0 | — | — | — | — | — | — | 28,200 | 30,000 | 0.940 |
| | R70SF0 | — | — | — | — | — | — | 24,000 | 21,400 | 1.121 |
| | R100SF0 | — | — | — | — | — | — | 21,900 | 19,900 | 1.101 |
| | R30SF1 | — | — | — | — | — | — | 28,900 | 32,200 | 0.898 |
| | R70SF1 | — | — | — | — | — | — | 25,300 | 23,700 | 1.068 |
| | R100SF1 | — | — | — | — | — | — | 22,700 | 20,800 | 1.091 |
| | R30SF2 | — | — | — | — | — | — | 29,600 | 33,000 | 0.897 |
| | R70SF2 | — | — | — | — | — | — | 25,900 | 25,100 | 1.032 |
| | R100SF2 | — | — | — | — | — | — | 23,200 | 21,200 | 1.094 |
| | R30SF3 | — | — | — | — | — | — | 30,400 | 34,500 | 0.881 |
| | R70SF3 | — | — | — | — | — | — | 26,600 | 27,000 | 0.985 |
| | R100SF3 | — | — | — | — | — | — | 23,900 | 22,100 | 1.081 |

Note: No. represents name of test specimen in the literature.

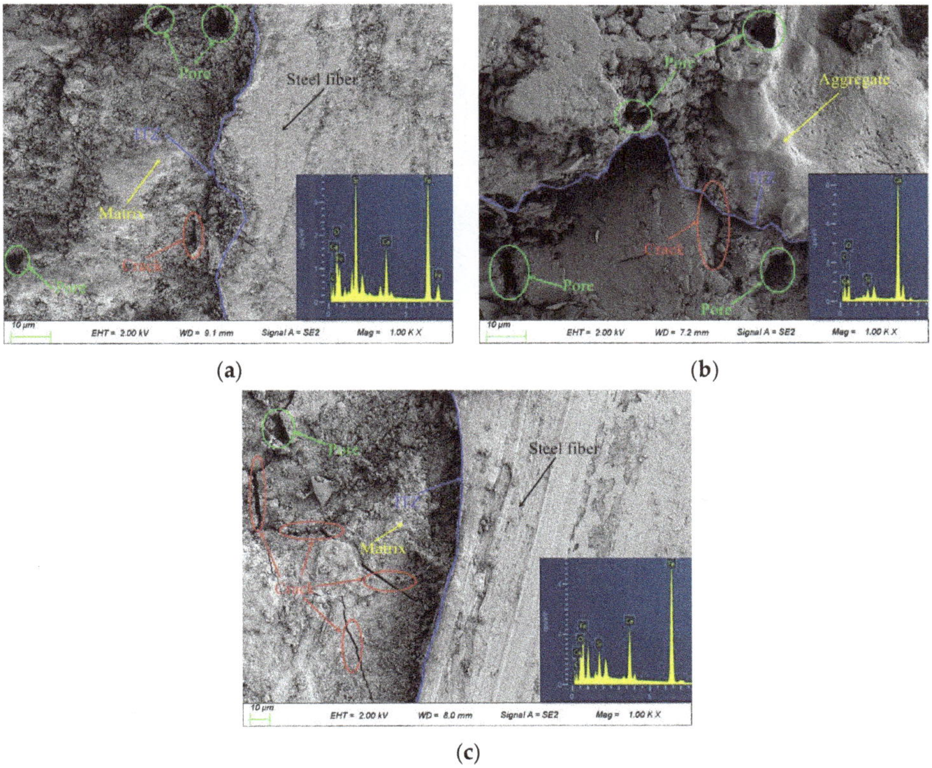


**Figure 15.** Morphologies of ITZs: (**a**) $v_{sf}$ = 1%; (**b**) $v_{sf}$ = 0%; (**c**) $v_{sf}$ = 2%.

It can be seen that, as the distance from the surfaces of the SFs increases, the concentration of $Fe^{2+}$ ions first decreases and subsequently increases. At 25–35 μm from the surfaces of the SFs, the $Fe^{2+}$ ion concentration is close to the lowest value. After 40 μm, the $Fe^{2+}$ ion concentration begins to level off, and at 100 μm, it is the inherent $Fe^{2+}$ ion concentration of the cement itself. The results show that a part of the $Fe^{2+}$ ions can be dissolved from the surfaces of the SFs, which is different from an ordinary aggregate interface. This part of the $Fe^{2+}$ ions enters the water film, and the superposition of the ions dissolved in the cement and the ions dissolved from the SFs makes the concentration of ions in the interface higher than that in the ordinary aggregate interface. $Fe^{2+}$ ions can reduce some of the adverse effects of the water film, and more CSH is formed close to the SFs. This increases the probability of the CSH contacting the SFs and causes it to fill the porous structure close to the surface of the SFs, therefore enhancing the compactness of the structure. The $Fe^{2+}$ ion concentration is lower in the A area slightly further from the fiber surface, and the needle-shaped AFt and hexagonal-shaped CH mainly accumulated in the area. Owing to the influence of solid particle shape, the sparse accumulation of particles around the side wall produces numerous pores, resulting in the formation of a loose structure in the ITZ around the aggregates, which can be considered to be a weaker area in the SF–matrix. However, excessive SFs can have a detrimental effect on the concrete, which requires a more detailed analysis of the microscopic pore structure parameters.

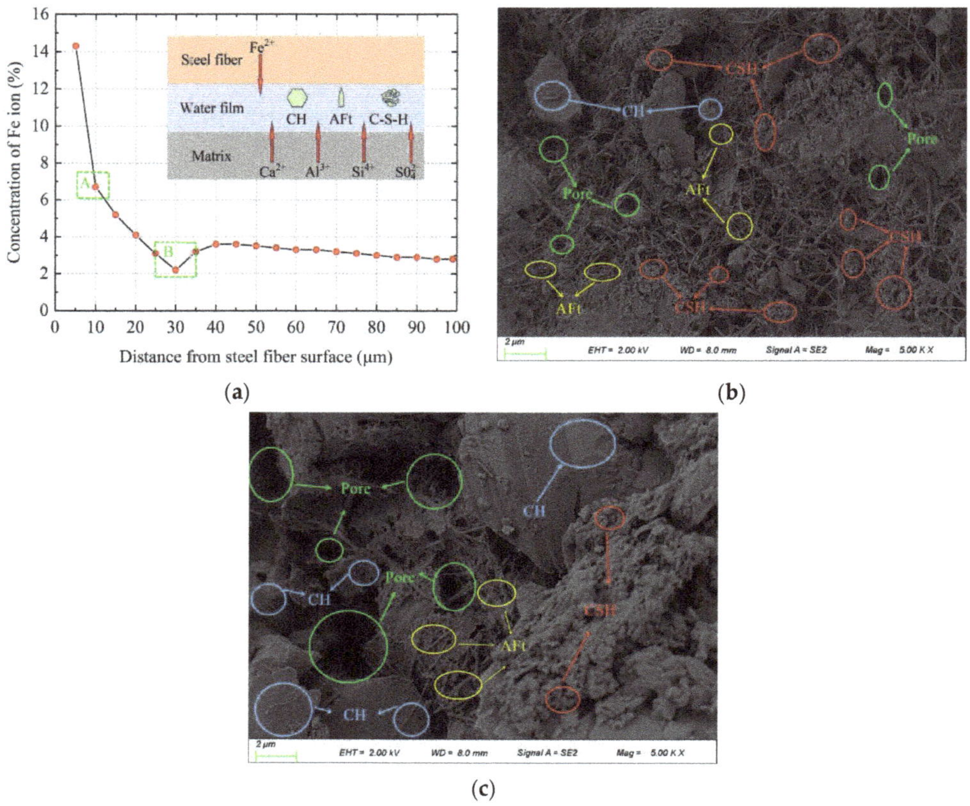


**Figure 16.** Micromorphology of SFRRC: (**a**) Element fraction by EDXA; (**b**) A area; (**c**) B area.

## 4. Internal Mechanisms of the Mechanical Properties in SFRRC

### *4.1. Microscopic Pore Structure Analysis*

The microscopic pore structures of the SFRRCs were determined using the MIP method, and the pore structure parameters of each specimen are listed in Table 8 and shown in Figure 17. The pore structures can be divided into four levels according to the pore size [39]: harmless pores (diameter < 20 nm), less-harmful pores (diameter 20–50 nm), harmful pores (diameter 50–200 nm), and more-harmful pores (diameter > 200 nm).

**Table 8.** Pore characteristic parameters and strength of SFRRCs.

| Groups | Orthogonal Design | Total Porosity (%) | Total Pore Volume (mL/g) | Average Pore Diameter (nm) | $f_{cu}$ (MPa) | $f_{ts}$ (MPa) |
|---|---|---|---|---|---|---|
| SFRRC-1 | A1B1C1D1 | 15.56 | 0.0727 | 32.56 | 48.4 | 3.74 |
| SFRRC-2 | A1B2C2D2 | 11.88 | 0.0591 | 15.88 | 51.9 | 4.46 |
| SFRRC-3 | A1B3C3D3 | 15.12 | 0.0716 | 30.73 | 47.1 | 5.10 |
| SFRRC-4 | A2B1C2D3 | 13.88 | 0.0712 | 17.32 | 39.2 | 4.10 |
| SFRRC-5 | A2B2C3D1 | 18.13 | 0.0863 | 30.17 | 37.0 | 4.29 |
| SFRRC-6 | A2B3C1D2 | 18.77 | 0.0993 | 36.61 | 35.1 | 2.87 |
| SFRRC-7 | A3B1C3D2 | 19.17 | 0.0879 | 33.69 | 36.0 | 4.39 |
| SFRRC-8 | A3B2C1D3 | 20.32 | 0.1031 | 39.81 | 33.8 | 2.61 |
| SFRRC-9 | A3B3C2D1 | 18.88 | 0.0921 | 19.7 | 32.7 | 3.29 |

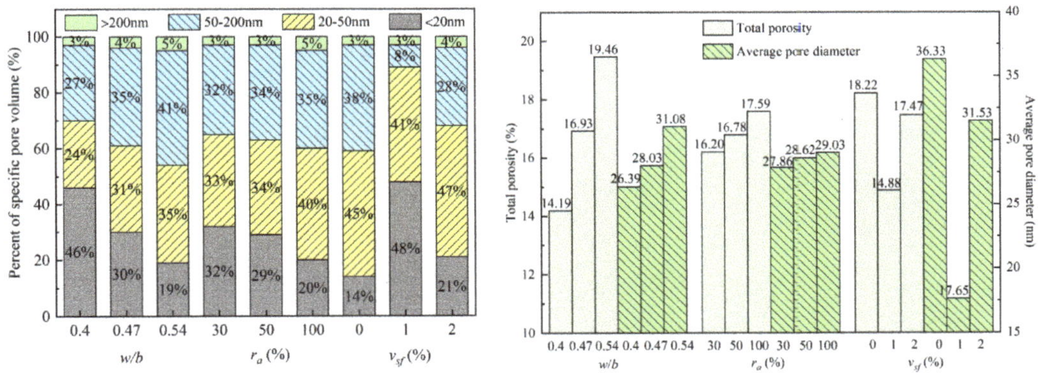


(**a**) (**b**)

**Figure 17.** Pore characteristic parameters: (**a**) Pore size distribution; (**b**) Total porosity and pore diameter.

From Table 8 and Figure 17, it can be found that the water–cement ratio, replacement rate of the RCAs, and SFs have little effect on the content of the harmful pores in the SFRRCs. With the increase in the water-binder ratio and the RCA replacement rate, the porosity and volume percentages of the less-harmful, harmful, and more-harmful pores of the specimens increase, whereas the volume percentage of the harmless pores decreases. With the increase in the SF volume content, the porosity and pore size of the SFRRCs show first a decreasing and subsequently an increasing trend. When the SF volume content is 1%, the total porosity and average pore diameter of the SFRRCs are the smallest, which are 18.33% and 51.42% lower than those of the specimens without SFs, respectively. When the SF volume content is 2%, all pore structure parameters of the SFRRCs show an increasing trend. Compared to the SFRRC with 1% SF volume content, the total porosity and pore size of the former increase by 17.41% and 78.64%, respectively. The results show that mixing an appropriate amount of the SFs into the concrete can effectively prevent the generation of the concrete microcracks caused by water loss and reduce the size and number of microcracks. These therefore reduce the total porosity of concrete and make the distribution of the pore structure more reasonable. When the amount of the SFs is extremely large, their agglomeration occurs in the concrete, resulting in an uneven distribution of the pore structure, and increasing the total porosity and pore size of the concrete. It is generally believed that the $f_{cu}$ of an SFRRC mainly depends on the compactness of its matrix, and the $f_{ts}$ of an SFRRC mainly depends on the bonding strength of the SF–matrix interface. The SFs can bond with the matrix well and have the effect of strengthening and toughening the concrete; therefore, the $f_{ts}$ of the concrete is significantly improved with the increase in the SFs. However, excessive addition of the SFs adversely affects the pore structure of the concrete, resulting in a decrease in $f_{cu}$.

### *4.2. Relationship between Strength and Pore Structure of SFRRCs*

The static pore structure of concrete as a porous medium is generally analyzed using the Menger sponge model, and this method has been recognized by many researchers [40]. As shown in Figure 18, each side of the initial element with side length $R$ is equally divided into m parts, following which m$^3$-small cube elements are obtained and n small cube elements are randomly eliminated. After k times, the number of remaining cubes is $(m^3 - n)^k$ and the side length of the smallest cube unit is $r_k = R/m^k$. The fractal dimension, $D_m$, of the sponge model is calculated using the box-counting method, as expressed below:

$$D_m = \lim_{k \to +\infty} \frac{\ln(m^3 - n)^k}{\ln(m^k/R)} = \ln\left(m^3 - n\right)/\ln m \quad (10)$$

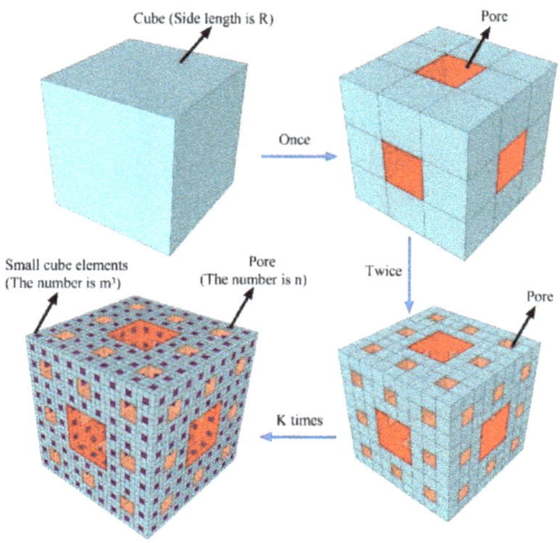


**Figure 18.** Menger sponge model.

According to the fractal theory, the remaining cube volume, $V_s$, of the sponge model has a power relationship with the length, $r_k$, of the smallest cube elements, and their derivatives can be obtained as follows:

$$\frac{\ln V_s}{\ln r_k} = \frac{\ln\left[(m^3 - n)^k \times (R/m^k)^3\right]}{\ln(R/m^k)} \propto (3 - D_m) \tag{11}$$

The remaining cube volume of the sponge model is:

$$V_s \propto {r_k}^{3-D_m} \tag{12}$$

The pore volume of the remaining cube, $V_p$, is:

$$V_p = R^3 - V_s \tag{13}$$

Based on Equations (12) and (13), the following formula is obtained:

$$\lg(-dV_p/dr_k) \propto (2 - D_m)\lg r_k \tag{14}$$

It can be seen from Figure 19 that the variation law of the total porosity of the SFRRCs is opposite to the variation law of the fractal dimension, which indicates that porosity can reflect the variations in the pore structure composition and distribution in the concrete to a certain extent. A large fractal dimension implies a highly complex pore structure and high composition in the concrete, and a high tortuosity of the capillary pores.

$f_{cu}$ and $f_{ts}$ are the most basic mechanical indicators of the macroscopic properties of concrete, and the magnitude of the strength is closely related to the microscopic pore structure. Research results show that the mechanical strength of concrete is related to its porosity [41], pore size distribution, and fractal dimension, and researchers commonly use porosity or fractal dimension to calculate the strength of concrete. The results of the quadratic polynomial fitting of the relations between the porosity, fractal dimension, and strength of the concrete are shown in Figure 20.

It can be seen from Equation (14) that the logarithm of $-dV_p/dr_k$ and $r_k$ have a linear relationship, and the fractal dimension, $D_m$, can be obtained by fitting of the curve, and the result is shown in Figure 19.

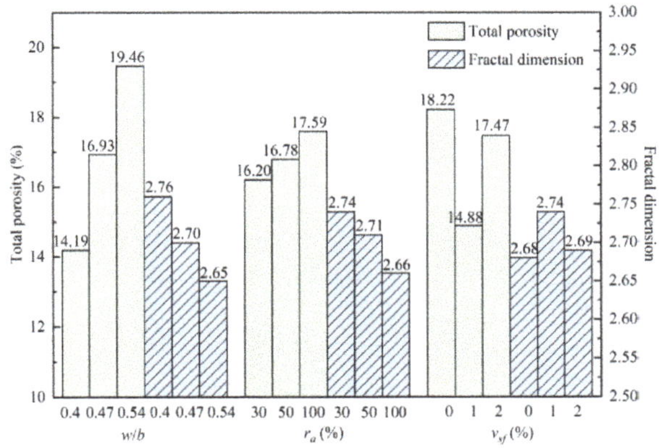


**Figure 19.** Total porosity and fractal dimension.

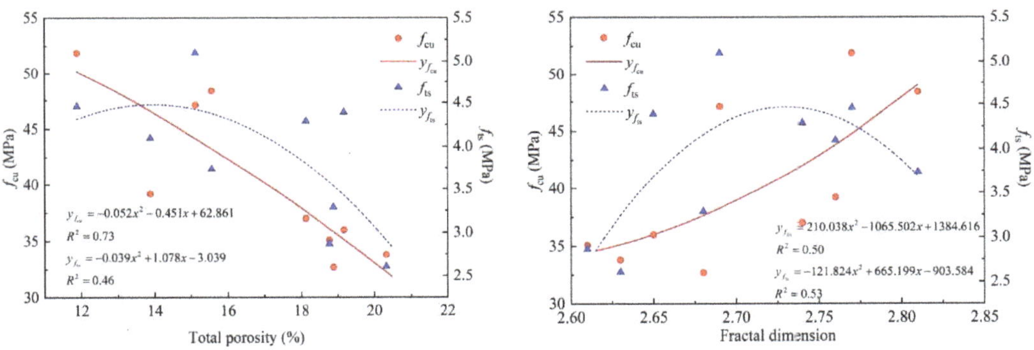


(a) (b)

**Figure 20.** Fitting of SFRRC strength and pore characteristic parameters: (**a**) Total porosity fitting curve; (**b**) Fractal dimension fitting curve.

It can be found that the fitting between the concrete strength and the porosity or the fractal dimension needs to be improved. To obtain a higher precision calculation model, it is necessary to consider the effects of both the porosity and fractal dimension, and the influence coefficients of the SF volume content and the replacement rate of the RCAs should be added. Based on the above analysis results, a calculation model of SFRRC strength based on the following formula is proposed:

$$f=(e_1P_t^2+e_2D_m^2+e_3P_tD_m+e_4P_t+e_5D_m+e_6)(1+e_7\lambda_m)(1+e_8r_a) \quad (15)$$

where $f$ is the SFRRC strength, $e_i$ (I = 1–8) are the regression coefficients of strength, $D_m$ is the fractal dimension, $\lambda_m$ is the product of the SF volume content and the aspect ratio of the SFs, and $r_a$ is the replacement rate of the RCAs.

Based on Equation (15), the relationships between the concrete strength and the pore structure parameters, SF volume content, and RCA ratio were obtained by multiple regression analysis, and $f_{cu}$ and $f_{ts}$ are calculated as shown in Figure 21. It can be found that the r-squared of the multifactor calculation models for $f_{cu}$ and $f_{ts}$ are 0.89 and 0.97, respectively. Moreover, the maximum errors of the compressive strength and splitting tensile strength do not exceed 10% and 5%, respectively, indicating that the regression effect is significant. Therefore, the calculation model results are in good agreement with the experimental results and can accurately describe the quantitative relationships between the macromechanical properties of SFRRCs and the microscopic pore structure parameters.

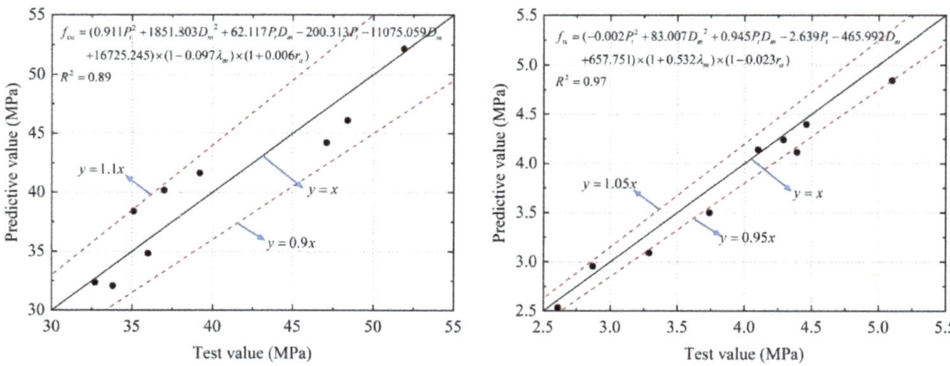


(**a**) (**b**)

**Figure 21.** Multiple regression calculation of SFRRC strength: (**a**) Compressive strength; (**b**) Splitting tensile strength.

## 5. Conclusions

This study investigated the mechanical properties of SFRRCs with different water–cement ratios, RCA replacement rates, and SF volume contents by orthogonal tests. Combined with an SFRRC microstructure analysis, the following conclusions were drawn:

- The SFs can inhibit the development of SFRRC cracks, increase the strength and ductility of the SFRRC, and significantly improve the failure mode of the SFRRC, changing it from brittle failure to plastic failure.
- With the increase in the SF volume content, the $f_{ts}$ of the SFRRCs increase significantly. When the SFs volume content is 1% and 2%, the $f_{ts}$ increases by 28.66% and 49.51%. At the same time, the $f_{cu}$ increased by 5.55% and 2.45%, indicating that too much SFs dosing will instead reduce the $f_{cu}$ of SFRRC, which is related to the microscopic pore structure of SFRRCs.
- The formula for calculating concrete strength in the code is not applicable to SFRRCs. By a multifactor analysis, a mechanical calculation model for SFRRCs is obtained using Equations (7)–(9), and the calculation and test results have a high degree of fit.
- The $Fe^{2+}$ ions dissolved on the surface of the SFs can improve the structure of the ITZ and increase the bonding strength of the interface. The weak region of the SF–matrix ITZ is approximately 25–35 μm from the SF surfaces, which is related to the concentration of $Fe^{2+}$ ions.
- When the SF volume content is 1%, the total porosity and average pore diameter of the SFRRCs are 18.33% and 51.42% lower than those of the specimens without SFs. Compared to the SFRRC with 1% SF volume content, the total porosity and pore size of SFRRCs increased by 17.41% and 78.64%, respectively, when the SF content was 2%. This phenomenon is because adding the appropriate amount of SFs reduces the size and number of cracks in the ITZ and makes the pore structure distributions of the SFRRCs more reasonable. An excessive amount of SFs causes an uneven distribution of the pore structure, increasing the total pore space and the pore size, therefore reducing the compressive strength of the SFRRCs.
- It is inaccurate to use the porosity or the fractal dimension alone to calculate the strength of an SFRRC. The regression of a multifactor strength calculation model using the fractal dimension, porosity, SF volume content, and RCA replacement rate is significant, and the model can accurately describe the quantitative relationships between the strength of the SFRRCs and the pore structure parameters.

**Author Contributions:** Conceptualization, H.Y. (Hanquan Yuan) and L.Z.; methodology, H.Y. (Hanquan Yuan); software, H.Y. (Hanquan Yuan); validation, H.Y. (Hanquan Yuan), L.Z. and X.W.; formal analysis, H.Y. (Hanquan Yuan); investigation, H.Y. (Hanquan Yuan) and X.W.; resources, L.Z.; data curation, H.Y. (Hanquan Yuan) and X.W.; writing—original draft preparation, H.Y. (Hanquan Yuan); writing—review and editing, H.Y. (Hanquan Yuan); visualization, H.Y. (Hanquan Yuan) and L.Z.; supervision, L.Z. and H.Y. (Hongtao Yang); project administration, L.Z.; funding acquisition, L.Z. and H.Y. (Hongtao Yang). All authors have read and agreed to the published version of the manuscript.

**Funding:** This research was supported by the Shaanxi Province Key Research and Development Program on Industry Innovation Chain (grant number. 2018ZDCXL-SF-03-03-01).

**Institutional Review Board Statement:** Not applicable.

**Informed Consent Statement:** Not applicable.

**Data Availability Statement:** Not applicable.

**Conflicts of Interest:** The authors declare no conflict of interest.

## References

1. Xiao, J.; Li, W.; Fan, Y.; Huang, X. An overview of study on recycled aggregate concrete in China (1996–2011). *Constr. Build. Mater.* **2012**, *31*, 364–383. [CrossRef]
2. Reis, G.; Quattrone, M.; Ambros, W.M.; Cazacliu, B.G.; Sampaio, C.H. Current Applications of Recycled Aggregates from Construction and Demolition: A Review. *Materials* **2021**, *14*, 1700. [CrossRef] [PubMed]
3. García-Sanz-Calcedo, J.; Al-Kassir, A.; Yusaf, T. Economic and environmental impact of energy saving in healthcare buildings. *Appl. Sci.* **2018**, *8*, 440. [CrossRef]
4. Lv, Z.; Liu, C.; Zhu, C.; Bai, G.; Qi, H. Experimental study on a prediction model of the shrinkage and creep of recycled aggregate concrete. *Appl. Sci.* **2019**, *9*, 4322. [CrossRef]
5. López-Uceda, A.; Fernández-Ledesma, E.; Salas-Morera, L.; Jiménez, J.R.; Suescum-Morales, D. Effect of the Composition of Mixed Recycled Aggregates on Physical–Mechanical Properties. *Crystals* **2021**, *11*, 1518. [CrossRef]
6. Akbarnezhad, A.; Ong, K.C.G.; Zhang, M.H.; Tam, C.T.; Foo, T.W.J. Microwave assisted beneficiation of recycled concrete aggregates. *Constr. Build. Mater.* **2011**, *25*, 3469–3479. [CrossRef]
7. Katz, A. Treatments for the improvement of recycled aggregate. *J. Mater. Civil. Eng.* **2004**, *16*, 597–603. [CrossRef]
8. Dilbas, H.; Cakir, O.; Atis, C.D. Experimental investigation on properties of recycled aggregate concrete with optimized Ball Milling Method. Constr. *Build. Mater.* **2019**, *212*, 716–726. [CrossRef]
9. Zhao, Y.; Zeng, W.; Zhang, H. Properties of recycled aggregate concrete with different water control methods. *Constr. Build. Mater.* **2017**, *152*, 539–546. [CrossRef]
10. Wang, J.; Zhang, J.; Cao, D.; Dang, H.; Ding, B. Comparison of recycled aggregate treatment methods on the performance for recycled concrete. *Constr. Build. Mater.* **2020**, *234*, 117366. [CrossRef]
11. Zhang, H.; Liu, W.; Lin, X.; Su, S.; Zhao, B. To ameliorate the performance of recycled aggregate concrete (RAC) by pre-treating aggregate in sulfoaluminate cement slurry and water glass solution. *J. Build. Eng.* **2021**, *44*, 103364. [CrossRef]
12. Chen, A.; Han, X.; Chen, M.; Wang, X.; Guo, T. Mechanical and stress-strain behavior of basalt fiber reinforced rubberized recycled coarse aggregate concrete. *Constr. Build. Mater.* **2020**, *260*, 119888. [CrossRef]
13. Yin, S.; Tuladhar, R.; Collister, T.; Combe, M.; Sivakugan, N.; Deng, Z. Post-cracking performance of recycled polypropylene fibre in concrete. *Constr. Build. Mater.* **2015**, *101*, 1069–1077. [CrossRef]
14. Meesala, C.R. Influence of different types of fiber on the properties of recycled aggregate concrete. *Struct. Concr.* **2019**, *20*, 1656–1669. [CrossRef]
15. Shah, A.A.; Ribakov, Y. Recent trends in steel fibered high-strength concrete. *Mater. Des.* **2011**, *32*, 4122–4151. [CrossRef]
16. Qin, Y.; Zhang, X.; Chai, J.; Xu, Z.; Li, S. Experimental study of compressive behavior of polypropylene-fiber-reinforced and polypropylene-fiber-fabric-reinforced concrete. *Constr. Build. Mater.* **2019**, *194*, 216–225. [CrossRef]
17. Ndruru, H.; Simanjuntak, R.M.; Tampubolon, S.P. Utilization of copper fiber waste to increase compressive strength and split tensile strength of rigid pavement. *IOP Conf. Ser. Earth Environ. Sci.* **2021**, *878*, 012052. [CrossRef]
18. Jenifer, J.V.; Brindha, D. Development of hybrid steel-basalt fiber reinforced concrete—in aspects of flexure, fracture and microstructure. *Rev. Constr.* **2021**, *20*, 62–90.
19. Lee, S.C.; Oh, J.H.; Cho, J.Y. Compressive behavior of fiber-reinforced concrete with end-hooked steel fibers. *Materials* **2015**, *8*, 1442–1458. [CrossRef]
20. Bentur, A.; Diamond, S.; Mindess, S. The microstructure of the steel fibre-cement interface. *J. Mater. Sci.* **1985**, *20*, 3610–3620. [CrossRef]
21. Hannawi, K.; Bian, H.; Prince-Agbodjan, W.; Raghavan, B. Effect of different types of fibers on the microstructure and the mechanical behavior of Ultra-High Performance Fiber-Reinforced Concretes. *Compos. Part B Eng.* **2016**, *20*, 214–220. [CrossRef]
22. *ASTM-C33*; Standard Specification for Concrete Aggregates. ASTM International: West Conshohocken, MI, USA, 2013.

23. *GB/T14685-2011*; Pebble and Crushed Stone Code for Buildings. China Standard Press: Beijing, China, 2010.
24. Zhu, L.; Zhao, C.; Dai, J. Prediction of compressive strength of recycled aggregate concrete based on gray correlation analysis. *Constr. Build. Mater.* **2021**, *273*, 121750. [CrossRef]
25. *GB/T50081*; Standard for Test Method of Mechanical Properties on Ordinary Concrete. China Architecture & Building Press: Beijing, China, 2002.
26. Bai, G.; Zhu, C.; Liu, C.; Liu, B. An evaluation of the recycled aggregate characteristics and the recycled aggregate concrete mechanical properties. *Constr. Build. Mater.* **2020**, *240*, 117978. [CrossRef]
27. Kou, S.C.; Poon, C.S.; Etxeberria, M. Influence of recycled aggregates on long term mechanical properties and pore size distribution of concrete. *Cement. Concr. Compos.* **2011**, *33*, 286–291. [CrossRef]
28. Belén, G.F.; Fernando, M.A.; Diego, C.L.; Sindy, S.P. Stress-strain relationship in axial compression for concrete using recycled saturated coarse aggregate. *Constr. Build. Mater.* **2011**, *25*, 2335–2342. [CrossRef]
29. Zhu, H.; Gao, D.; Tang, J. Strength indexes of SF reinforced high strength concrete and their interrelations. *J. Build. Mater.* **2009**, *12*, 323–327.
30. Arezoumandi, M.; Smith, A.; Volz, J.S.; Khayat, K.H. An experimental study on flexural strength of reinforced concrete beams with 100% recycled concrete aggregate. *Eng. Struct.* **2015**, *88*, 154–162. [CrossRef]
31. *GB50010-2010*; Code for Design of Concrete Structures. China Architecture & Building Press: Beijing, China, 2010.
32. *ACI318-14*; Building Code Requirements for Structural Concrete and Commentary. American Concrete Institute: Farmington Hills, MI, USA, 2014.
33. Gao, D.; Zhang, L.; Nokken, M. Compressive behavior of steel fiber reinforced recycled coarse aggregate concrete designed with equivalent cubic compressive strength. *Constr. Build. Mater.* **2017**, *141*, 235–244. [CrossRef]
34. Gao, D.; Yan, Y.; Pang, Y.; Tang, J.; Yang, L.; Gu, Z. Effects of groove and steel fiber on shear properties of concrete with recycled coarse aggregate. *Materials.* **2020**, *13*, 4537. [CrossRef]
35. Carneiro, J.A.; Lima, P.R.; Leite, M.B.; Toledo, R.D. Compressive stress-strain behavior of steel fiber reinforced-recycled aggregate concrete. *Cement Concr. Compos.* **2014**, *46*, 65–72. [CrossRef]
36. Yazc, E.; Arel, H.A. The effect of steel fiber on the bond between concrete and deformed steel bar in SFRCs. *Constr. Build. Mater.* **2013**, *40*, 299–305.
37. Ahmadi, M.; Farzin, S.; Hassani, A.; Motamedi, M. Mechanical properties of the concrete containing recycled fibers and aggregates. *Constr. Build. Mater.* **2017**, *144*, 392–398. [CrossRef]
38. Kachouh, N.; El-Hassan, H.; El-Maaddawy, T. Influence of steel fibers on the flexural performance of concrete incorporating recycled concrete aggregates and dune sand. *J. Sustain. Cement Based Mater.* **2020**, *10*, 165–192. [CrossRef]
39. Niu, D.; Jiang, L.; Bai, M.; Miao, Y. Study of the performance of steel fiber reinforced concrete to water and salt freezing condition. *Mater. Des.* **2013**, *44*, 267–273. [CrossRef]
40. Zhang, B.; Liu, W.; Liu, X. Scale-dependent nature of the surface fractal dimension for bi- and multi-disperse porous solids by mercury porosimetry. *Appl. Surf. Sci.* **2006**, *253*, 1349–1355. [CrossRef]
41. Cui, S.; Liu, P.; Cui, E.; Su, J.; Huang, B. Experimental study on mechanical property and pore structure of concrete for shotcrete use in a hot-dry environment of high geothermal tunnels. *Constr. Build. Mater.* **2018**, *173*, 124–135. [CrossRef]

*materials*

MDPI

*Article*

# Research on the Mechanical and Physical Properties of Basalt Fiber-Reinforced Pervious Concrete

**Jian Wu [1,2], Qian Pang [1], Yuanyuan Lv [1], Jinpeng Zhang [1] and Shan Gao [3,*]**

1 Shaanxi Key Laboratory of Safety and Durability of Concrete Structures, Xijing University, Xi'an 710123, China
2 The Youth Innovation Teen of Shaanxi Universities, Xijing University, Xi'an 710123, China
3 School of Civil Engineering, Harbin Institute of Technology, Harbin 150090, China
* Correspondence: gaoshan@hit.edu.cn

**Abstract:** The aim of this study was to investigate the properties of fiber-reinforced pervious concrete. Ordinary cement, silica fume, coarse aggregate, and basalt fibers were used to produce the concrete mix. The fibers were mixed with pervious concrete at the levels of 0 kg/m$^3$, 2 kg/m$^3$, 4 kg/m$^3$, 6 kg/m$^3$, and 8 kg/m$^3$ to the investigate their influence on the mechanical and physical properties of pervious concrete. It could be observed that the cubic compressive strength, axial compressive strength, and flexural strength increased and then decreased as the content of basalt fiber increased, while the permeability and porosity of the pervious concrete decreased with the increase in the basalt fiber content. The mesostructure of pervious concrete was also studied through industrial computed tomography (ICT); the testing phenomenon showed that the fibers had a significant influence on the arrangement of the aggregate, cement paste, and the interfacial transition zone, and excessive basalt fiber resulted in poor characteristics of the interfacial transition zone (ITZ) and inferior strength properties. It was found that incorporating a basalt fiber content of 4 kg/m$^3$ could achieve a balance between the mechanical and physical properties of pervious concrete, which was suitable for structural applications.

**Keywords:** pervious concrete; basalt fiber; strength; permeability; mesostructure



**Citation:** Wu, J.; Pang, Q.; Lv, Y.; Zhang, J.; Gao, S. Research on the Mechanical and Physical Properties of Basalt Fiber-Reinforced Pervious Concrete. *Materials* **2022**, *15*, 6527. https://doi.org/10.3390/ma15196527

Academic Editor: Angelo Marcello Tarantino

Received: 19 August 2022
Accepted: 18 September 2022
Published: 20 September 2022



## 1. Introduction

Nowadays, various types of concrete gradually appear with the development of the economy, such as pervious concrete, fiber-reinforced concrete, and self-healing concrete [1–4]. Among these new types of concrete, pervious concrete has gradually become a research hotspot. Pervious concrete is a material consisting of cement, water, and coarse aggregates, which could protect the environment by replenishing the groundwater and reducing storm water runoff [1,5]. This could help to maintain ecological balance and alleviate the urban heat island effect. Therefore, although the construction cost has increased, pervious concrete is widely used all over the world in road paving, sidewalks, and pathways due to its high porosity and good thermal properties [6]. However, in order to obtain high porosity and permeability, pervious concrete often contains little or no fine aggregate [7], which results in an obvious reduction in strength and durability. Therefore, although pervious concrete can better promote the protection of the environment and the management of rainwater, its application is limited due to low strength [8]. Based on the previous research, the parameters that influence the properties are the cement content, water-to-cement, and the ratio of cement-to-aggregate [9–11].

In order to protect the environment, recycled aggregate is gradually being introduced as a partial replacement for natural coarse aggregate [12–14]. The reason for this is that a huge amount of concrete waste is produced every year, which not only requires much space for its disposal, but also pollutes the environment [15]. Although the addition of recycled aggregate can increase the permeability, due to its defects, excessive recycled aggregate will adversely affect the mechanical properties [16–18]. Kachouh et al. [19]

found that the optimum recycled aggregate replacement ratio was 30%, while the negative effect on the compressive strength could be seen when the recycled aggregate content was higher. Yap et al. [17] showed that the compressive strength of pervious concrete with 60–100% recycled aggregate as a partial replacement for natural aggregate was 20–40% lower than that of the pervious concrete with 100% natural aggregate. The study of El-Hassan [20] showed that the compressive strength of samples with 100% recycled aggregate was 78% lower than the control group. Given these deficiencies, recycled aggregate was mainly used in low-to-medium strength concrete and pavement applications [21,22]. Therefore, natural aggregate is used in this paper to improve the mechanical performance of pervious concrete.

Supplementary cementitious materials could improve the mechanical and physical properties of concrete due to the effects of microfilling and pozzolanic reactivity [23,24]. Amongst these materials, silica fume is more commonly used because it could refine the microstructure (pores and gaps) of the concrete matrix, and also produce additional gel due to the fact that the reactive silica reacts with the $Ca(OH)_2$ of the cement [25]. Chen et al. [26] studied the influence of silica fume on the strength, fracture, and fatigue of pervious concrete. The results showed that replacing the cement with silica fume could enhance the bonding effect between the aggregate and cement paste. Adil et al. [27] found that the mechanical properties of pervious concrete could be improved with a 5% content of silica fume, while the silica fume was not affected by the addition of silica fume. Amornsrivilai et al. [28] observed that the microstructure of pervious concrete could be improved due to the pore refinement by silica fume. Therefore, silica fume is added into the concrete mix to enhance the mechanical and physical properties of pervious concrete.

Besides the silica fume, another technique that is being widely used is fiber reinforcement. The reinforcing fibers could not only improve the seismic behavior of the structure [29,30], but also enhance the mechanical performance of concrete due to the bridging action of the fibers [31–33]. Industrial fibers have been successfully added into concrete to improve the mechanical properties, especially the tensile flexural strength. Basalt fibers are widely used in the production of fiber-reinforced concrete due to their low cost, simple production process, and high elastic modulus [34,35]. Mahapara et al. [36] concluded that the mechanical strength of basalt fiber-reinforced concrete increased by 88%, 89%, and 92% at 7 days, 28 days, and 90 days, respectively. Abbass et al. [36] found that concrete with an appropriate basalt fiber content could reduce the weight loss rate through acid corrosion by 27.43%, and the penetration depth of chloride ions was 2 mm. Xu et al. [37] studied basalt fiber reinforcement for geopolymer concrete. The results showed that the compressive strength and flexural strength of geopolymer concrete increased with the increase in the basalt fiber content, and the basalt fiber could refine the pores. Saloni et al. [38] found that the addition of basalt fibers was beneficial to the fiber–matrix transition zone, and the appropriate content of basalt fibers led to an increase in the setting time and mechanical strength of the concrete. Şahin et al. [39] evaluated the influence of basalt fiber on geopolymer mortars. The results showed that the compressive strength and flexural strength of the fiber-reinforced concrete increased by 25% and 50%, respectively, compared to that of the control samples. Yu et al. [40] investigated the effect of basalt fiber powder on the mechanical properties and microstructure of concrete. The results showed that 0.20 vol% basalt fiber could improve the antiabrasion property significantly, and the incorporation of basalt fiber would refine the pores and promote the densification of the concrete. Wang et al. [41] used MIMICS software to reconstruct 3D CT images of pervious concrete, and then the effect of the fiber length on the mechanical property of pervious concrete was studied. The results showed that the numerical simulation could better reflect the failure process of the specimens. Although many investigations have been carried out on basalt fiber-reinforced concrete, there are few works on the effect of acidic corrosion on the mechanical properties of pervious concrete, and SEM images cannot obtain the distribution of cement paste and aggregate. Therefore, the influence of acidic corrosion and the proportion of basalt fiber on the mechanical property of pervious concrete was studied in this paper, and ICT was used to analyze the internal structure of fiber-reinforced pervious concrete.

In this paper, we further investigated the effect of basalt fiber with different levels of content (2 kg/m$^3$, 4 kg/m$^3$, 6 kg/m$^3$, and 8 kg/m$^3$) on the mechanical and physical performance of pervious concrete. This paper is an attempt to establish a balance between the strength and permeability of pervious concrete. The cubic compressive strength, axial compressive strength, flexural tensile strength, permeability, porosity, acidic corrosion behavior, and mesostructure of pervious concrete were studied. The compressive strength and flexural tensile strength increased and then decreased with higher amounts of basalt fiber, while the porosity and permeability decreased as the basalt fiber content increased. Therefore, basalt fibers could achieve enhanced improvement in the physical and mechanical performance of pervious concrete. The research methodology is given in Figure 1.

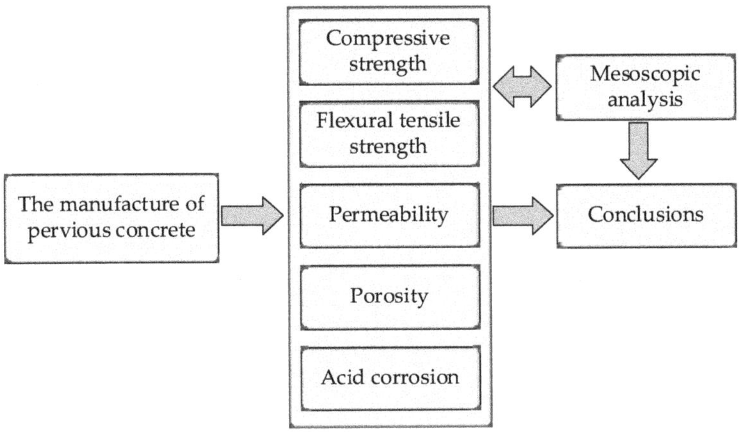


**Figure 1.** The flow chart of methodology.

## 2. Materials and Methods

### *2.1. Materials*

#### 2.1.1. Cement

In order to ensure the mechanical property of the pervious concrete, the strength grade of the Portland cement was established as 42.5. This study used the ordinary Portland cement purchased from Shenzhen Yuanheng Building Materials Co., Ltd. (Shenzhen, China), and the performance index provided by this company is given in Table 1.

**Table 1.** The performance indexes of cement.

| Initial Setting Time (min) | Final Setting Time (min) | Flexural Strength (MPa) | | Compressive Strength (MPa) | |
|---|---|---|---|---|---|
| | | 3 d | 28 d | 3 d | 28 d |
| 255 | 290 | 5.2 | 8.85 | 24.6 | 48.4 |

#### 2.1.2. Silica Fume

Silicon fume is collected and treated from the smoke and dust escaping from the waste gas in the process of melting industrial silicon and ferrosilicon at high temperature in an electric furnace. Adding an appropriate amount of silica fume can improve the fluidity, compactness, and strength of pervious concrete. The silica fume was purchased from Henan Yuanheng Environmental Protection Engineering Co., Ltd. (Zhengzhou, China), and the technical indexes given by the company are presented in Table 2.

**Table 2.** The main technical indexes of silica fume.

| Fire-Resistance Temperature (°C) | Volumetric Weight (kg/m$^3$) | Mean Diameter (μm) | Specific Surface Area (m$^2$) |
|---|---|---|---|
| 1600 | 1650 | 0.2 | 24 |

### 2.1.3. Aggregate

The performance of pervious concrete is affected by the type, size, and porosity of the aggregate. In this paper, limestone was selected as the aggregate. The physical properties and grain size distribution of the aggregate were investigated, according to Chinese Code GB/T 14685-2011 [42], as presented in Tables 3 and 4.

**Table 3.** Physical properties of aggregate.

| Aggregate Size (mm) | Bulk Density (kg/m$^3$) | Apparent Density (kg/m$^3$) | Porosity (%) |
|---|---|---|---|
| 5–10 | 1580 | 2750 | 40.2 |

**Table 4.** Sieve analysis of aggregate.

| Sieve Size (mm) | 2.36 | 4.75 | 9.5 |
|---|---|---|---|
| Accumulated retained percentage (%) | 99 | 95 | 11 |

### 2.1.4. Basalt Fiber

Basalt fiber could improve the mechanical and physical performance of concrete. The basalt fibers used in this paper were provided by Zhejiang Jinshi Co., Ltd. (Hangzhou, China), and the material characteristics were tested and provided by the company, as shown in Table 5.

**Table 5.** The technical indexes of basalt fiber.

| Density (kg/m$^3$) | Tensile Strength (MPa) | Elastic Modulus (GPa) | Length (mm) | Diameter (μm) | Elongation (%) |
|---|---|---|---|---|---|
| 2650 | 4800 | 110 | 20 | 15 | 3.1 |

### 2.1.5. Other Materials

Polycarboxylate high-performance water reducer is used to reduce the water consumption in the preparation of pervious concrete, so as to improve the fluidity of the concrete. The Q8081PCA liquid equilibrium polycarboxylate acid high-performance water reducer was used in this paper, and the dosage of the water reducer was determined by the trial matching of the specimen. The water used in the pervious concrete was running water.

## 2.2. *Mixture Proportion*

The impact of material additives was studied by adding basalt fiber into the concrete mix at the levels of 2 kg/m$^3$, 4 kg/m$^3$, 6 kg/m$^3$, and 8 kg/m$^3$. The detailed mix design for the experiments is given in Table 6.

## 2.3. *Test Method*

The pervious concrete was designed to meet the requirements of Chinese Code CJJ/T 135-2009 [43]. The cubic compressive strength, axial compressive strength, and flexural tensile strength of the specimens were tested, according to Chinese Code GB/T 50081-2019 [44]. The testing of the porosity and permeability coefficient was based on Chinese Code CJJ/T 253-2016 [45] and Chinese Code CJJ/T 135-2009 [43]. The acidic corrosion behavior was selected to study the durability of the pervious concrete, according to Chinese

Code GB/T 50476-2019 [46]. Three specimens were prepared to determine the individual mechanical and physical performance by average values. All the testing was conducted after the specimens had been cured in a concrete standard curing box for 28 days, with the condition of 22 °C and 95% humidity.

**Table 6.** Mix proportion of the pervious concrete.

| Number | Material Consumption (kg/m$^3$) | | | | | |
|---|---|---|---|---|---|---|
| | Aggregate | Cement | Water | Silica Fume | Water-Reducing Agent | Basalt Fiber |
| BF-0 | 1617 | 388 | 121 | 25 | 4.13 | 0 |
| BF-2 | 1617 | 388 | 121 | 25 | 4.13 | 2 |
| BF-4 | 1617 | 388 | 121 | 25 | 4.13 | 4 |
| BF-6 | 1617 | 388 | 121 | 25 | 4.13 | 6 |
| BF-8 | 1617 | 388 | 121 | 25 | 4.13 | 8 |

## 3. Experiment and Discussion

### *3.1. Mechanical Properties of Fiber-Reinforced Pervious Concrete*

To fully investigate the compressive strength of the pervious concrete, two sizes of specimens were designed: (i). 100 mm × 100 mm × 100 mm, (ii). 100 mm × 100 mm × 400 mm. The groups (i) and (ii) were used to obtain the cubic compressive strength, axial compressive strength, and flexural tensile strength of the pervious concrete, respectively.

#### 3.1.1. Cubic Compressive Strength

The failure patterns of the cubic specimens are given in Figure 2. The specimens without basalt fiber were completely destroyed after reaching the maximum bearing capacity, the aggregate was dispersed, and the crushing phenomenon was obvious. Although for the specimens with basalt fiber, some fragments fell off at the corner of the block at the end of testing, but the integrity of the pervious concrete was maintained well.

(a) (b)

**Figure 2.** The failure patterns of (**a**) cubic specimens without basalt fiber and (**b**) cubic specimens with basalt fiber.

Figure 3 shows the effect of basalt fiber on the cubic compressive strength of the pervious concrete. Although the cubic compressive strength increases and then decreases as the basalt fiber content increases, the cubic compressive strength value of the pervious concrete with basalt fiber is higher than that of the pervious concrete without basalt fiber (control specimen). The peak value of cubic compressive strength is shown with a basalt fiber content of 4 kg/m$^3$, which increases by 24% compared to the control specimen. The red dashed line in Figure 3 indicates that the compressive strength of the pervious concrete can meet the requirements of CJJ/T 135-2009 [43] (20 MPa, the red dashed line in Figure 3).

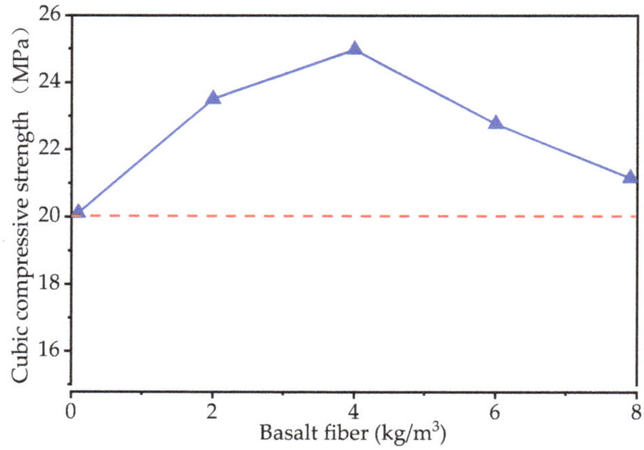


**Figure 3.** The cubic compressive strength of the specimens with different levels of basalt fiber content.

### 3.1.2. Axial Compressive Strength

The compressive strength of concrete is related to the shape of the specimen. The prism specimen can better reflect the actual compressive strength of concrete than that of cube samples. Therefore, the axial compressive strength of the pervious concrete was tested in this paper.

The failure patterns of concrete prisms under the action of axial load are given in Figure 4. For the specimens without basalt fiber, a wide vertical crack appeared in the middle of the prism. Although the prism was not completely destroyed, it had completely lost the bearing capacity. Although for the specimen with basalt fiber, small cracks occurred on the side of the specimen and a small amount of the aggregate fell off, the specimen as a whole remained intact.

(**a**) (**b**)

**Figure 4.** The failure patterns of (**a**) concrete prism without basalt fiber and (**b**) concrete prism with basalt fiber under axial compression.

The effect of basalt fiber on the axial compressive strength of the pervious concrete is presented in Figure 5. It can be seen that with an increase in the basalt fiber content, there is an increase and then a decrease in the axial compressive strength. The axial compressive strength of the pervious concrete increases from 14.84 MPa for the control specimen (i.e., the basalt fiber content is 0 kg/m$^3$) to 20.63 MPa for a basalt fiber content of 4 kg/m$^3$, and then decreases to 18.00 MPa for a basalt fiber content of 8 kg/m$^3$. The results also indicate that the addition of basalt fiber can improve the axial compressive strength of pervious concrete.

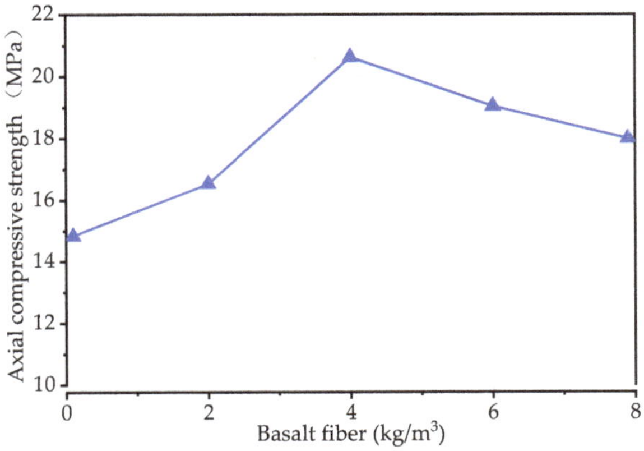


**Figure 5.** The axial compressive strength of specimens with different levels of basalt fiber content.

3.1.3. Flexural Tensile Strength

Pervious concrete is inevitably subjected to flexural and tensile loads in the practical usage process. Therefore, it is essential to study the flexural tensile strength to prevent the appearance of cracks in pervious concrete.

Figure 6 shows the failure pattern of a concrete prism under the action of a lateral load. Whether or not the basalt fiber was added to the pervious concrete, the cracks mainly appeared in the middle of the specimens. The difference between the specimens with basalt fiber and control specimens is that without the connection of the fiber, once the crack appeared at the bottom, it would develop rapidly to the top of the specimen.

**Figure 6.** The flexural tensile failure patterns of (**a**) concrete prism without basalt fiber and (**b**) concrete prism with basalt fiber.

The effect of basalt fiber on the flexural tensile strength of the pervious concrete is presented in Figure 7. The flexural tensile strength increases and then decreases with the increasing of the basalt fiber content. The maximum strength value is 4.82 MPa when the basalt fiber content is 2 kg/m$^3$, and then the strength value decreases to 2.84 MPa for the basalt fiber content of 8 kg/m$^3$. In CJJ/T 135-2009 [43], it is stated that the flexural tensile strength should not be lower than 2.5 MPa (the red dashed line in Figure 7), meaning that all the specimens met the application requirement.

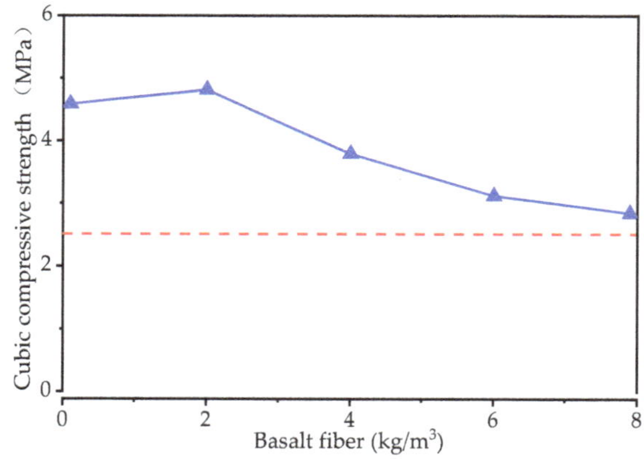


**Figure 7.** The flexural tensile strength of the specimens with different levels of basalt fiber content.

Combining the results of the cubic compressive strength, axial compressive strength, and flexural tensile strength, it could be concluded that the mechanical properties of the fiber-reinforced pervious concrete can meet the requirements of CJJ/T 135-2009 [43], and that the appropriate addition of basalt fiber could improve the mechanical properties.

### *3.2. Permeability and Porosity of Fiber-Reinforced Pervious Concrete*

#### 3.2.1. Permeability

Cylinder specimens were prepared to obtain the permeability coefficient, as given in Figure 8. The cylinder specimen should be connected to the cylinder wall using gasket cement, to ensure that the water can only flow through the specimen. The permeability coefficient could be obtained from Equation (1).

$$k_{\mathrm{T}} = \frac{QL}{AHt} \tag{1}$$

in which $k_{\mathrm{T}}$ is the permeability coefficient when the water temperature is $T$ °C, $Q$ is the amount of water seeping out within $t$ seconds, $L$ is the thickness of the specimen, $A$ is the upper surface area of the specimen, $H$ is the water level difference, and $t$ is the time. The results of the calculation are shown in Figure 9.

**Figure 8.** The testing of permeability coefficient.

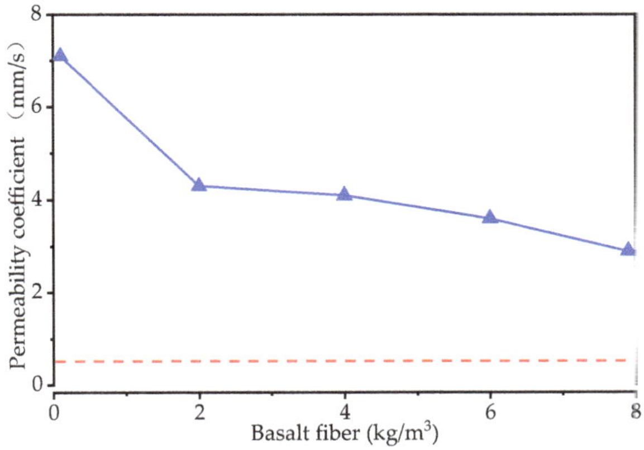


**Figure 9.** The permeability coefficient of pervious concrete with different levels of basalt fiber content.

As shown in Figure 9, the permeability coefficient (mm/s) decreases as the basalt fiber content increases. Compared to the specimen without basalt fiber, the permeability coefficient decreases rapidly (0–2 kg/m$^3$) and then slowly (2–8 kg/m$^3$) with the increasing of the basalt fiber content. As can be seen, the permeability coefficient is roughly 2.9–7.1 mm/s, which is higher than the regulation of CJJ/T 135-2009 [43] (0.5 mm/s, the red dashed line in Figure 9).

3.2.2. Porosity

As shown in Figure 10, the cubic specimens were immersed in water for one day, and then the specimens were taken out to be dried until the surfaces were dried. The porosity could be obtained through the quality difference before and after drying. The porosity can be calculated by Equation (2).

$$P = \left[1 - \left(\frac{m_2 - m_1}{\rho_W V_0}\right)\right] \times 100\% \quad (2)$$

in which $P$ is porosity, $m_1$ is the quality of the specimen under water, $m_2$ is the quality of the specimen after air drying for 24 h, $\rho_W$ is the density of water at normal temperature, and $V_0$ is the volume of the specimen.

**Figure 10.** The testing of porosity.

The porosity of the specimens is given in Figure 11. The porosity of the specimens, when compared with the control specimens, showed a percentage decrease of 28.2%, 35.5%, 40.6%, and 42.7% for the basalt fiber contents of 2 kg/m$^3$, 4 kg/m$^3$, 6 kg/m$^3$, and 8 kg/m$^3$, respectively. However, the porosity could meet the requirements of CJJ/T 135-2009 [43] (10%, the red dashed line in Figure 11). It is clear that the porosity has a relationship with the permeability, and the addition of basalt fibers did not change this trend.

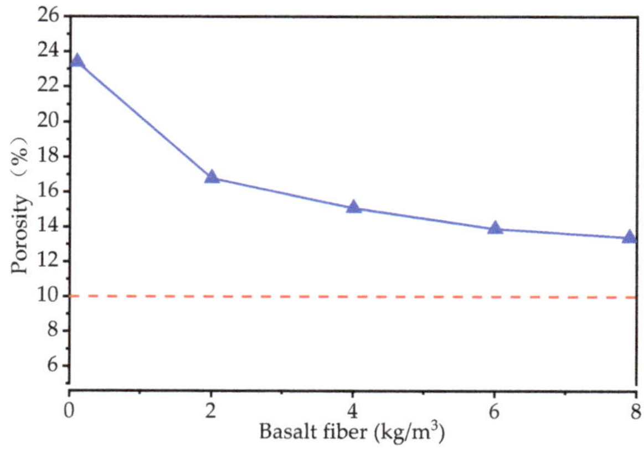


**Figure 11.** The porosity of the specimens with different levels of basalt fiber content.

3.2.3. The Relationship between the Permeability and Porosity

It is well known that the porosity is vital to the permeability. Therefore, this paper presented the relationship between the permeability coefficient and porosity in Figure 12, in which the red line (fitting curve) is the result of the regression analysis. The formula of the fitting curve can be expressed as Equation (3).

$$k_T = -2.04 + 0.39P \tag{3}$$

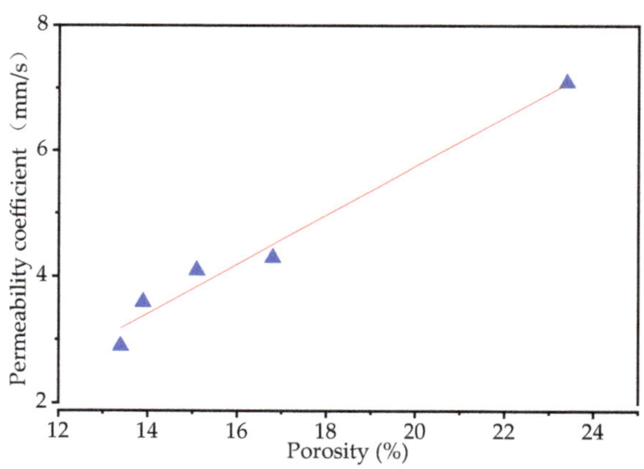


**Figure 12.** The relationship between permeability and porosity.

It can be seen from Figure 12 that the permeability coefficient is linearly dependent on porosity, and the permeability coefficient increases with the increasing of porosity.

*3.3. The Acidic Corrosion Behavior of Fiber-Reinforced Pervious Concrete*

Due to the large number of pores, corrosive substances can easily enter the interior of pervious concrete, which will destroy the internal structure of pervious concrete and reduce the performance of pervious concrete. In this paper, the specimens were put into acid solution (hydrochloric acid solution with pH = 1), whose surface was at least 20 mm higher than the upper surface of the specimens, as shown in Figure 13. The specimens were soaked in the acid solution for 20 days, 40 days, and 60 days.

**Figure 13.** The acidic corrosion of the specimens.

The corrosion phenomenon of the specimens (with a basalt fiber content of 2 $kg/m^3$) is shown in Figure 14, and the specimens with the other levels of basalt fiber content exhibited the same experimental phenomenon. It could be seen that when the corrosion time was 20 days, the aggregate did not fall off, and a white colloid appeared, almost filling the surface gap of the specimen; when the corrosion time lasted for 40 days, the aggregate was obviously exposed and fell off slightly, the white colloid gradually disappeared, and the soaking solution became turbid; when the corrosion time was 60 days, the surface of the specimens was seriously corroded and became dark brown, and the specimens showed an uneven shape due to the aggregate falling off (at the corners of the specimens).

(a) (b) (c) (d)

**Figure 14.** The testing phenomenon of specimens (with basalt fiber content of 4 $kg/m^3$) for acidic corrosion: (**a**) 0 day, (**b**) 20 days, (**c**) 40 days, and (**d**) 60 days.

Figure 15 shows the effect of the basalt fiber content and acidic corrosion time on the cubic compressive strength of the specimens. As shown in Figure 15, for the pervious concrete specimen, which had a different level of basalt fiber content, the cubic compressive strength increases and then decreases as the acidic corrosion time increases. The reason for increasing the compressive strength is the production of acidic corrosion. $CaSO_4{\cdot}2H_2O$, $CaAl_2Si_2O_8$, and $Ca_3Al_6O_{12}{\cdot}CaSO_4$ are produced due to acidic corrosion. When the corrosion time is short, these expansive substances could block the pores and reduce the corrosion speed of the pervious concrete. However, for the acidic corrosion of 40 days and 60 days, the expansive substance gradually decomposes in the acid solution, which can accelerate the corrosion process of the pervious concrete [47]. It can also be seen from Figure 15 that the cubic compressive strength cannot meet the requirements of CJJ/T 135-2009 [43] with acidic corrosion of 60 days, indicating that the acidic corrosion will seriously affect the mechanical properties of pervious concrete.

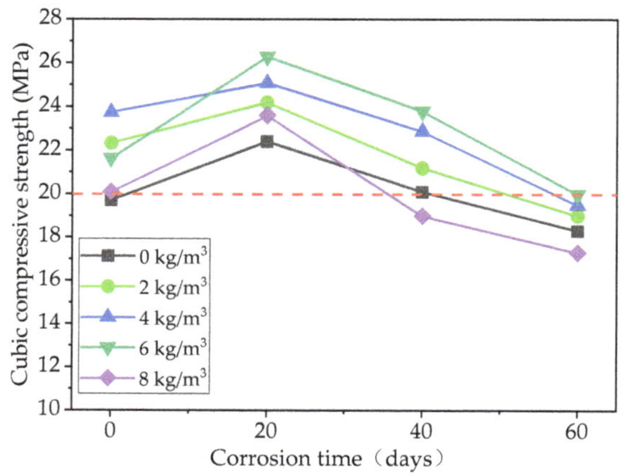


**Figure 15.** The cubic compressive strength of specimens with different levels of basalt fiber content and different corrosion times.

### 3.4. The Mesostructure of Fiber-Reinforced Pervious Concrete

In order to obtain the mesostructure of the pervious concrete, industrial computed tomography (ICT) was used to scan the specimen. Figures 16 and 17, which can be obtained after processing with Avizo software, are the 3D images and 2D images of the cross section respectively. The blue-scale in Figure 16 and the grey-scale in Figure 17 are the aggregates and the cement paste. As shown in Figure 16, the proper addition of basalt fiber is beneficial to uniform dispersion of aggregate and cement paste (Figure 16a–c), while too much basalt fiber causes more cement paste to bond to the fiber, meaning that the bond between the aggregates and the cement paste will be weakened due to the reduction in cement paste on the aggregates (Figure 16d,e).

The same trend can be seen from Figure 17. For the control specimen (Figure 17a), the cement paste is mainly bonding to the aggregate, and there are lots of pores in the pervious concrete. With the increase in the basalt fiber content, more and more cement paste is bonded to the fiber, which reduces the pores in the pervious concrete (Figure 17b–e).

### 3.5. Analysis of the Relation between Mechanical Properties and Physical Performance

As described in [48], pervious concrete could be divided into three parts: aggregate, cement paste, and the interfacial transition zone (ITZ), in which the ITZ represents the point/path at which there is debonding between the aggregates and the cement paste. The same phenomenon can also be seen in Figures 16 and 17. Therefore, three different failure modes could be observed—aggregate failure, cement paste failure, and ITZ failure. The failures are given in Figure 18 with the yellow marks. In the testing of cubic compressive strength, for the specimens without or with less basalt fiber content (0 kg/m$^3$, 2 kg/m$^3$, and 4 kg/m$^3$), three failure modes could be observed, while for the specimens with a higher basalt fiber content (6 kg/m$^3$ and 8 kg/m$^3$), the failure was mainly along the ITZ. The same phenomenon could also be seen in the testing of axial compressive strength. Considering the changes in the strength value of mechanical performance, it could be concluded that the proper addition of basalt fiber helps to achieve the strength balance of the aggregate, cement paste, and the ITZ, which will ensure the strength of the specimen. Although excessive basalt fiber results in poor characteristics of the ITZ and inferior strength properties.

It is widely accepted that porosity is vital to the mechanical performance and physical properties of pervious concrete. As described in the testing results of permeability and porosity, and taking the mesostructure of the fiber-reinforced pervious concrete into account, the addition of basalt fiber causes more cement to bond to the fibers, which blocks the pores in pervious concrete. It is also clear that porosity has a positive relationship with permeability, and the addition of basalt fibers does not change this trend.

In order to improve the physical and mechanical properties of pervious concrete and promote the application of pervious concrete in actual construction, the optimum basalt fiber content should be established. Comparing the samples with a basalt fiber content of 0 kg/m$^3$ (control samples) and 4 kg/m$^3$, the cubic compressive strength and axial compressive strength increased by 24% and 39%, respectively, while the flexural tensile strength, permeability, and porosity were reduced by 17%, 42%, and 35%, respectively, with the 4 kg/m$^3$ basalt fiber content. For the specimens for acidic corrosion of 0 day, 20 days, 40 days, and 60 days, the cubic compressive strength of the samples with a basalt fiber content of 4 kg/m$^3$ was 21%, 12%, 14%, and 7% higher than that of the control specimens, respectively. All the testing results meet the requirements of specification, indicating that the optimum basalt fiber content should be determined as 4 kg/m$^3$.

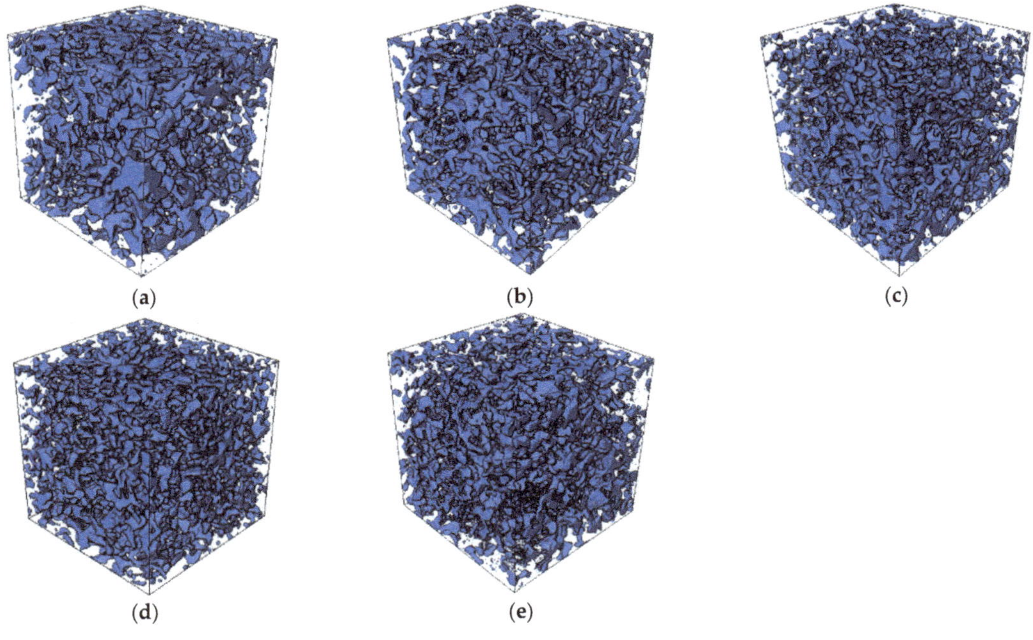


**Figure 16.** The 3D images of the specimens with different levels of content of basalt fiber: (**a**) 0 kg/m$^3$, (**b**) 2 kg/m$^3$, (**c**) 4 kg/m$^3$, (**d**) 6 kg/m$^3$, and (**e**) 8 kg/m$^3$.

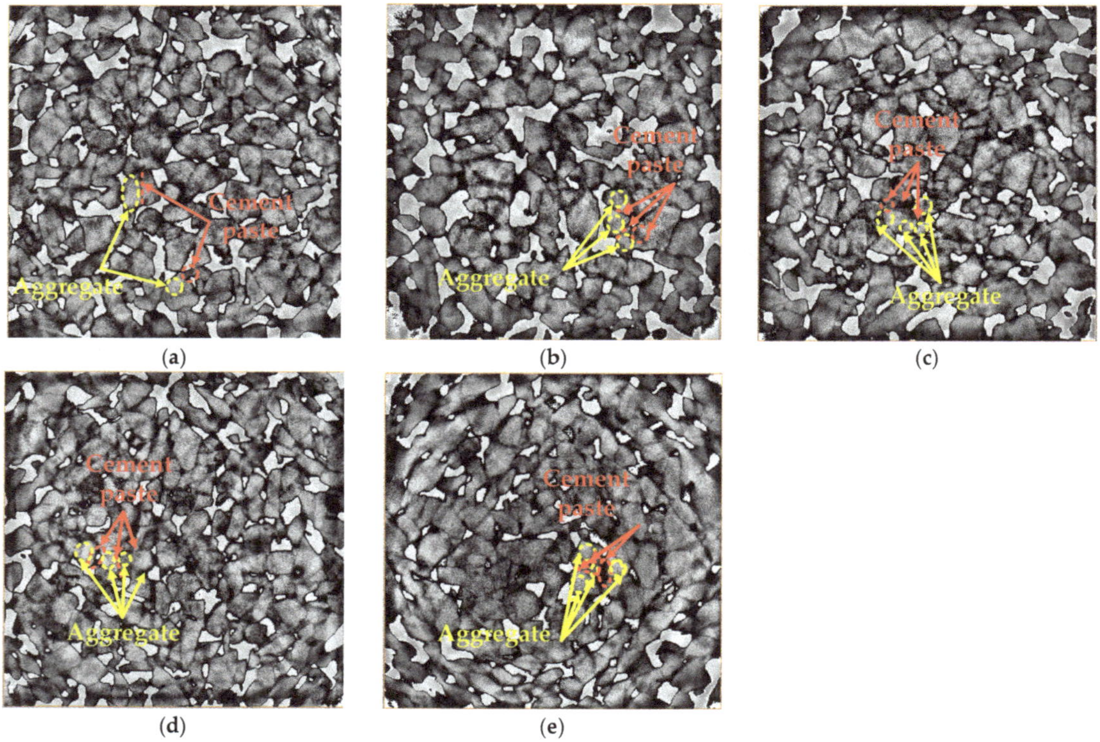


**Figure 17.** The 2D images of cross sections of the specimens with different levels of content of basalt fiber: (**a**) 0 kg/m$^3$, (**b**) 2 kg/m$^3$, (**c**) 4 kg/m$^3$, (**d**) 6 kg/m$^3$, and (**e**) 8 kg/m$^3$.

**Figure 18.** The failure of the specimen: (**a**) sample without basalt fiber and (**b**) sample with basalt fiber.

## 4. Conclusions

This study investigated the effect of basalt fiber on the physical and mechanical properties of pervious concrete. To achieve the aim of this paper, the pervious concrete matrix was mixed incorporating basalt fiber at the level of 2 kg/m$^3$, 4 kg/m$^3$, 6 kg/m$^3$, and 8 kg/m$^3$. The strength testing, porosity testing, permeability testing, and acid corrosion testing were conducted on the testing specimen. The following conclusions were made:

1. The incorporation of basalt fiber enhances the cubic and axial compressive strength, and the compressive strength increases and then decreases with the increasing of the basalt fiber content. Compared to the matrix without basalt fiber, the cubic and axial compressive strength of the matrix with a basalt content of 4 $kg/m^3$ increased by 24% and 39%, respectively.
2. The contribution of basalt fiber to the flexural strength of the concrete matrix reduces with the increasing of the basalt fiber content. Compared to the control samples, the flexural tensile strength of the matrix with a basalt fiber content of 4 $kg/m^3$ reduces by 17%.
3. A pervious concrete matrix incorporating basalt fiber showed a lower permeability and porosity than that of the mix without fiber. Moreover, the permeability coefficient and the porosity of the pervious concrete with a basalt fiber content of 4 $kg/m^3$ were 42% and 35% lower than those of the control samples.
4. For the concrete matrix with the basalt fiber content of 4 $kg/m^3$, the cubic compressive strengths of the matrix with acidic corrosion of 0 day, 20 days, 40 days, and 60 days were 21%, 12%, 14%, and 7% higher than those of the control group, respectively. The compressive strength increased and then decreased with the increase in the corrosion time, and the addition of basalt fibers does not change this trend.
5. The testing results of ICT scanning indicated that the basalt fiber obviously affects the distribution of the cement paste and aggregate, hence affecting the physical and mechanical properties of the pervious concrete. The optimum basalt fiber content can realize the balance between the cement paste, aggregate, and the ITZ.

The physical and mechanical properties of fiber-reinforced pervious concrete could satisfy the requirements of CJJ/T 135-2009. The obtained results indicated that it is possible to investigate whether pervious concrete is suitable for structural application by the addition of basalt fiber, which can maximize the application of pervious concrete.

**Author Contributions:** Conceptualization, S.G. and J.W.; methodology, J.W.; software, Y.L.; validation, S.G. and J.W.; formal analysis, J.Z.; investigation, Q.P.; resources, Q.P.; data curation, Y.L.; writing—original draft preparation, J.W.; writing—review and editing, S.G.; visualization, Q.P.; supervision, S.G.; project administration, J.W.; funding acquisition, J.W. All authors have read and agreed to the published version of the manuscript.

**Funding:** This research was supported by the Natural Science Foundation of Shaanxi Province (2021JQ-876) and the Scientific Research Foundation for High-level Talents (XJ17T08).

**Institutional Review Board Statement:** Not applicable.

**Informed Consent Statement:** Not applicable.

**Data Availability Statement:** Data sharing is not applicable to this article.

**Acknowledgments:** The authors would like to thank the Shaanxi Key Laboratory of Safety and Durability of Concrete Structures for the project testing.

**Conflicts of Interest:** The authors declare no conflict of interest.

## References


1. Deo, O.; Neithalath, N. Compressive behavior of pervious concretes and a quantification of the influence of random pore structure features. *Mater. Sci. Eng. A* **2010**, *528*, 402–412. [CrossRef]
2. Aslam, F.; Zaid, O.; Althoey, F.; Alyami, S.H.; Qaidi, S.M.; de Prado Gil, J.; Martínez-García, R. Evaluating the influence of fly ash and waste glass on the characteristics of coconut fibers reinforced concrete. *Struct. Concr.* **2022**. *early view*. [CrossRef]
3. Akeed, M.H.; Qaidi, S.; Ahmed, H.U.; Faraj, R.H.; Mohammed, A.S.; Emad, W.; Tayeh, B.A.; Azevedo, A.R.G. Ultra-high-performance fiber-reinforced concrete. Part IV: Durability properties, cost assessment, applications, and challenges. *Case Stud. Constr. Mat.* **2022**, *17*, e01271. [CrossRef]
4. Onyelowe, K.C.; Ebid, A.M.; Riofrio, A.; Baykara, H.; Soleymani, A.; Mahdi, H.A.; Jahangir, H.; Ibe, K. Multi-Objective Prediction of the Mechanical Properties and Environmental Impact Appraisals of Self-Healing Concrete for Sustainable Structures. *Sustainability* **2022**, *14*, 9573. [CrossRef]

5. Ho, H.-L.; Huang, R.; Hwang, L.-C.; Lin, W.-T.; Hsu, H.-M. Waste-Based Pervious Concrete for Climate-Resilient Pavements. *Materials* **2018**, *11*, 900. [CrossRef] [PubMed]
6. Huang, J.; Luo, Z.; Khan, M.B.E. Impact of aggregate type and size and mineral admixtures on the properties of pervious concrete: An experimental investigation. *Constr. Build. Mater.* **2020**, *265*, 120759. [CrossRef]
7. Chandrappa, A.K.; Biligiri, K.P. Pervious concrete as a sustainable pavement material—Research findings and future prospects: A state-of-the-art review. *Constr. Build. Mater.* **2016**, *111*, 262–274. [CrossRef]
8. Mahalingam, R.; Mahalingam, S.V. Analysis of pervious concrete properties. *J. Croat. Assoc. Civ. Eng.* **2016**, *68*, 493–501.
9. Joshi, T.; Dave, U. Evaluation of strength, permeability and void ratio of pervious concrete with changing W/C ratio and aggregate size. *Int. J. Civ. Eng. Technol.* **2016**, *7*, 276–284.
10. Yu, F.; Sun, D.; Wang, J.; Hu, M. Influence of aggregate size on compressive strength of pervious concrete. *Constr. Build. Mater.* **2019**, *209*, 463–475. [CrossRef]
11. Wang, H.; Li, H.; Liang, X.; Zhou, H.; Xie, N.; Dai, Z. Investigation on the mechanical properties and environmental impacts of pervious concrete containing fly ash based on the cement-aggregate ratio. *Constr. Build. Mater.* **2019**, *202*, 387–395. [CrossRef]
12. Lori, A.R.; Hassani, A.; Sedghi, R. Investigating the mechanical and hydraulic characteristics of pervious concrete containing copper slag as coarse aggregate. *Constr. Build. Mater.* **2019**, *197*, 130–142. [CrossRef]
13. Lu, G.; Liu, P.; Wang, Y.; Faßbender, S.; Wang, D.; Oeser, M. Development of a sustainable pervious pavement material using recycled ceramic aggregate and bio-based polyurethane binder. *J. Clean. Prod.* **2019**, *220*, 1052–1060. [CrossRef]
14. Zareei, S.A.; Ameri, F.; Bahrami, N.; Shoaei, P.; Moosaei, H.R.; Salemi, N. Performance of sustainable high strength concrete with basic oxygen steel-making (BOS) slag and nano-silica. *J. Build. Eng.* **2019**, *25*, 100791. [CrossRef]
15. Tam, V.W.Y.; Soomro, M.; Evangelista, A.C.J. A review of recycled aggregate in concrete applications (2000–2017). *Constr. Build. Mater.* **2018**, *172*, 272–292. [CrossRef]
16. Wongsa, A.; Zaetang, Y.; Sata, V.; Chindaprasirt, P. Properties of lightweight fly ash geopolymer concrete containing bottom ash as aggregates. *Constr. Build. Mater.* **2016**, *111*, 637–643. [CrossRef]
17. Yap, S.P.; Chen, P.Z.C.; Goh, Y.; Ibrahim, H.; Mo, K.H.; Yuen, C.W. Characterization of pervious concrete with blended natural aggregate and recycled concrete aggregates. *J. Clean. Prod.* **2018**, *181*, 155–165. [CrossRef]
18. Lang, L.; Duan, H.; Chen, B. Properties of pervious concrete made from steel slag and magnesium phosphate cement. *Constr. Build. Mater.* **2019**, *209*, 95–104. [CrossRef]
19. Kachouh, N.; El-Hassan, H.; El Maaddawy, T. Effect of steel fibers on the performance of concrete made with recycled concrete aggregates and dune sand. *Constr. Build. Mater.* **2019**, *213*, 348–359. [CrossRef]
20. El-Hassan, H.; Kianmehr, P.; Zouaoui, S. Properties of pervious concrete incorporating recycled concrete aggregates and slag. *Constr. Build. Mater.* **2019**, *212*, 164–175. [CrossRef]
21. Saberian, M.; Li, J.; Nguyen, B.; Wang, G. Permanent deformation behaviour of pavement base and subbase containing recycle concrete aggregate, coarse and fine crumb rubber. *Constr. Build. Mater.* **2018**, *178*, 51–58. [CrossRef]
22. Yaowarat, T.; Horpibulsuk, S.; Arulrajah, A.; Mohammadinia, A.; Chinkulkijniwat, A. Recycled Concrete Aggregate Modified with Polyvinyl Alcohol and Fly Ash for Concrete Pavement Applications. *J. Mater. Civ. Eng.* **2019**, *31*, 4019103. [CrossRef]
23. Lu, J.-X.; Yan, X.; He, P.; Poon, C.S. Sustainable design of pervious concrete using waste glass and recycled concrete aggregate. *J. Clean. Prod.* **2019**, *234*, 1102–1112. [CrossRef]
24. Shilar, F.A.; Ganachari, S.V.; Patil, V.B.; Javed, S.; Khan, T.M.Y.; Baig, R.U. Assessment of Destructive and Nondestructive Analysis for GGBS Based Geopolymer Concrete and Its Statistical Analysis. *Polymers* **2022**, *14*, 3132. [CrossRef]
25. Zamanabadi, S.N.; Zareei, S.A.; Shoaei, P.; Ameri, F. Ambient-cured alkali-activated slag paste incorporating micro-silica as repair material: Effects of alkali activator solution on physical and mechanical properties. *Constr. Build. Mater.* **2019**, *229*, 116911. [CrossRef]
26. Chen, Y.; Wang, K.; Wang, X.; Zhou, W. Strength, fracture and fatigue of pervious concrete. *Constr. Build. Mater.* **2013**, *42*, 97–104. [CrossRef]
27. Adil, G.; Kevern, J.T.; Mann, D. Influence of silica fume on mechanical and durability of pervious concrete. *Constr. Build. Mater.* **2020**, *247*, 118453. [CrossRef]
28. Amornsrivilai, P.; Tia, M.; Lee, M.-G.; Su, Y.-M. Effects of Fly Ash and Silica Fume on Permeability of Concrete Made With Porous Limestone and Non-Porous Aggregates. *J. Test. Eval.* **2016**, *45*, 335–345. [CrossRef]
29. Ztürk, B.; Yilmaz, C. Analytical investigation of effect of retrofit application using FRP on seismic behavior of a monumental building at historical Cappadocia region of Turkey. In Proceedings of the 9th U.S. National and 10th Canadian Conference on Earthquake Engineering, Toronto, ON, Canada, 25–29 July 2010.
30. Ztürk, B.; Yilmaz, C.; Senturk, T. Effect of FRP retrofitting application on seismic behavior of a historical building at Nigde, Turkey. In Proceedings of the 14th European Conference on Earthquake Engineering, Ohrid, Republic of Macedonia, 30 August–3 September 2010.
31. Akand, L.; Yang, M.; Wang, X. Effectiveness of chemical treatment on polypropylene fibers as reinforcement in pervious concrete. *Constr. Build. Mater.* **2018**, *163*, 32–39. [CrossRef]
32. Al-Hadithi, A.; Noaman, A.T.; Mosleh, W.K. Mechanical properties and impact behavior of PET fiber reinforced self-compacting concrete (SCC). *Compos. Struct.* **2019**, *224*, 111021. [CrossRef]

33. Mastali, M.; Dalvand, A. Use of silica fume and recycled steel fibers in self-compacting concrete (SCC). *Constr. Build. Mater.* **2016**, *125*, 196–209. [CrossRef]
34. Fiore, V.; Scalici, T.; Di Bella, G.; Valenza, A. A review on basalt fibre and its composites. *Compos. Part B Eng.* **2015**, *74*, 74–94. [CrossRef]
35. Kumutha, R.; Fathima, I.S.; Vijai, D.K. Experimental investigation on properties of basalt fiber reinforced geopolymer concrete. *J. Mech. Civil Eng.* **2017**, *14*, 105–109. [CrossRef]
36. Abbass, M.; Singh, G. Impact strength of rice husk ash and basalt fibre based sustainable geopolymer concrete in rigid pavements. *Mater. Today Proc.* **2021**, *61*, 250–257. [CrossRef]
37. Xu, J.; Kang, A.; Wu, Z.; Xiao, P.; Gong, Y. Effect of high-calcium basalt fiber on the workability, mechanical properties and microstructure of slag-fly ash geopolymer grouting material. *Constr. Build. Mater.* **2021**, *302*, 124089. [CrossRef]
38. Saloni; Parveen; Pham, T.M. Enhanced properties of high-silica rice husk ash-based geopolymer paste by incorporating basalt fibers. *Constr. Build. Mater.* **2020**, *245*, 118422. [CrossRef]
39. Şahin, F.; Uysal, M.; Canpolat, O.; Aygörmez, Y.; Cosgun, T.; Dehghanpour, H. Effect of basalt fiber on metakaolin-based geopolymer mortars containing rilem, basalt and recycled waste concrete aggregates. *Constr. Build. Mater.* **2021**, *301*, 124113. [CrossRef]
40. Yu, H.Y.; Meng, T.; Zhao, Y.X.; Liao, J.P.; Ying, K.J. Effects of basalt fiber powder on mechanical properties and microstructure of concrete. *Case Stud. Constr. Mater.* **2022**, *17*, e01286. [CrossRef]
41. Wang, W.; Zhu, J.; Cheng, X.; Liu, S.; Jiang, D.; Wang, W. Numerical simulation of strength of basalt fiber permeable concrete based on CT technology. *Case Stud. Constr. Mater.* **2022**, *17*, e01348. [CrossRef]
42. *GB/T 14685-2011*; Pebble and Crushed Stone for Construction. China Construction Industry Press: Beijing, China, 2011.
43. *CJJ/T 135-2009*; Technical Specification for Pervious Cement Concrete Pavement. China Construction Industry Press: Beijing, China, 2009.
44. *GB/T 50081-2019*; Standard for Test Methods of Concrete Physical and Mechanical Properties. China Construction Industry Press: Beijing, China, 2019.
45. *CJJ/T 253-2016*; Technical Specification for Application of Pervious Recycled Aggregate Concrete. China Construction Industry Press: Beijing, China, 2016.
46. *GB/T 50476-2019*; Standard for Design of Concrete Structure Durability. China Construction Industry Press: Beijing, China, 2019.
47. Chen, M.C.; Kai, W.; Li, X. Deterioration mechanism of cementitious materials under acid rain attack. *Eng. Fail Anal.* **2013**, *27*, 272–285. [CrossRef]
48. Oni, B.C.; Xia, J.; Liu, M. Mechanical properties of pressure moulded fibre reinforced pervious concrete pavement brick. *Case Stud. Constr. Mater.* **2020**, *13*, e00431. [CrossRef]

materials

MDPI

Article

# An Investigation of the Properties of Expanded Polystyrene Concrete with Fibers Based on an Orthogonal Experimental Design

Yi Sun [1,2], Chenxi Li [1,2], Junjie You [3,*], Changming Bu [1,2,*], Linwen Yu [4,*], Zhitao Yan [1,2], Xinpeng Liu [1,2], Yi Zhang [5] and Xianrui Chen [6]

1 School of Civil Engineering and Architecture, Chongqing University of Science & Technology, Chongqing 401331, China; sunyi@cqust.edu.cn (Y.S.); lichenxi@cqust.edu.cn (C.L.); yzt@cqust.edu.cn (Z.Y.); liu_simple@cqust.edu.cn (X.L.)
2 Chongqing Key Laboratory of Energy Engineering Mechanics & Disaster Prevention and Mitigation, Chongqing 401331, China
3 School of Civil Engineering, Southwest Jiaotong University, Chengdu 610031, China
4 College of Materials Science and Engineering, Chongqing University, Chongqing 400044, China
5 Chongqing Construction Residential Engineering Co., Ltd., Chongqing 400015, China; kzb@cqzj.com.cn
6 Chongqing Tidy Green New Material Co., Ltd., Chongqing 401221, China; 2012026@cqust.edu.cn
* Correspondence: junjie_you@my.swjtu.edu.cn (J.Y.); buchangming@cqust.edu.cn (C.B.); linwen.yu@cqu.edu.cn (L.Y.)

**Abstract:** Expanded polystyrene (EPS) concrete is commonly used as the core material of commercial sandwich panels (CSPs). It is environmentally friendly and lightweight but has poor strength. Adding fibers can improve the microstructure of EPS concrete and reduce the weakening effect of EPS beads on the mechanical properties of concrete. An orthogonal experimental design (OED) was used in this paper to analyze the influence of length and content of polypropylene fiber (PF), glass fiber (GF), and carbon fiber (CF) on the physical and mechanical properties and micromorphology of EPS concrete. Among them, CFs have the most apparent impact on concrete and produce the most significant improvements in all properties. According to the requirements of the flexural performance of CSPs, the splitting tensile strength was taken as the optimization index, and the predicted optimal combination (OC) of EPS concrete with fibers was selected. The variations in the material properties, mechanical properties, and microstructure with age were analyzed. The results show that with increasing age, the dry density, compressive strength, and splitting tensile strength of concrete are markedly improved relative to those of the CSP core material and the control case (CC), and even the degree of hydration is improved.

**Keywords:** EPS concrete; fiber-reinforced concrete; orthogonal experimental design; mechanical properties; microstructures



**Citation:** Sun, Y.; Li, C.; You, J.; Bu, C.; Yu, L.; Yan, Z.; Liu, X.; Zhang, Y.; Chen, X. An Investigation of the Properties of Expanded Polystyrene Concrete with Fibers Based on an Orthogonal Experimental Design. *Materials* **2022**, *15*, 1228. https://doi.org/10.3390/ma15031228

Academic Editor: Roman Fediuk

Received: 28 December 2021
Accepted: 1 February 2022
Published: 7 February 2022



## 1. Introduction

Expanded polystyrene (EPS) beads are byproducts of petroleum engineering and have the advantages of facility acquisition low densities, economical prices, low water absorption, and great thermal insulation performance. However, they still have the following disadvantages: they are nondegradable, difficult to recycle and treat, and they have poor mechanical strength. EPS beads can be used for the partial replacement of raw aggregates and can be made into EPS concrete, which not only utilizes industrial waste but also saves the usage of aggregates and reduces the density and weight of concrete. It is worth noting that EPS concrete is fairly environmentally friendly.

Owing to the low density of EPS beads, which is generally lower than 20 kg/m$^3$, the density of concrete can be sharply reduced by adding EPS beads. As the proportion of EPS increases, the density, compressive strength, splitting tensile strength, and elastic modulus of concrete present decreasing trends [1–5]. Moreover, the size of the EPS beads also has an obvious effect on the material properties [6] under the same density of EPS concrete. As the size decreases, the compressive strength performance of concrete and the binding force between cement and EPS beads are both enhanced [1,7].

Fly ash (FA) is used as supplementary cementitious material, and it improves the water absorption rate [8] and reduces the dry shrinkage of concrete. Silica fume (SF) is an industrial byproduct that is suitable for filling cement pores [9] due to its small diameter, and by this way, the durability could be improved [10]. In addition, the mechanical properties of concrete can be properly improved by the gel that is formed by rehydration with cement hydration products and SF. Similarly, the addition of SF not only improves the durability of the concrete [11] but also enhances the compressive strength and reduces the porosity. In particular, it can tightly connect the interfacial transition zone (ITZ) between EPS beads and the interface of concrete, which improves the segregation of materials [12].

EPS beads have minimal mechanical strength. Thus, the strength of EPS concrete is mainly determined by the strength of concrete without EPS beads and its microstructure. The ordinary method for enhancing the mechanical strength of EPS concrete without affecting its low density and lightweight design is to add fibers. Fibers have a significant enhancement effect on the mechanical properties of concrete [13–16]. Typically, PF [17–19], GF [20,21], and CF [22,23] are commonly used as reinforcing materials to enhance the various properties of concrete. The addition of fibers has improved the anti-shrinkage, splitting tensile strength, compressive strength, and flexural properties of EPS concrete to varying degrees. As the proportion of fibers increases, the compressive strength also increases. However, when the proportion of fibers exceeds 1% [24], the compressive strength exhibits a downward trend with an increase in the proportion of fibers [25–29]. For EPS concrete, the peak value of the compressive strength is affected by the different proportions of fibers, but the overall variation trend is the same [19]. Correspondingly, fibers can upgrade the microstructure of concrete to a certain extent and improve the mechanical properties accordingly. Therefore, it is a major tendency to use fibers to compensate for the weakening effect of EPS on concrete.

In summary, the mechanical properties of EPS concrete are severely weakened due to the poor mechanical properties of the EPS beads added to concrete. Nevertheless, recent research on fiber-modified EPS concrete is relatively limited, especially research on the various properties at different ages. Therefore, it is necessary to investigate the influence of various factors of fibers on the material and mechanical properties of concrete.

In this paper, the orthogonal test method is mainly used to investigate the influence of the three factors of fibers, which are the type, length, and proportion, on the various properties of EPS concrete. The physical properties, mechanical properties, and micromorphology of the specimens are investigated. On this basis, the application of sandwich panels is taken as an example, and the mechanical properties and microscopic characteristics of EPS concrete at various ages are discussed according to the requirements of improving the splitting tensile strength.

## 2. Materials and Methods

### 2.1. Materials

In this paper, ordinary Portland cement (OPC) (manufactured by Yangchun Cement Co., Ltd. from Weifang, China) according to the requirements of P. O. 42.5 in GB175-2007 [30] and Type I in ASTM C150-20 [31] was used. The chemical composition and performance indicators are shown in Table 1. SF with a particle size of 0.1~0.3 μm was supplied, which met the requirements of GB/T 27690-2011 [32] and ASTM C1240-20 [33] The chemical composition is shown in Table 2. Natural river sands (NRSs) were used as fine aggregates and tested by the screening method and moisture proportion measurement method specified in GB/T 14684-2011 [34]. The results are shown in Table 3. The NRSs belonged to natural sand grade type I with a fineness modulus of 1.73, moisture content of

9.61%, and a bulk density of 1664 kg/m$^3$. A polycarboxylate-based superplasticizer (SP) (S04B type) was adopted.

**Table 1.** Chemical composition and performance indicators.

| $SO_3$ | MgO | $Cl^-$ | Slag | Gypsum | Ignition Loss | Specific Surface Area |
|---|---|---|---|---|---|---|
| 2.20% | 3% | 0.04% | 12% | 6% | 4% | 358 m$^2$/kg |

**Table 2.** Chemical composition of silica fume.

| Composition | $SiO_2$ | $Al_2O_3$ | $Fe_2O_3$ | MgO | CaO | $Na_2O$ |
|---|---|---|---|---|---|---|
| w/% | 96.74% | 0.32 | 0.008 | 0.1 | 0.11 | 0.09 |

**Table 3.** River sand sieving analysis.

| Sieve Hole (mm) | 9.5 | 4.75 | 2.36 | 1.18 | 0.6 | 0.3 | 0.15 | <0.15 |
|---|---|---|---|---|---|---|---|---|
| sand sieve weight (g) | 691.0 | 705.0 | 681.2 | 572.0 | 561.2 | 554.0 | 555.0 | 464.8 |
| record weight (g) | 691.0 | 705.0 | 688.4 | 625.8 | 643.2 | 673.0 | 684.4 | 572.0 |
| sand sieve allowance (g) | 0.0 | 0.2 | 7.2 | 53.8 | 82.0 | 19.0 | 129.4 | 107.2 |
| sieve allowance percentage (%) | 0.0 | 0.0 | 1.4 | 10.8 | 16.5 | 23.9 | 25.9 | 21.5 |
| cumulative sieve remainder percentage (%) | 0.0 | 0.0 | 1.0 | 12.0 | 29.0 | 53.0 | 79.0 | 100.0 |
| through percentage | 100.0 | 100.0 | 99.0 | 88.0 | 71.0 | 47.0 | 21.00 | 0.00 |
| fineness modulus | 1.73 | | | | | | | |

EPS beads with an apparent density of 25.1 kg/m$^3$ and a bulk density of 11.31 kg/m$^3$ were adopted and tested under the screening method and bulk density measurement method specified in GB/T 17431-2010 [35] and ASTM C136-19 [36]. The fibers used in this article included PF, GF, and CF, which had lengths of 6 mm, 9 mm, and 12 mm, respectively. The macrostructure and main physical performance of the fibers are separately shown in Figure 1 and Table 4.

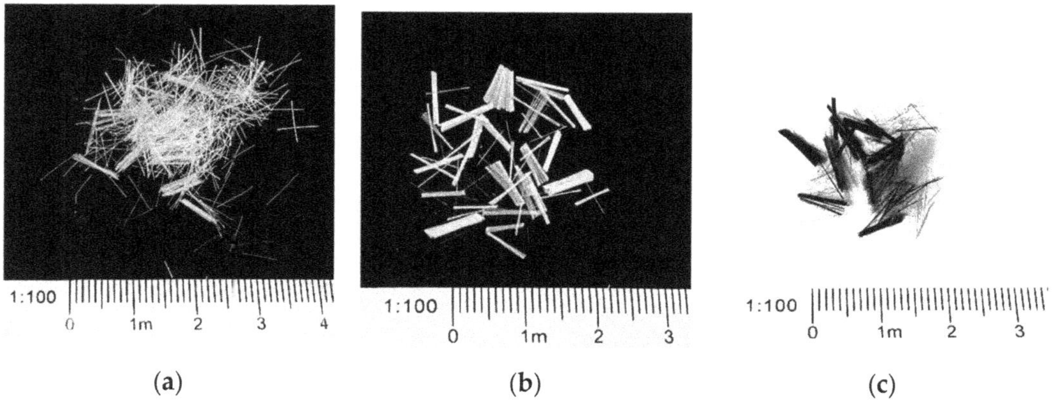


**Figure 1.** Macroscopic morphology of the fibers: (**a**) PF; (**b**) GF; (**c**) CF.

**Table 4.** Physical properties.

| Type | Density | Tensile Strength | Elastic Modulus | Breaking Elongation |
|---|---|---|---|---|
| PF | 0.91 g/cm$^3$ | 360 MPa | 4236 MPa | 28.4% |
| GF | 2.5 g/cm$^3$ | 469 MPa | 4286 MPa | 21.6% |
| CF | 1.80 g/cm$^3$ | 4900 MPa | 240,000 MPa | 2.1% |

### 2.2. Mix Design

An orthogonal experimental design (OED) was adopted to determine the optimal indexes of the fibers, with three factors and three levels, as shown in Table 5. Thereinto, the three factors are fiber type, fiber length, and fiber dosage, and the three levels correspond to corresponding factors, respectively.

**Table 5.** OED table.

| Level | Factor | | |
|---|---|---|---|
| | A-Type | B-Length (mm) | C-Proportion (%) |
| 1 | PF | 6 | 0.5 |
| 2 | GF | 9 | 1.0 |
| 3 | CF | 12 | 1.5 |

EPS concrete without fiber was added as the control case (CC) based on the OED. The naming rules are as follows: the first letter is fiber type, the first two digits are fiber length, and the last two digits are fiber content. For instance, P0605 means choosing PF with a length of 6 mm and a dosage of 0.5%. The mixed number of other contents such as cement, SF, water, NRS, SP [37], and EPS, was decided by the previous test study. The mix design is shown in Table 6. The concrete mixing process is as follows. First of all, the cement, silica fume (SF), fly ash (FA), and river sand are dry mixed for 1 min. After the solid mixture is mixed evenly, the liquid superplasticizer (SP) and water are added stirring for 5 min; then the expanded polystyrene (EPS) beads are added to the mixture continue stirring for 2 min, forming a stable and uniform fresh concrete mixture. Finally, adding various types of fibers are stirred for another 2 min, and loaded into the concrete standard mold.

**Table 6.** Mix design.

| Mix Designation | Cement (kg/m$^3$) | SF (kg/m$^3$) | Water (kg/m$^3$) | NRS (kg/m$^3$) | SP (kg/m$^3$) | EPS (kg/m$^3$) | Fiber | | |
|---|---|---|---|---|---|---|---|---|---|
| | | | | | | | Type | Length (mm) | Proportion (kg/m$^3$) |
| CC | 508.91 | 127.23 | 152.67 | 407.12 | 4.07 | 10 | - | - | - |
| P0605 | 508.91 | 127.23 | 152.67 | 407.12 | 4.07 | 10 | PF(A1) | 6 (B1) | 4.55 (C1) |
| G0905 | 508.91 | 127.23 | 152.67 | 407.12 | 4.07 | 10 | GF(A2) | 9 (B2) | 12.50(C1) |
| C1205 | 508.91 | 127.23 | 152.67 | 407.12 | 4.07 | 10 | CF(A3) | 12(B3) | 9.10 (C1) |
| C0910 | 508.91 | 127.23 | 152.67 | 407.12 | 4.07 | 10 | CF(A3) | 9 (B2) | 18.20(C2) |
| G0610 | 508.91 | 127.23 | 152.67 | 407.12 | 4.07 | 10 | GF(A2) | 6 (B1) | 25.00(C2) |
| P1210 | 508.91 | 127.23 | 152.67 | 407.12 | 4.07 | 10 | PF(A1) | 12(B3) | 9.10 (C2) |
| P0915 | 508.91 | 127.23 | 152.67 | 407.12 | 4.07 | 10 | PF(A1) | 9 (B2) | 13.65(C3) |
| G1215 | 508.91 | 127.23 | 152.67 | 407.12 | 4.07 | 10 | GF(A2) | 12(B3) | 37.50(C3) |
| C0615 | 508.91 | 127.23 | 152.67 | 407.12 | 4.07 | 10 | CF(A3) | 6 (B1) | 27.30(C3) |

### 2.3. Experimental Methods

Slump and dry density tests were performed according to GB/T 50080-2016 [38], and the water absorption test was related to JG/T 266-2011 [39]. To prevent the EPS particles in the concrete from being in a molten state due to high temperatures, a thermostatic drying oven at 60 °C was used. The dry shrinkage test was measured by a concrete shrinkage expansion machine HSP 540 (manufactured by ZKLD from Beijing, China), according to

GB/T 50082-2009 [40], for the ages of 1 day, 3 days, 7 days, 14 days, 28 days, 45 days, 60 days, 90 days and 120 days, and the size of the specimen was 100 mm × 100 mm × 515 mm. In addition, the ultrasonic pulse velocity (UPV) test was carried out using a ZBL-U520 nonmetallic ultrasonic detector according to ASTM C597-16 [41] and ACI 228.1R-03 [42]. The compressive strength, splitting tensile strength, uniaxial compressive strength, and static elastic modulus tests were by GB/T 50081-2019 [43].

A KYKY-EM6200 (manufactured by KYKY from Beijing, China) scanning electron microscope (SEM) and SmartLab-9 (manufactured by Rigaku from Japan) diffractometer for X-ray diffraction (XRD) were used to analyze the microstructure and various diffraction phases of concrete. The scanning voltage, electric current, speed, and scanning 2θ range were 45 kV, 200 mA, 12°/min (step size was 0.02°), and 5–60°, respectively.

## 3. Results and Discussion

The test results of the material and the mechanical properties with the curing age of 28 days of concrete are shown in Table 7.

**Table 7.** Material and mechanical properties.

| Mix Designation | Slump (mm) | Dry Density (kg/m$^3$) | Water Absorption (%) | UPV (km/s) | Compressive Strength (MPa) | Splitting Tensile Strength (MPa) | Uniaxial Compressive Strength (MPa) | Static Elastic Modulus (GPa) |
|---|---|---|---|---|---|---|---|---|
| CC | 196 | 1036.60 | 5.5 | 2.86 | 8.58 | 0.91 | 7.68 | 6.97 |
| P0605 | 115 | 1043.71 | 6.1 | 2.83 | 9.41 | 0.88 | 6.63 | 7.51 |
| G0905 | 10 | 1083.38 | 3.5 | 3.02 | 9.19 | 1.17 | 9.98 | 8.48 |
| C1205 | 7 | 1045.44 | 5.4 | 2.91 | 12.35 | 2.04 | 11.11 | 3.92 |
| C0910 | 1 | 1151.20 | 4.5 | 3.02 | 12.16 | 2.04 | 8.86 | 11.32 |
| G0610 | 12 | 1069.16 | 4.4 | 3.01 | 10.08 | 1.23 | 10.09 | 9.91 |
| P1210 | 72 | 1118.93 | 4.4 | 2.90 | 8.41 | 1.30 | 8.78 | 8.09 |
| P0915 | 11 | 986.42 | 5.9 | 2.74 | 7.53 | 1.06 | 10.61 | 9.17 |
| G1215 | 2 | 1085.07 | 9.5 | 3.00 | 6.12 | 0.89 | 3.24 | 4.35 |
| C0615 | 1 | 1097.71 | 5.7 | 2.92 | 8.43 | 2.35 | 4.06 | 9.29 |

### *3.1. Properties of Fresh Concrete*

Table 6 shows that the slump ranges from 1 mm to 115 mm. The minimum value of C0615 and C0910 was 1, the maximum value of P0605 was 115, and the value of CC was 196 mm.

The results show that the addition of fibers can greatly enhance the consistency and reduce the flowability of concrete, resulting in a decrease in workability. The mean response analysis of the slump values is shown in Figure 2.

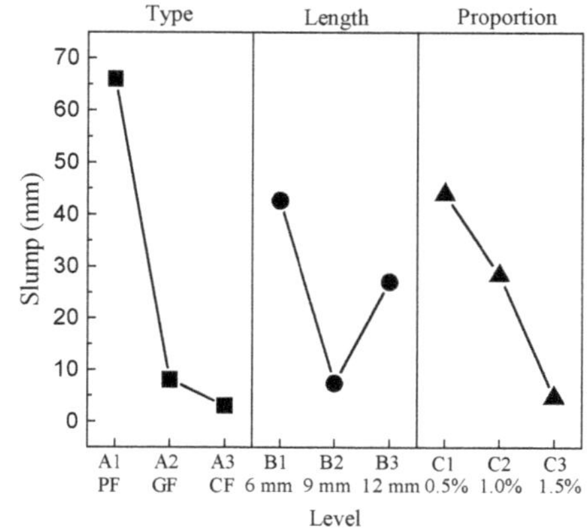


**Figure 2.** Mean slump response.

For the type of fiber factor, the maximum is level 1, and the minimum is level 3. For the length factor, the minimum value appears at level 2, and the maximum value appears at level 1. The slump has strong negative correlations with the proportion of fibers; namely, the minimum value is at level 3 and the maximum value is at level 1.

### 3.2. Dry Density

The dry density ranges from 986.42 kg/m$^3$ to 1151.2 kg/m$^3$. P0915 has the minimum dry density value, P1210 has the maximum dry density value, and the dry density of CC is 1036.6 kg/m$^3$. The results show that the addition of fibers can influence the dry density to some degree. The mean response analysis of the dry density is shown in Figure 3.

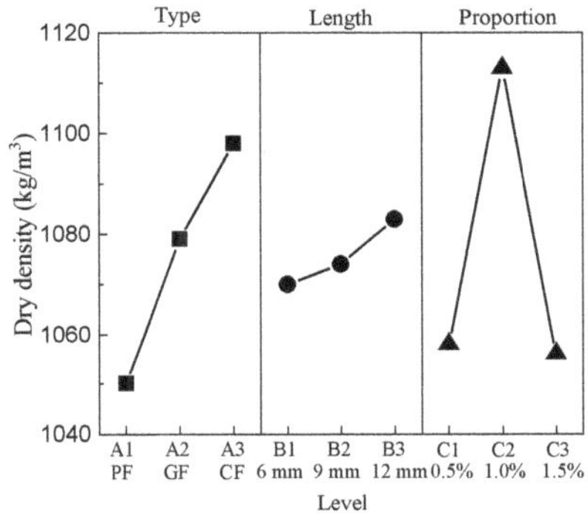


**Figure 3.** Mean dry density response.

For the type of fiber factor, the minimum is level 1, and the maximum is level 3. The dry density has significant positive correlations with the lengths of fibers; namely, the minimum value is at level 1 and the maximum value is at level 3. For the proportion factor, the minimum value appears at level 3, and the maximum value appears at level 2.

### 3.3. Water Absorption

The water absorption rate ranges from 3.5% to 9.5%. G0905 has the minimum value, G1215 has the maximum value, and the water absorption of CC is 5.5%. The results show that the addition of fibers has a certain effect on water absorption. The mean response analysis of water absorption is shown in Figure 4.

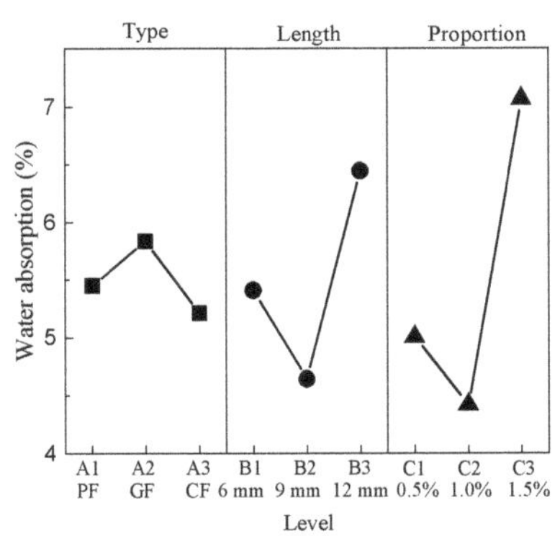


**Figure 4.** Mean water absorption response.

The minimum value of the type factor appears at level 3, and the maximum value appears at level 2. For the length factor, the minimum value appears at level 2, and the maximum value appears at level 3. The minimum value of the proportion factor appears at level 2, and the maximum value appears at level 3.

### 3.4. Ultrasonic Pulse Velocity

The UPV ranges from 2.74 km/s to 3.02 km/s. P0515 has the minimum value, G0905 and G0910 have the maximum values, and the UPV of CC is 2.86 km/s. The results show that the addition of fibers has an influence on the UPV to a certain extent. The mean response analysis of the UPV is shown in Figure 5.

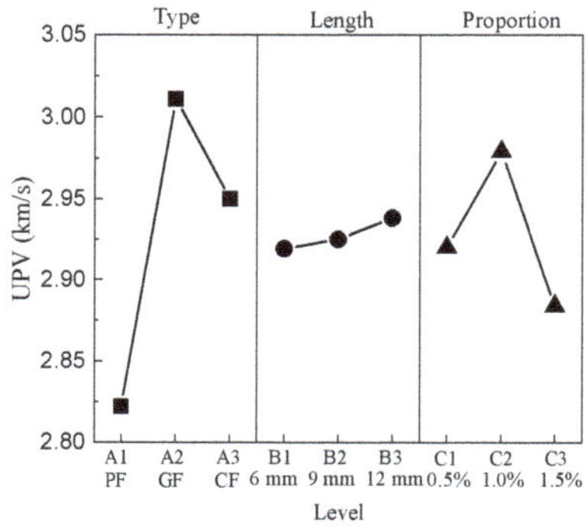


**Figure 5.** Mean UPV response.

For the type of fiber, the minimum value of the type factor appears at level 1, and the maximum value appears at level 2. The UPV is positively correlated with the lengths of fibers, where the minimum value is at level 1 and the maximum value is at level 3. The minimum value of the proportion factor appears at level 3, and the maximum value appears at level 2.

### 3.5. Dry Shrinkage

The drying shrinkage values of CC and other specimens were tested, and the curing ages were from 1 day to 120 days (9 age stages). Correspondingly, the changing trend of CC was used as the reference baseline to evaluate the improvement degree of the fiber for dry shrinkage, and the relationship between the change rate of dry shrinkage and curing time was drawn, as shown in Figure 6.

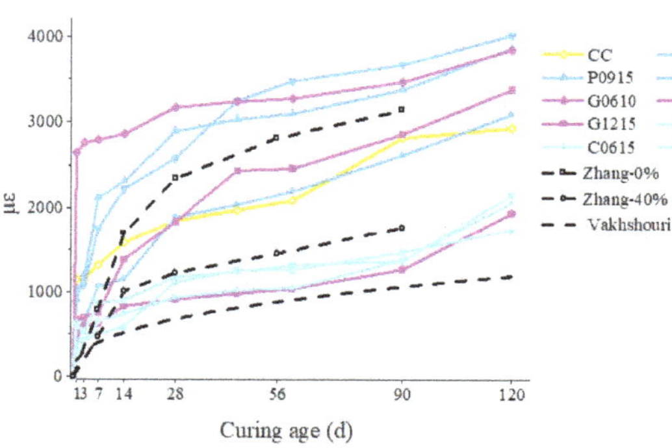


**Figure 6.** The trend of the dry shrinkage values of the modified core material with curing age.

It can be seen from the figure that the dry shrinkage value increases with curing age, and the curve of CC is in the middle of the overall ranges for all cases. The results indicate that the addition of fibers has a significant effect on dry shrinkage. In particular, the dry shrinkage values of C1205, C0615, C0910, and G1215 were improved significantly. Correspondingly, CF has a greater influence on dry shrinkage than GF and PF.

The mean response of dry shrinkage at the standard age (28 days) is shown in Figure 7. For the type of fiber, the minimum and maximum values appear at level 3 and level 1, respectively. The dry shrinkage value has a negative correlation with the length factor; namely, the minimum value is at level 3 and the maximum value is at level 1. The minimum value of the proportion factor is level 3, and the maximum value is level 2.

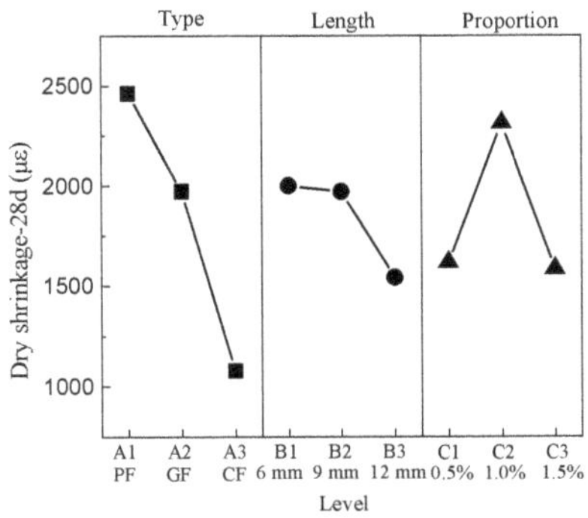


**Figure 7.** Mean dry shrinkage response.

### 3.6. Compressive Strength

The compressive strength ranges from 6.12 MPa to 12.35 MPa. G1215 has the minimum value, C1205 has the maximum value, and the compressive strength of CC is 8.58 MPa. The results show that the addition of fibers has an apparent effect on compressive strength. The mean response analysis of the compressive strength is shown in Figure 8.

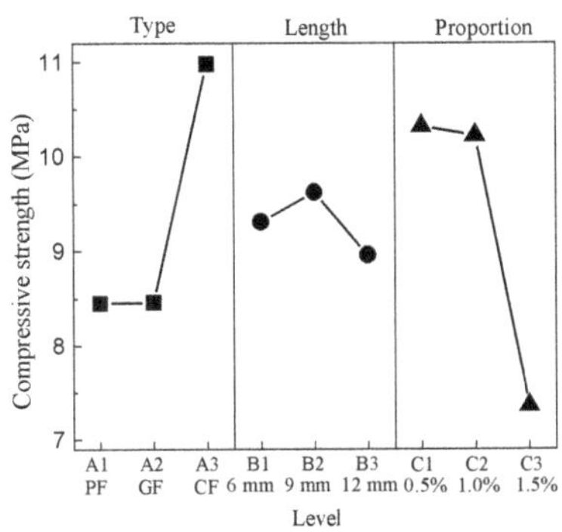


**Figure 8.** Mean compressive strength response.

For the type of fiber factor, the minimum value is level 1, and the maximum value is level 3. For the length factor, the minimum value appears at level 3, and the maximum value appears at level 2. There is a negative correlation between the compressive strength and the proportion of fibers, where the minimum is at level 3 and the maximum is at level 1.

### 3.7. Splitting Tensile Strength

The splitting tensile strength ranges from 0.88 MPa to 2.35 MPa. P0605 has the minimum value, C0615 has the maximum value, and the splitting tensile strength of CC is 0.91 MPa. The results show that the addition of fibers has a great influence on the splitting tensile strength. The mean response analysis of the splitting tensile strength is shown in Figure 9.

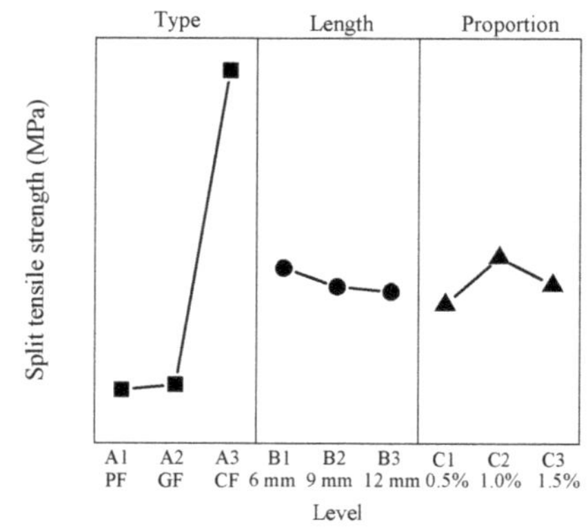


**Figure 9.** Mean splitting tensile strength response.

The minimum value of the type factor appears at level 1, and the maximum value appears at level 3. The fiber length factor is negatively correlated with the fiber length factor, where the minimum appears at level 3 and the maximum appears at level 1. For the proportion factor, the minimum value appears at level 1, and the maximum value appears at level 2.

### 3.8. Uniaxial Compressive Strength

The uniaxial compressive strength ranges from 3.24 MPa to 11.11 MPa, G1215 has the minimum value, C1205 has the maximum value, and the axial compressive strength of CC is 7.68 The results show that the addition of fibers significantly influences the axial compressive strength. The mean response analysis of the uniaxial compressive strength is shown in Figure 10.

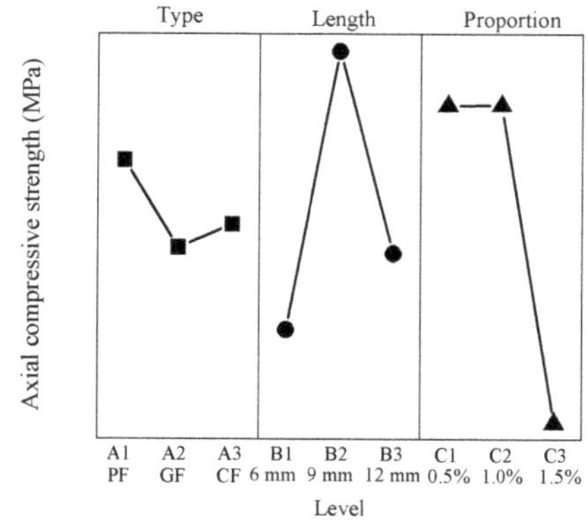


**Figure 10.** Mean axial compressive strength response.

For the axial compressive strength, the minimum value of the type factor appears at level 2, and the maximum value appears at level 1. The minimum value of the length factor appears at level 1, and the maximum value appears at level 2. The uniaxial compressive strength is negatively correlated with the proportion, where the minimum value appears at level 3 and the maximum value appears at level 1.

### 3.9. Static Elastic Modulus

The static elastic modulus ranges from 3.92 GPa to 11.32 GPa. C1205 has the minimum value, C0910 has the maximum value, and the static elastic modulus of CC is 6.79 GPa. The results show that the addition of fibers has an obvious effect on the static elastic modulus. The mean response analysis is shown in Figure 11.

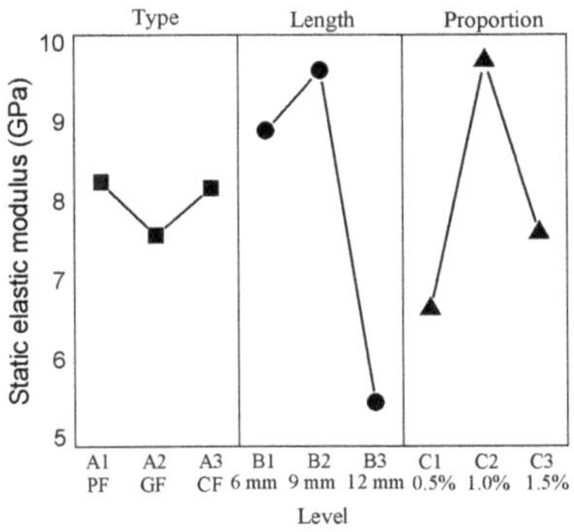


**Figure 11.** Mean static elastic modulus response.

The minimum value of the type factor appears at level 2, and the maximum value appears at level 1. The minimum value of the length factor appears at level 3, and the maximum value appears at level 1. For the proportion factor, the minimum value appears at level 1, and the maximum value appears at level 2.

### *3.10. Scanning Electron Microscope*

The situation of the surface may be alterative during the mixing and forming processes of concrete. Hence, a rougher fiber surface is favorable for combining the concrete matrix and fibers [44]. Figure 12a,c,e show the apparent morphology of the fiber before being mixed into EPS concrete, and Figure 12b,d,f show the surfaces of the fibers added into EPS concrete.

As shown in Figure 12a,b, the surface of PF is quite smooth, but its surface presents relatively regular vertical microcracks after being mixed in concrete.

Although a small number of hydration crystals is visibly attached to its surface, it looks smoother compared with the surface of GF and CF in EPS concrete. Therefore, the boundary between PF and concrete is inferior to that of the others, which accounts for the general lower mechanical properties of EPS concrete mixed with PF.

As shown in Figure 12c,d, a relatively smooth GF surface was observed. However, the surface of GF became extremely rough, and lamellar spalling damage appeared, resembling fish scales after it was mixed in EPS concrete. This probably accounts for the alkaline environment of the cement matrix and the chemical etching caused by $Ca(OH)_2$ [45,46]. However, the mechanical properties of GF are weakened by damage to its surface. The roughness of the surface is conducive to the attachment of massively hydrated crystals and the synergistic bonding between fibers and the cement matrix. The mechanical properties of EPS concrete mixed with GF are generally better than those of concrete mixed with PF, but the overall improvement is not obvious.

As shown in Figure 12e,f, some regular fine strips on the surface of CF emerged. The diameter and contact area with the cement matrix of CF are smaller and larger, respectively than those of PF and GF. The formation and adhesion of the hydration crystals are promoted by the increase in the fine strips on the surface of CF after it is mixed in concrete. Furthermore, it can be observed that massive crystals are attached to the surface of CF, making it more efficient for the bonding and adhesion of hydration crystals. Moreover, the mechanical properties of EPS concrete are favorably improved, which is confirmed by the compressive strength and splitting tensile strength results in the OED.

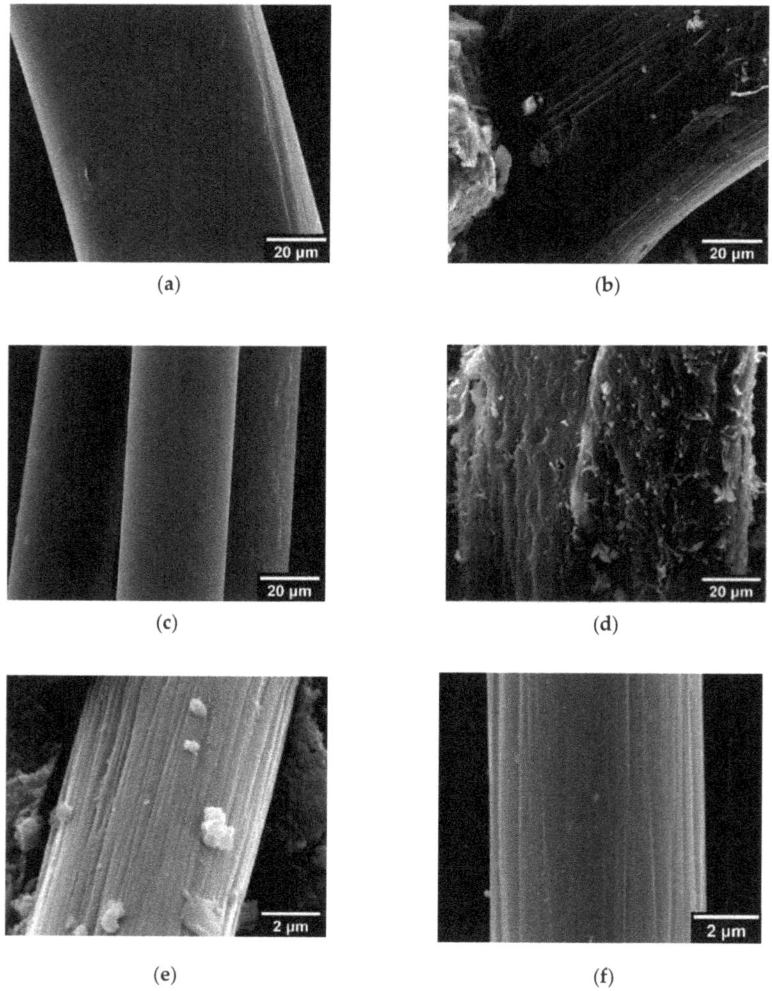


**Figure 12.** Micrographs of fiber surfaces: (**a**) original surface of PF; (**b**) surface of PF mixed in P0605; (**c**) original surface of GF; (**d**) surface of GF mixed in G0905; (**e**) original surface of CF; (**f**) surface of CF mixed in C1205.

## 4. Application

Commercial sandwich panels (CSPs) are a common type of prefabricated concrete panels that commonly use EPS concrete as their core material. The tensile strength of EPS concrete becomes the controlling factor [47–49] for the safety guarantee of sandwich panels when they encounter moments of flexure. Hence, as the core material of CSPs, the optimal combination (OC) of EPS concrete was selected according to the splitting tensile strength.

In addition, other performance tests were performed on OC and compared with CC and the CSP core [50] material. The results are shown in Figure 13. It can be seen from the test results that the dry density, splitting tensile strength, and compressive strength of OC compared to the CSP material increased by 31%, 591%, and 536%, respectively. For CC, the dry density, splitting tensile strength, and compressive strength are higher than those of OC and are 14%, 172%, and 13%, respectively.

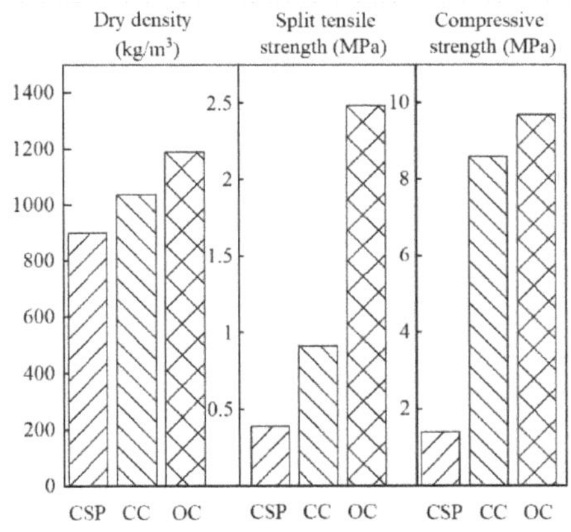


**Figure 13.** Comparison of properties.

Hence, the mechanical properties of CSPs are predominantly improved after modification by adding fibers. The density of the core material is increased by approximately 30%, but there is a 600% increase in the mechanical properties.

Tests of dry density, compressive strength, and splitting tensile strength under OC at different curing ages and the influence of age on its properties were obtained. The age stages of this test were 7, 14, 28, 56, and 90 days. Three samples of each age stage were tested to obtain the dry density, compressive strength, and splitting tensile strength. The average value of each test result is shown in Table 8.

**Table 8.** Effect of age (7 to 90 days) on OC.

| Name | Age (days) | Dry Shrinkage (kg/m$^3$) | Compressive Strength (MPa) | Splitting Tensile Strength (MPa) |
|---|---|---|---|---|
| OC-7 d | 7 | 1161.28 | 9.43 | 1.73 |
| OC-14 d | 14 | 1173.80 | 9.44 | 1.80 |
| OC-28 d | 28 | 1187.78 | 9.68 | 2.48 |
| OC-56 d | 56 | 1197.55 | 11.90 | 2.54 |
| OC-90 d | 90 | 1200.95 | 11.98 | 2.63 |

## 4.1. Dry Density

As shown in Table 8, the dry density of OC at different ages ranges from 1161.28 kg/m$^3$ to 1200.95 kg/m$^3$. The curve fitting of the dry density is shown in Figure 14.

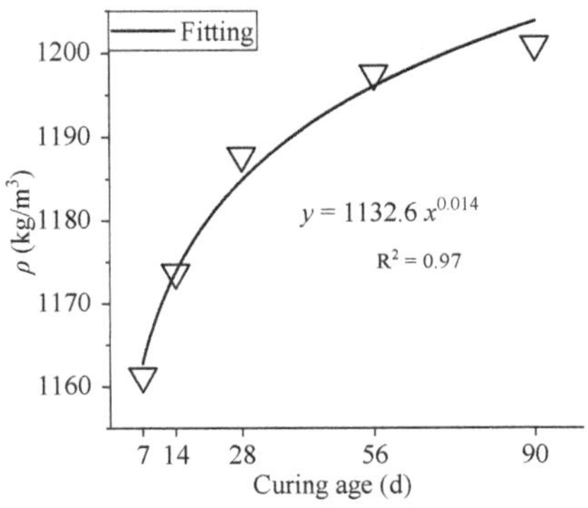


**Figure 14.** Curve fitting of dry density.

The dry density continues to improve with increasing curing time, and the rate of increase gradually slows down as the age increases. When the age increases from 7 days to 28 days, 14 days to 28 days, 28 days to 56 days, and 28 days to 90 days, the dry density increases by 1%, 2%, 0.8%, and 1%, respectively. Although increasing age has an impact on the dry density, the improvement is not significant compared to that at the standard age; the improvement is only approximately 1% to 2%.

### 4.2. Compressive Strength

The compressive strength of OC at different ages ranges from 9.43 MPa to 11.98 MPa. The curve fitting of dry density is shown in Figure 15.

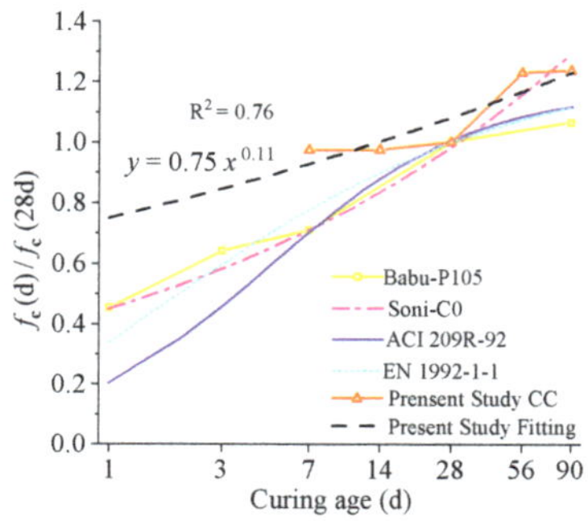


**Figure 15.** Curve fitting of the compressive strength.

The compressive strength continues to improve with increasing curing time, and the rate of increase gradually slows down as the age increases. When the age increases from 7 days to 28 days, 14 days to 28 days, 28 days to 56 days, and 28 days to 90 days, the dry density increases by 2.5%, 2.5%, 23%, and 24%, respectively. It shows that the compressive strength of OC at an early age is approximately 98% of that at the standard age and is higher than the compressive strength of CC at the standard age. The compressive properties of OC increase significantly after reaching the standard age, but the growth rate slows down after 56 days. As shown in Figure 15, compared with the previous study [51,52] and the prediction curve of ordinary concrete [53,54], it was found that the compressive strength increased with increasing age. At an early age (within 14 days), the compressive strength of OC concrete is markedly superior to that of ordinary concrete. For instance, compared with P105, the compressive strength of OC at 7 days increased by nearly 110%. The ITZs between CF and the cement matrix significantly improved at the early age stage due to the high hydration degree, which enhanced the mechanical properties of the concrete material.

### 4.3. Splitting Tensile Strength

The splitting tensile strength of OC at different ages ranges from 1.73 MPa to 2.63 MPa. The fitting curve of the dry density is shown in Figure 16.

When the age increases from 7 days to 28 days and from 14 days to 28 days, the splitting tensile strength increases by 0.75 MPa and 0.68 MPa, namely, 1% and 2%, respectively. When the age increases from 28 days to 56 days and 90 days, respectively, the tensile strength increases by 0.06 MPa and 0.15 MPa, which are improvements of 2% and 6%, respectively. However, after reaching the standard age, the splitting tensile strength continues to increase, but the improvement is not substantial. The splitting tensile strength continues to improve with increasing curing time, and the rate of increase gradually slows down as the age increases. When the age increases from 7 days to 28 days, 14 days to 28 days, 28 days to 56 days, and 28 days to 90 days, the dry density increases by 1%, 2%, 2%, and 6%, respectively. After reaching the standard age, the splitting tensile strength continues to increase, but the rate of increase is not sharp.

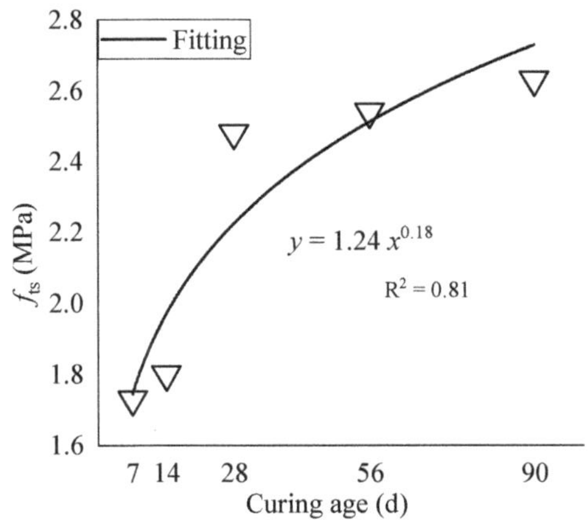


**Figure 16.** Curve fitting of the splitting tensile strength.

### *4.4. Scanning Electron Microscope*

The OC material at four ages (7, 14, 28, and 120 days) was analyzed by SEM. As shown in Figure 17a–d, there is no obvious gap near the ITZ between the cement matrix and the roots of CF at all ages. The ITZs at the roots of PF and GF are shown in Figure 17g,h and are more apparent than those in CF. Compared with the curing conditions at the standard age, the bonding strength between the fiber and matrix is already tight at an early age, which explains why the mechanical properties of OC at an early age are significantly higher than those of CC at a standard age.

Clustered and thickened hydrated crystals are attached to the matrix and fibers of the surface and increase with age. Moreover, the overall mechanical properties are further enhanced, which may be attributed to the large number of hydration crystals observed on the surface of CF in the SEM micrographs, making the connection between the fiber root and the cement matrix closer.

As shown in Figure 17e,f, partial micropores attached to denser acicular ettringite hydrated crystals are visible in the cement matrix at a prolonged age (120 days) and account for the enhancement of the mechanical properties.

### *4.5. X-ray Diffraction*

The OC material at ages 7, 14, 28, and 120 days was characterized by XRD, as shown in Figure 18.

The various phases of the different ages are roughly the same. The main phases are the calcite and portlandite phases, as well as the nonahydrate alite and belite phases.

Similarly, the diffraction peaks of alite and belite ($Ca_3SiO_3$ and $Ca_2SiO_4$) at 7 days and 14 days are more significant than those at 28 days, especially the peaks at 7 days. The above shows that the longer the curing period is, the more sufficient the hydration degree is $SiO_2$ continuously reacts with the cement matrix for secondary hydration, which is the main component of SF mixed in concrete. $Ca(OH)_2$ is consumed continuously, which sufficiently promotes the secondary hydration of OC as the curing age increases. Accordingly, this is corroborated by the phenomenon observed by SEM.

The diffraction peak of Aft, namely, ettringite, was observed in the XRD pattern at the age of 120 days, accounting for the long curing period and the greater the degree of hydration. In fact, the adhesion between the fibers and the cement matrix is also improved when the hydration crystals are increased. Consequently, the mechanical properties of concrete are more effectively improved by the above two effects, hydration, and adhesion, whose effects are amplified when they are used together.

Figure 17. Microscopic morphological analysis: (**a**) OC in 7 days; (**b**) OC in 14 days; (**c**) OC in 28 days; (**d**) OC in 120 days; (**e**) Aft in 120 days; (**f**) Aft in 120 days; (**g**) P0605 in 28 days; (**h**) G0905 in 28 days.

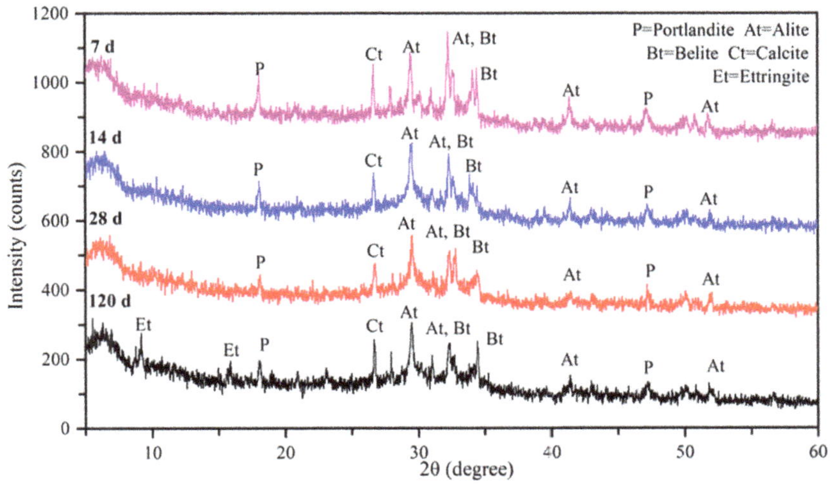


**Figure 18.** XRD diffraction pattern.

## 5. Conclusions

To investigate the effect of fibers on EPS concrete under the factors of the fiber type, length, and proportion, an OED with three factors and three levels was designed. The material, mechanical properties, and microstructure of EPS concrete with fibers were tested. Furthermore, based on the CSP in the practice project, taking the splitting tensile strength as the selection criterion, OC was predicted. The dry density, the splitting tensile strength, the compressive strength, SEM, and XRD of the specimens at various ages were analyzed, and the following conclusions were obtained.

The addition of fibers reduced the fluidity and workability of EPS concrete to a great extent. For the dry density, water absorption rate, and UPV, the addition of fibers had little effect. However, the dry shrinkage rate was considerably altered after the fibers were mixed, and the overall modification effect of CF was better than that of PF and GF. The compressive strength, splitting tensile strength, uniaxial compressive strength, and static elastic modulus of EPS concrete with fibers were significantly improved. Among them, the improvement in CF was the most remarkable. The microscopic analysis showed that the mixing of fibers promotes the hydration of concrete. Correspondingly, with increasing curing time, the hydration effect is also better. Furthermore, the hydration crystals considerably enhance the connection between the fibers and cement matrix, thereby improving the overall mechanical properties of the concrete. CF has the most marked improvement in mechanical properties due to its large contact area with the cement matrix and the large number of hydration crystals attached to its surface.

Compared with CC, the dry density, splitting tensile strength, and compressive strength of OC increased by 14%, 172%, and 13%, respectively. Similarly, compared with the CSP, the dry density, splitting tensile strength, and compressive strength of OC increased by 31%, 591%, and 536%, respectively. That is, the mechanical properties are notably enhanced. The compressive strength increased rapidly at an early age and reached 98% of the compressive strength at the standard age. There was no obvious gap near the ITZ between CF and the cement matrix at all ages, which was observed by SEM and XRD analyses. Moreover, with increasing curing age, the abundant hydrated crystals on the surface of CF also multiplied, which is conducive to the further enhancement of mechanical properties. Moreover, the appearance of ettringite was observed at 120 days, which also proved that the addition of fibers can promote the hydration of concrete.

**Author Contributions:** Y.S.: Conceptualization, Methodology, and Investigation. C.L.: Validation, Formal analysis, Visualization, and Writing—original draft. J.Y.: Conceptualization, Data curation, and Writing—review and editing. C.B.: Validation and Investigation. L.Y.: Methodology and Software. Z.Y.: Project administration. X.L.: Resources. Y.Z.: Resources. X.C.: Resources. All authors have read and agreed to the published version of the manuscript.

**Funding:** This work was funded by the Natural Science Foundation of China (grant numbers 51708060 and 51808088) and Graduate Research and Innovation Projects of Chongqing University of Science & Technology (grant numbers YKJCX2120618).

**Institutional Review Board Statement:** Not applicable.

**Informed Consent Statement:** Not applicable.

**Data Availability Statement:** The data presented in this study are available on request from the corresponding author.

**Conflicts of Interest:** The authors declare that they have no known competing financial interests or personal relationships that could have appeared to influence the work reported in this paper.

## Abbreviations

| | |
|---|---|
| EPS | Expanded polystyrene |
| FA | Fly ash |
| SF | Silica fume |
| ITZ | Interfacial transition zone |
| OPC | Ordinary Portland cement |
| NRS | Natural river sand |
| PF | Polypropylene fiber |
| GF | Glass fiber |
| CF | Carbon fiber |
| SP | Superplasticizer |
| CC | Control case |
| OED | Orthogonal experimental design |
| UPV | Ultrasonic pulse velocity |
| SEM | Scanning electron microscope |
| XRD | X-ray diffraction |
| CSP | Commercial sandwich panel |
| OC | Optimal combination |

## References

1. Saradhi Babu, D.; Ganesh Babu, K.; Wee, T.H. Properties of lightweight expanded polystyrene aggregate concrete containing fly ash. *Cem. Concr. Res.* **2005**, *35*, 1218–1223. [CrossRef]
2. Chung, S.-Y.; Abd Elrahman, M.; Stephan, D. Effects of expanded polystyrene (EPS) sizes and arrangements on the properties of lightweight concrete. *Mater. Struct.* **2018**, *51*, 57. [CrossRef]
3. Nikbin, I.M.; Golshekan, M. The effect of expanded polystyrene synthetic particles on the fracture parameters, brittleness and mechanical properties of concrete. *Theor. Appl. Fract. Mech.* **2018**, *94*, 160–172. [CrossRef]
4. Vakhshouri, B.; Nejadi, S. Size effect and age factor in mechanical properties of BST Light Weight Concrete. *Constr. Build. Mater.* **2018**, *177*, 63–71. [CrossRef]
5. Zhang, J.; Chen, B.; Yu, F. Preparation of EPS-Based Thermal Insulation Mortar with Improved Thermal and Mechanical Properties. *J. Mater. Civ. Eng.* **2019**, *31*, 04019183. [CrossRef]
6. Le Roy, R.; Parant, E.; Boulay, C. Taking into account the inclusions' size in lightweight concrete compressive strength prediction. *Cem. Concr. Res.* **2005**, *35*, 770–775. [CrossRef]
7. Babu, K.G.; Babu, D.S. Behaviour of lightweight expanded polystyrene concrete containing silica fume. *Cem. Concr. Res.* **2003**, *33*, 755–762. [CrossRef]
8. Ganesh Babu, K.; Saradhi Babu, D. Performance of fly ash concretes containing lightweight EPS aggregates. *Cem. Concr. Compos.* **2004**, *26*, 605–611. [CrossRef]
9. Fathi, M.; Yousefipour, A.; Farokhy, E.H. Mechanical and physical properties of expanded polystyrene structural concretes containing Micro-silica and Nano-silica. *Constr. Build. Mater.* **2017**, *136*, 590–597. [CrossRef]
10. Lesovik, V.S.; Zagorodnyuk, L.K.h.; Babaev, Z.K.; Dzhumaniyazov, Z.B. Analysis of the Causes of Brickwork Efflorescence in the Aral Sea Region. *Glass Ceram.* **2020**, *77*, 277–279. [CrossRef]

11. Sadrmomtazi, A.; Sobhani, J.; Mirgozar, M.A.; Najimi, M. Properties of multi-strength grade EPS concrete containing silica fume and rice husk ash. *Constr. Build. Mater.* **2012**, *35*, 211–219. [CrossRef]
12. Madandoust, R.; Ranjbar, M.M.; Yasin Mousavi, S. An investigation on the fresh properties of self-compacted lightweight concrete containing expanded polystyrene. *Constr. Build. Mater.* **2011**, *25*, 3721–3731. [CrossRef]
13. Gu, L.; Ozbakkaloglu, T. Use of recycled plastics in concrete: A critical review. *Waste Manag.* **2016**, *51*, 19–42. [CrossRef]
14. Rostami, R.; Zarrebini, M.; Mandegari, M.; Mostofinejad, D.; Abtahi, S.M. A review on performance of polyester fibers in alkaline and cementitious composites environments. *Constr. Build. Mater.* **2020**, *241*, 117998. [CrossRef]
15. Salari, Z.; Vakhshouri, B.; Nejadi, S. Analytical review of the mix design of fiber reinforced high strength self-compacting concrete. *J. Build. Eng.* **2018**, *20*, 264–276. [CrossRef]
16. Murali, G.; Abid, S.R.; Amran, Y.H.M.; Abdelgader, H.S.; Fediuk, R.; Susrutha, A.; Poonguzhali, K. Impact performance of novel multi-layered prepacked aggregate fibrous composites under compression and bending. *Structures* **2020**, *28*, 1502–1515. [CrossRef]
17. Chen, B.; Liu, J.; Chen, L.-Z. Experimental study of lightweight expanded polystyrene aggregate concrete containing silica fume and polypropylene fibers. *J. Shanghai Jiaotong Univ.* **2010**, *15*, 129–137. [CrossRef]
18. Koksal, F.; Mutlulay, E.; Gencel, O. Characteristics of isolation mortars produced with expanded vermiculite and waste expanded polystyrene. *Constr. Build. Mater.* **2020**, *236*, 117789. [CrossRef]
19. Babavalian, A.; Ranjbaran, A.H.; Shahbeyk, S. Uniaxial and triaxial failure strength of fiber reinforced EPS concrete. *Constr. Build. Mater.* **2020**, *247*, 118617. [CrossRef]
20. Ali, Y.A.; Fahmy, E.H.; AbouZeid, M.N.; Shaheen, Y.B.; Mooty, M.N. Use of expanded polystyrene wastes in developing hollow block masonry units. *Constr. Build. Mater.* **2020**, *241*, 118149. [CrossRef]
21. Cheng, C.; Hong, S.; Zhang, Y.; He, J. Effect of expanded polystyrene on the flexural behavior of lightweight glass fiber reinforced cement. *Constr. Build. Mater.* **2020**, *265*, 120328. [CrossRef]
22. Shaheen, E.; Shrive, N.G. Optimization of mechanical properties and durability of reactive powder concrete. *ACI Mater. J.* **2006**, *103*, 444.
23. Jiang, C.X.; Li, R.; Mo, J.Y.; Zhong, D.W. Effect of fiber surface treatment on interfacial mechano-electric properties of carbon fiber reinforced concrete. *Adv. Mater. Res.* **2011**, *211*, 1087–1090. [CrossRef]
24. Ochi, T.; Okubo, S.; Fukui, K. Development of recycled PET fiber and its application as concrete-reinforcing fiber. *Cem. Concr. Compos.* **2007**, *29*, 448–455. [CrossRef]
25. Kayali, O.; Haque, M.; Zhu, B. Drying shrinkage of fibre-reinforced lightweight aggregate concrete containing fly ash. *Cem. Concr. Res.* **1999**, *29*, 1835–1840. [CrossRef]
26. Ramadevi, K.; Manju, R. Experimental investigation on the properties of concrete with plastic PET (bottle) fibres as fine aggregates. *Int. J. Emerg. Technol. Adv. Eng.* **2012**, *2*, 42–46.
27. Nibudey, R.N.; Nagarnaik, P.B.; Parbat, D.K.; Pande, A.M. Strength and fracture properties of post consumed waste plastic fiber reinforced concrete. *Int. J. Civ. Struct. Environ. Infrastruct. Eng. Res. Dev.* **2013**, *3*, 9–16.
28. Karahan, O.; Atiş, C.D. The durability properties of polypropylene fiber reinforced fly ash concrete. *Mater. Des.* **2011**, *32*, 1044–1049. [CrossRef]
29. Bagherzadeh, R.; Sadeghi, A.-H.; Latifi, M. Utilizing polypropylene fibers to improve physical and mechanical properties of concrete. *Text. Res. J.* **2012**, *82*, 88–96. [CrossRef]
30. *GB 175-2007*; Common Portland Cement. China Architecture & Building Press: Beijing, China, 2007.
31. *C150/C150M−20*; Standard Specification for Portland Cement. ASTM International: West Conshohocken, PA, USA, 2020.
32. *GB/T 27690-2011*; Silica Fume for Cement Mortar and Concrete. Standards Press of China: Beijing, China, 2011.
33. *C1240−20*; Standard Specification for Silica Fume Used in Cementitious Mixtures. ASTM International: West Conshohocken, PA, USA, 2020.
34. *GB/T 14684-2011*; Sand for Construction. Standards Press of China: Beijing, China, 2011.
35. *GB/T 17431.2-201*; Lightweight Aggregates and Its Test Methods-Part 2: Test Methods for Lightweight Aggregates. Standards Press of China: Beijing, China, 2010.
36. *C136/C136M−19*; Standard Test Method for Sieve Analysis of Fine and Coarse Aggregates. ASTM International: West Conshohocken, PA, USA, 2019.
37. Smirnova, O.M.; Potyomkin, D.A. Influence of ground granulated blast furnace slag properties on the superplasticizers effect. *Int. J. Civ. Eng. Technol.* **2018**, *9*, 874–880.
38. *GB/T 50080-2016*; Standard for Test Method of Performance on Ordinary Fresh Concrete. China Architecture & Building Press: Beijing, China, 2016.
39. *JG/T 266-2011*; Foamed Concrete;. Standards Press of China: Beijing, China, 2011.
40. *GB/T 50082-2009*; Standard for Test Methods of Long-Term Performance and Durability of Ordinary Concrete. China Architecture & Building Press: Beijing, China, 2009.
41. *C597-16*; Standard Test Method for Pulse Velocity Through Concrete. ASTM International: West Conshohocken, PA, USA, 2016.
42. *228.1R-03*; In-Place Methods to Estimate Concrete Strength. American Concrete Institute: Farmington Hills, MI, USA, 2003.
43. *GB/T 50081-2019*; Standard for Test Methods of Concrete Physical and Mechanical Properties. China Architecture & Building Press: Beijing, China, 2019.

44. López-Buendía, A.M.; Romero-Sánchez, M.D.; Climent, V.; Guillem, C. Surface treated polypropylene (PP) fibres for reinforced concrete. *Cem. Concr. Res.* **2013**, *54*, 29–35. [CrossRef]
45. Litherland, K.; Oakley, D.; Proctor, B. The use of accelerated ageing procedures to predict the long term strength of GRC composites. *Cem. Concr. Res.* **1981**, *11*, 455–466. [CrossRef]
46. Mills, R. Preferential precipitation of calcium hydroxide on alkali-resistant glass fibres. *Cem. Concr. Res.* **1981**, *11*, 689–697. [CrossRef]
47. Fahmy, E.H.; Shaheen, Y.B.; Abou Zeid, M.N.; Gaafar, H.M. Ferrocement sandwich and hollow core panels for floor construction. *Can. J. Civ. Eng.* **2012**, *39*, 1297–1310. [CrossRef]
48. Fernando, P.; Jayasinghe, M.; Jayasinghe, C. Structural feasibility of Expanded Polystyrene (EPS) based lightweight concrete sandwich wall panels. *Constr. Build. Mater.* **2017**, *139*, 45–51. [CrossRef]
49. Portal, N.W.; Flansbjer, M.; Zandi, K.; Wlasak, L.; Malaga, K. Bending behaviour of novel Textile Reinforced Concrete-foamed concrete (TRC-FC) sandwich elements. *Compos. Struct.* **2017**, *177*, 104–118. [CrossRef]
50. Sun, Y.; You, J.; Zhou, J.; Liu, X.; Yu, L.; Bu, C.; Yan, Z.; Chen, X. Quantified research on the nonuniform distribution of expanded polystyrene beads in sandwich panels. *Constr. Build. Mater.* **2020**, *263*, 120672. [CrossRef]
51. Babu, D.S.; Ganesh Babu, K.; Tiong-Huan, W. Effect of polystyrene aggregate size on strength and moisture migration characteristics of lightweight concrete. *Cem. Concr. Compos.* **2006**, *28*, 520–527. [CrossRef]
52. Soni, N.; Shukla, D.K. Analytical study on mechanical properties of concrete containing crushed recycled coarse aggregate as an alternative of natural sand. *Constr. Build. Mater.* **2021**, *266*, 120595. [CrossRef]
53. *ACI 209R-92*; Prediction of Creep, Shrinkage, and Temperature Effects in Concrete Structures; American Concrete Institute: Farmington Hills, MI, USA, 1992.
54. *BS EN 1992-1-1*; Eurocode 2. Design of Concrete Structures-General Rules and Rules for Buildings; British Standards Institution: London, UK, 2004.

materials

MDPI

Article

# Study on Compression Deformation and Damage Characteristics of Pine Needle Fiber-Reinforced Concrete Using DIC

Yonggang Wang [1], Shan Gao [1,2,*] and Wei Li [1]

1 Shaanxi Key Laboratory of Safety and Durability of Concrete Structures, Xijing University, Xi'an 710123, China; 20180004@xijing.edu.cn (Y.W.); lw18894006385@163.com (W.L.)
2 School of Civil Engineering, Harbin Institute of Technology, Harbin 150090, China
* Correspondence: gaoshan@hit.edu.cn

**Abstract:** Natural fiber-reinforced concrete (NFRC) has the advantages of environmental protection, energy conservation and regeneration. However, studies conducted to improve the macro mechanical properties of concrete by pine needle fiber have achieved good results. In this paper, the deformation and compression damage of pine needle fiber-reinforced concrete (PNFRC) are analyzed by digital image correlation; a fractal dimension is used to quantify the shape of PNFRC after compression damage; and the results of scanning electron microscopy confirm the effect of fiber treatment on deformation and damage of concrete. The results showed that the horizontal strain field of PNFRC has strain concentration zones in the elastic deformation stage, indicating that the fiber enhances the deformation ability of concrete. The defined damage factor can reflect the damage of fiber-reinforced concrete (FRC). The damage curve of natural fiber concrete increases evenly and slowly compared to ordinary concrete.

**Keywords:** digital image correlation (DIC); natural fiber-reinforced concrete; pine needle fiber; crack detection and quantification; damage evolution factor; compression damage process

check for updates

**Citation:** Wang, Y.; Gao, S.; Li, W. Study on Compression Deformation and Damage Characteristics of Pine Needle Fiber-Reinforced Concrete Using DIC. *Materials* **2022**, *15*, 1654. https://doi.org/10.3390/ma15051654

Academic Editor: Dario De Domenico

Received: 14 January 2022
Accepted: 14 February 2022
Published: 23 February 2022



## 1. Introduction

Fiber-reinforced concrete has better flexibility and strength than ordinary concrete. However, natural fiber-reinforced concrete (NFRC) has the advantages of environmental protection, energy saving and regeneration. In previous studies, natural fiber dramatically improved the macroscopic mechanical properties of concrete. The researchers observed that palm fiber [1] reduced the early drying shrinkage of concrete; it is found that the addition of natural fibers creates a good thermal insulation performance for cement mortar, hemp shives [2], flax straw, granular cork, and palm fiber [3]. Especially in terms of mechanical properties, the researchers reported that natural fibers (such as sisal fiber [4], coir [5,6], coconut [7,8], açaí fiber [9] and human hair fiber [10]) improve the compressive strength and flexural strength of NFRC. Using sisal fiber could enhance the impact resistance [11,12] of NFRC. The addition of natural fiber (coconut shell fiber) [13] improves the ductility of concrete, compared with traditional concrete. The hornification of vegetable fibers [14] has a good effect on the durability of cement mortar composites. Meanwhile, the mechanical properties of cement mortar can be significantly improved by the pretreatment of natural fibers [15,16]. Zhou et al. [17] reported that the mechanical properties of hemp fiber concrete treated by boiling and alkalization are at least 10% higher than those of plain concrete. After heat treatment and $Na_2CO_3$ treatment on the surface of sisal fiber, the durability of sisal fiber in concrete improved, as demonstrated by Wei and Meyer [18].

In particular, previous studies found that pine needle fibers [19] significantly improved the macroscopic mechanical properties of concrete. Wang, Y. and Long, W. [20] found that the descending section of the stress–strain curve of NFRC was significantly changed by pine needle fiber. However, the overall damage and deformation process of NFRC has not been considered in current research. At the same time, the failure mechanisms and processes involving concrete are the basis of concrete mechanics research and have important guiding significance for the treatment and prevention of practical engineering. For this problem, the whole process of damage and deformation of the NFRC was analyzed by digital image correlation technique.

Digital image correlation (DIC) technology has the advantage of measuring objects without contact, which can obtain higher measurement accuracy than the traditional method [21] and can analyze the deformation information of any time and area during the loading process. Guo, Q., Wang, H. and Gao, Y. et al. [22] analyzed the effects of three kinds of mixed fibers on the deformation and strain of an asphalt mixture by DIC technology in low-temperature and crack resistance tests. The researchers measured the displacement and deformation of FRC (ultra-high performance fiber-reinforced concrete [23–25] and fabric-reinforced cement-based mortar [26]) and monitored the evolution of the crack strain field under direct tensile tests and bending. The changes in the crack number and crack opening of cement-based composite samples under direct tensile action at different curing ages [27] were monitored; the crack width distribution of UHPFRC and the stress crack width relationship during tension were obtained by an inverse calculation [28]. In addition, using DIC as a fracture mechanics tool, the bonding properties of carbon fiber-reinforced polymer [29] and fabric-reinforced cement-based mortar (FRCM) sheets [30] to concrete matrix were studied. Golewski, G.L. [31] proposed an analysis of the fracture process of type II concrete. Allami, K. and Colombi, P. et al. [32] studied and calculated the tensile properties of epoxy resin and CFRP sheets and the fracture energy of the CFRP concrete interface. In particular, Francisco, V. and Moro, C. et al. [33] evaluated the flexural strength, toughness, and adhesion of recycled polypropylene fiber-reinforced cementitious materials under high-temperature exposure, mainly using DIC and a scanning electron microscope (SEM). At the same time, some researchers analyzed the damage performance of concrete and rocks based on the data acquisition results of DIC [34–38].

In this paper, the deformation field and damage characteristics of PNFRC under compression are analyzed by DIC. All samples in this study contained a volume of 0–2% pine needle fibers treated with tap water soaking, boiling water, and diluted alkali, respectively. The time-varying evolution law of relative deformation field on the specimen surface, the initial cracking position of the specimen surface and the influence range of fiber on the deformation field were studied. Meanwhile, the failure mode and fractal characteristics of PNFRC were calculated. The influence of pine needle fibers on the deformation and damage of concrete is validated by the SEM results. Compared with different damage factor curves, the damage factor $D_{fa}$ accurately describes the damage of fiber-reinforced concrete in each stress stage.

## 2. Experimental Section

### 2.1. Test Materials and Specimens

#### 2.1.1. Materials and Mix Design

In this study, the coarse aggregate used for raw materials was 5–15 mm of continuously graded gravel, and natural river sand with fineness modulus of 2.95, conforming to the Chinese Standard GBT14685-2001 and GBT14684-2001, respectively. The cementitious material was 42.5 R ordinary Portland cement; The test water met the domestic water standard GB5749-2006 [39]. The concrete mix proportion created for this experiment was 0.49:1.0:1.615:2.636 (water:cement:sand:stone), and the strength grade was 30 MPa (C30). After the pine needle fiber was processed into 30 mm, it was pretreated in three different ways. The treated procedures and effects are shown in more detail in [19].

#### 2.1.2. Specimen Design

All test samples were prepared, poured, and maintained in strict accordance with GB50081-2019 [40]. The sample size was designed as a cube of 100 × 100 × 100 mm$^3$. The fiber contents added to the concrete were 0.5%, 1.0%, 1.5% and 2.0% of the volume of the test block, respectively. Table 1 shows the grouping information of test samples. In the table, the 'T', 'B' and 'A' represent three treatment methods (tap water soaking, boiling water, and alkalinized solution), respectively; the '30' means the fiber length, and the 'C' stands for the compressive test. The number after C is the fiber volume content. CC is plain concrete.

**Table 1.** Test sample details.

| W/B | Sample Size | Fiber Pretreated Method | Fiber Volume | | | | |
|---|---|---|---|---|---|---|---|
| | | | 0% | 0.5% | 1.0% | 1.5% | 2.0% |
| 0.49 | 100 × 100 × 100 mm$^3$ | Tap water soaking<br>Boiling water<br>Alkalinized solution | CC | T30C05<br>B30C05<br>A30C05 | T30C10<br>B30C10<br>A30C10 | T30C15<br>B30C15<br>A30C15 | T30C20<br>B30C20<br>A30C20 |

### 2.2. Test Materials and Specimens

The experimental system for studying the compression deformation and damage characteristics of pine needle fiber-reinforced concrete included a loading system and DIC acquisition system.

#### 2.2.1. Digital Image Correlation (DIC)

The DIC acquisition system is shown in Figure 1a. CCD cameras (Manta G-146B/C with a Sony ICX267 sensor, dpi: 2452 × 2056, Allied Vision, Puchheim, Germany) were used in this research, which ran at 10 fps(full resolution). The CCD cameras, using DIC technology, obtained the camera parameters through correction (Figure 1b) and collected the strain and deformation of speckle samples (Figure 1c) during the experiment. The strain region of the compressive damage process analysis and statistical damage analysis is shown in Figure 1d, and the linear strain region of this is displayed in Figure 1e. The software MATCHID-2D/STEREO Correlated 2018 (University of Leuven, Leuven, BE) was used to carry out the DIC analyses.

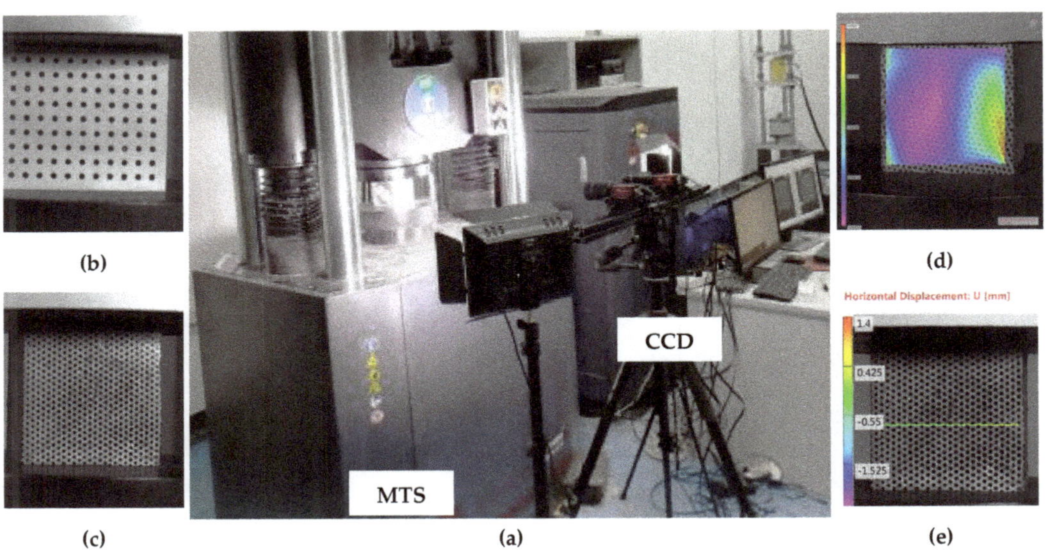


**Figure 1.** Test setup (**a**) MTS pressure testing machine and industrial camera test system (**b**) Camera parameter correction (**c**) Sample speckle (**d**) Area of surface analysis (**e**) Area of line analysis.

2.2.2. Compressive Strength Test of PNFRC

Hydraulic servo universal testing machine (MTS.e64.206, MTS Systems Corporation, Eden Prairie, MN, USA) was selected for the loading system, as shown in Figure 1a. According to GB50010-2010, all specimens were tested after 28 days of standard curing. The loading rate of the pressure testing machine was 0.5 MPa/s, and the data acquisition frequency of the pressure testing machine was set to 2 Hz.

## 3. Results and Analysis

### *3.1. Cracking Mode and Fractal Characteristics of PNFRC Samples*

Figure 2 illustrates the main cracks of the PNFRC samples after peak loading, and the blue color indicates the main cracks and the concrete shedding part of the concrete. Clearly, the PNFRC treated with alkali show many small cracks after rupture, which prove that the deformation of concrete is restrained and has a bridging effect between the concrete and pine needle fibers; the tap-water-treated pine needle fiber-reinforced concrete show mainly block shedding after damage, which indicates the inhibiting effect of tap-water-treated fibers on the cement hydration of concrete; however, the damage pattern of boiling-water-treated pine needle fiber-reinforced concrete is between the alkali-treated and tap-water-treated PNFRC.

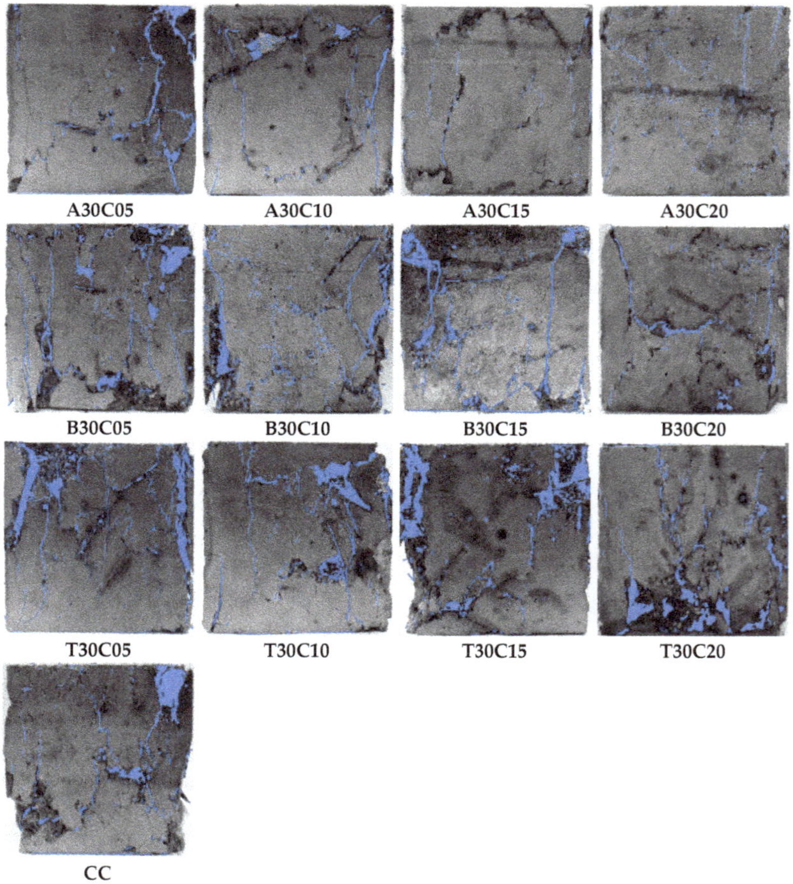


**Figure 2.** Failure characteristics of the samples.

The fractal dimension is used to evaluate the failure morphology of concrete with three treatments. Figure 3 shows the fractal dimension of cracks of pine needle fiber-reinforced concrete specimens. In general, the fractal dimension of alkali-treated PNFRC is slighter than that of the CC specimen, while the fractal dimension of boiling-water-treated and tap-water-treated fiber concrete is much greater than that of the plain concrete, which is consistent with the macro results of a damaged morphology.

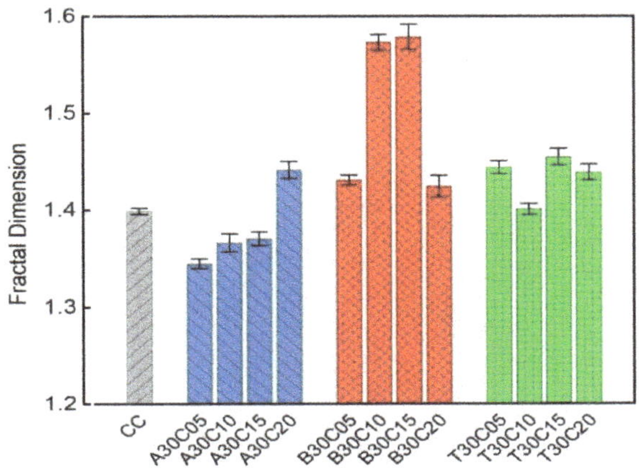


**Figure 3.** Failure characteristics of the samples.

### 3.2. Micro-Analysis of SEM

From the micro-level, the effects of different treatments of pine needle fibers on the deformation of concrete are demonstrated by SEM, as shown in Figure 4. In general, there are varying degrees of cement mortar on the fiber surface compared to the original pine needle fiber (Figure 4a). In particular, the surface of the fiber treated by dilute alkali is wrapped with the most cement mortar (Figure 4b), and the fiber adhere well to the cement matrix, cracking the cement matrix after destruction and particularly influencing the deformation and damage of the concrete.

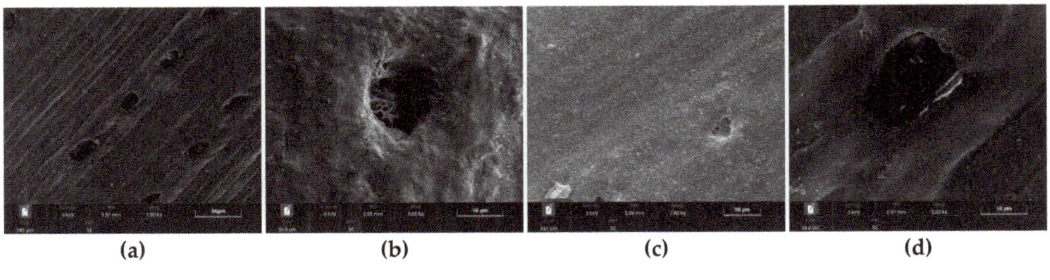


**Figure 4.** The morphology of fiber surface: (**a**) primitive fiber, (**b**) alkali-treated fiber, (**c**) boiling-water-treated fiber, and (**d**) tap-water-soaked fiber schemes follow the same formatting.

### 3.3. The Compressive Damage Process Analysis

Figure 5 displays the load-displacement curves of all samples. Clearly, the content and treatment of pine needle fibers have a direct influence on the macro mechanical properties and deformability of concrete. The effect of pine needle fibers treated with boiling water and diluted alkali on concrete is greater than PNFRC treated with tap water in strength, deformation peak value, and elastic modulus. Therefore, the horizontal strain field of pine needle fiber concrete treated with boiling water and dilute alkali was selected for analysis. The horizontal strain field corresponding to point A–E of the load-displacement curve of concrete samples was mainly analyzed. Point A–E is the corresponding load value of 10%, 50%, 70%, 90% and 100% of the peak value (Pmax ), respectively, as shown in Figures 6 and 7.

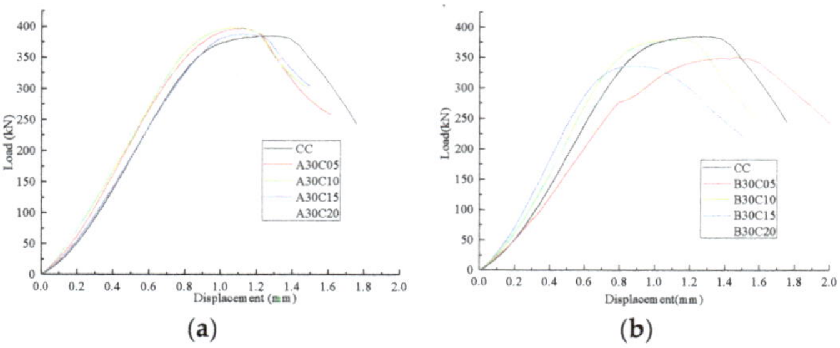


**Figure 5.** Load-displacement curves of all samples. (**a**) Alkalinized solution, (**b**) Boiling water, and (**c**) Tap water.

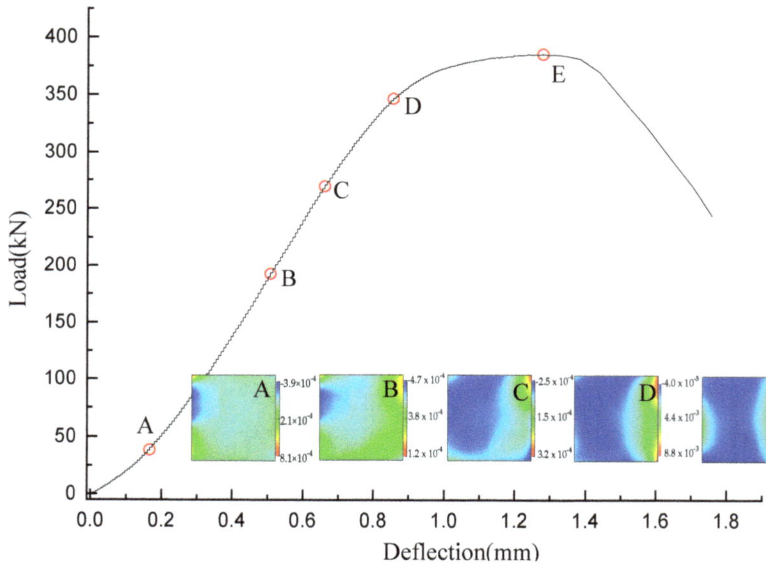


**Figure 6.** The load-displacement curve and horizontal strain field of ordinary concrete (CC) sample.

At the early phase of loading (before A), the coordinated deformation of the corner part of the specimen and the indenter resulted in local breakage, causing a weak strain concentration. In the early loading stage A–C, the strain field of the CC sample is relatively uniform, as the samples are in the linear stage of the load-displacement curve under compression. However, the strain fields on the surface of the PNFRC specimen present an uneven shape. At the elastic stage, the relative deformation fields of PNFRC were formed, and fiber and concrete bear forces together.

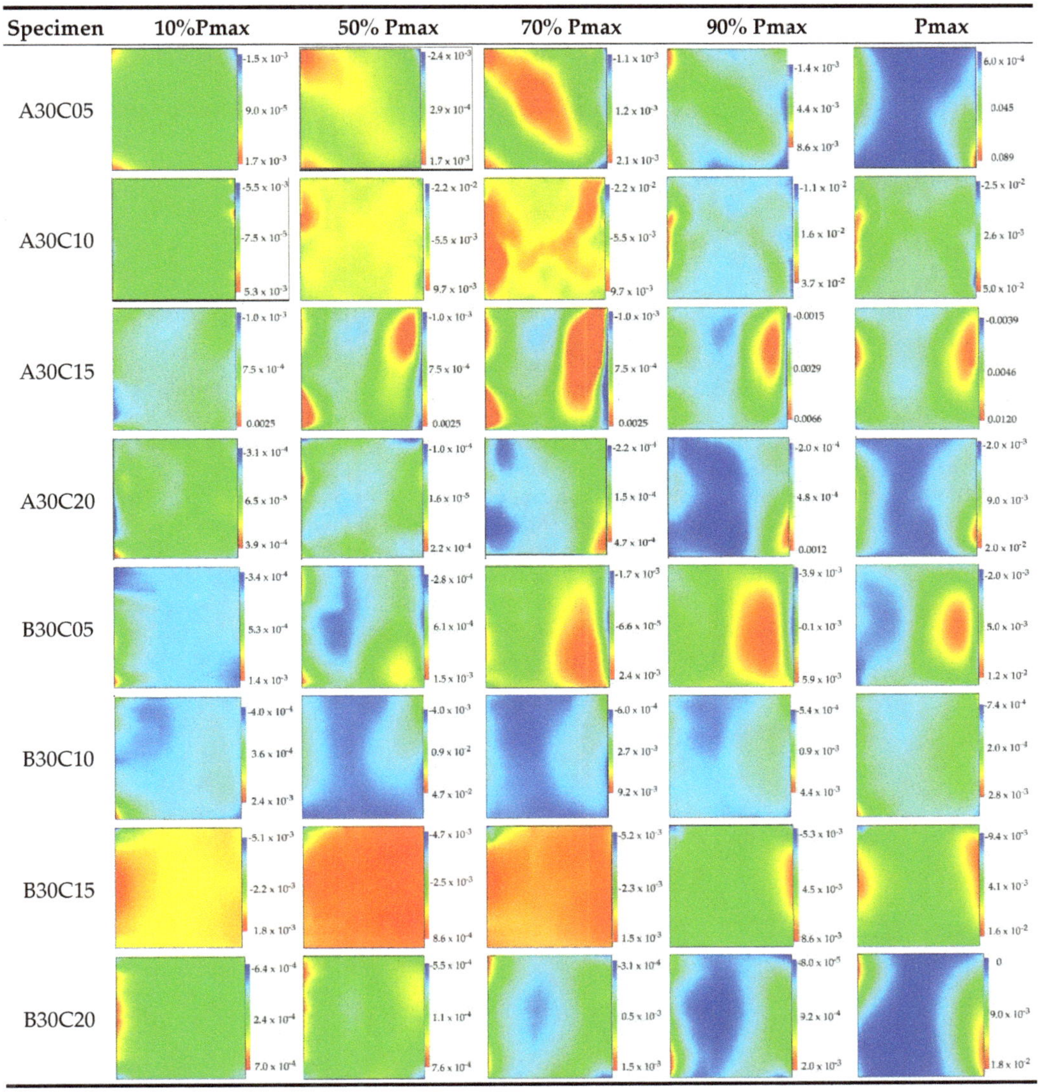


**Figure 7.** Evolution of horizontal strain field of PNFRC specimens.

In the C–D stage, there are strain concentration areas on the surface of the specimens in the nonlinear part of the load-displacement curve, and the strain field shows the clear heterogeneity of zoning. Especially for the specimens containing 0.5–1.5%, the impact range of the fibers on the relative deformation field of concrete increases, and a large strain concentration area appears on the sample surface.

### *3.4. Statistical Analysis of PNFRC Damage*

DIC is an important tool for quantifying sample damage. S.P. Ma [37] proposed to use the standard deviation of the horizontal strain field to represent strain field damage degree factor $D_{fa}$, as shown in Equations (1)–(2). Similarly, Y.R. Zhao et al. [35] and H. Zhang et al. [38] proposed the deviation of the strain field to calculate the damage degree factor.

According to a horizontal strain field analysis of compressive damage, five kinds of concrete samples with good modification effects were selected for statistical damage analysis. The damage factors, $D_{fm}$ and $D_{fl}$; in line strain region; and plane strain region (as shown in Figure 1d,e) were calculated by combining Equations (1) and (2). At the same time, the damage factor $D_{fa}$ was defined in the linear strain region as shown in Equation (3).

$$D_{f,j} = \frac{S_j}{S_{max}} \tag{1}$$

$$S_j = \sqrt{\frac{1}{n-1}\sum_{i=1}^{n}(\varepsilon_i - \overline{\varepsilon})^2} \tag{2}$$

$$D_{fa,j} = \sum_{i=1}^{n}\varepsilon_i / \sum_{i=1}^{n}\varepsilon_{max} \tag{3}$$

where n is the number of strain points in every DIC horizontal strain field analysis region; $\varepsilon_i$ stands for the strain at any point in the j-th horizontal strain field; $\overline{\varepsilon}$ indicates the strain mean of the j-th horizontal strain field; and $\varepsilon_{max}$ means the strain value in the strain field analysis region at the peak load.

Figure 8 lists the damage evolution curve of PNFRC samples with the changes in displacement. It is evident that the change of all damage factors ($D_{fl}$, $D_{fm}$ and $D_{fa}$) of ordinary concrete samples almost coincide. In general, the changes of $D_{fl}$ and $D_{fa}$ are more sensitive than those of $D_{fm}$ and can better reflect the details of the damage and deformation of samples. The $D_{fl}$ and $D_{fa}$ are not smooth in the whole change stage, which indicates that the damage of pine needle fiber-reinforced concrete is clearly affected. The $D_{fl}$ and $D_{fa}$ of the B30C05 samples also have a mutation at the point of load mutation.

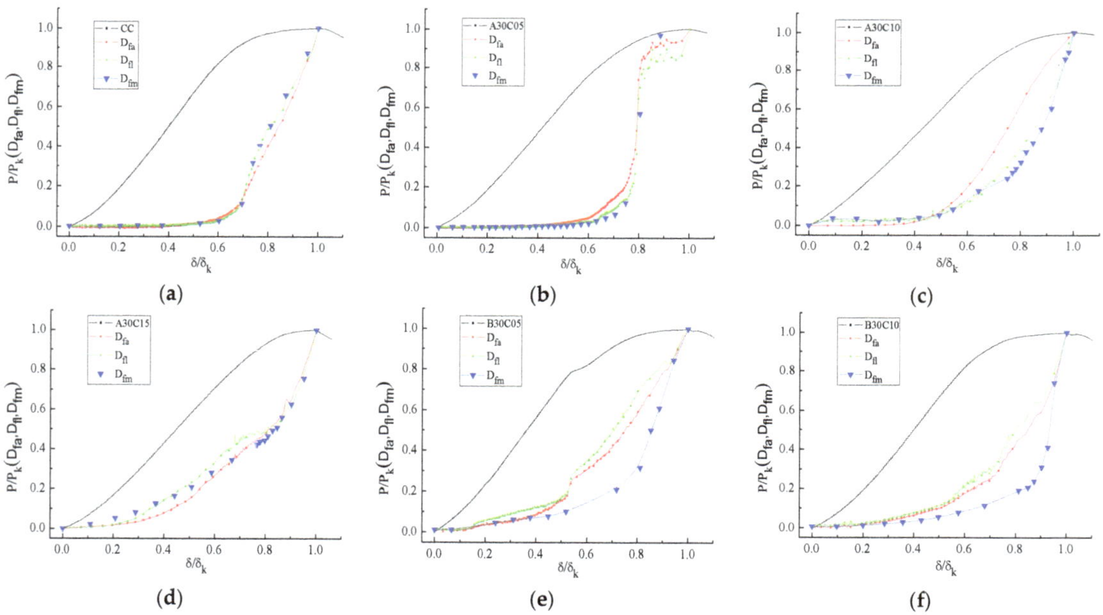


**Figure 8.** Damage evolution curve of PNFRC samples. (**a**) CC, (**b**) A30C05, (**c**) A30C10, (**d**) A30C15, (**e**) B30C05, and (**f**) B30C10.

Under a 10% load at the initial loading phase, the above three damage factors keep changing synchronously. Under the same deformation conditions, the more fiber content, the earlier the damage occurs to pine needle fiber-reinforced concrete specimens. The damage degree of diluted-alkali-treated samples is less than that of the boiling-water-treated sample when the load is in the elastic stage. The increase in the damage of PNFRC is relatively gentle compared to ordinary concrete.

## 4. Conclusions

In this study, DIC technology was used to study the compressive damage characteristics of PNFRC, and three damage evaluation methods were compared. The conclusions were as follows:

(1) The fractal dimension of fiber-reinforced concrete treated with tap water soaking, boiling water, and an alkali solution is consistent with the macroscopic results of damaged morphology. At the same time, it is proved that pine needle fiber can promote the deformation and damage of concrete from the micro point of view. The SEM results show that the surface of the alkali-treated fiber is wrapped with cement mortar, which produces cracks after the concrete is damaged.
(2) At 50% of the peak load, the strain concentration zone of the 0.5–1.5% PNFRC appears, and the strain concentration zone develops to the maximum at 90% of the peak load. The "X" type strain concentration zone of other samples is irregular at the peak load, compared with ordinary concrete and 2% pine needle fiber concrete.
(3) Adding natural fiber can change the damage growth rate of concrete and make the damage curve grow evenly and slowly. The samples A30C10 show the best deformation and damage performance. The damage degree of the samples treated with boiling water is greater than that of the samples treated with dilute alkali in the elastic deformation stage.
(4) The damage factor $D_{fa}$ could accurately reflect the compression damage of PNFRC. In the process of compression deformation, the defined damage factor can characterize the details of the damage and deformation of each specimen.

Therefore, it can be concluded that the damage of NFRC is different from that of ordinary concrete, and the damage factor evaluation method proposed in this study can accurately recognize the evolution mechanism of the plant fiber concrete instability precursor response crisis, and realize the damage evaluation of concrete. As a new type of natural reinforced composite material applied in buildings, PNFRC still needs further research. PNFRC can also be used to control the degree of damage by manual intervention.

**Author Contributions:** Conceptualization, writing—original draft preparation, Y.W.; methodology, Project administration, S.G.; Data curation, writing—review and editing, W.L. All authors have read and agreed to the published version of the manuscript.

**Funding:** This research received no external funding.

**Institutional Review Board Statement:** Not applicable.

**Informed Consent Statement:** Not applicable.

**Data Availability Statement:** The raw/processed data required to reproduce these findings cannot be shared at this time as they form part of an ongoing study.

**Conflicts of Interest:** The authors declare no conflict of interest.

## References

1. Tioua, T.; Kriker, A.; Barluenga, G.; Palomar, I. Influence of date palm fiber and shrinkage reducing admixture on self-compacting concrete performance at early age in hot-dry environment. *Constr. Build. Mater.* **2017**, *154*, 721–733. [CrossRef]
2. Rahim, M.; Douzane, O.; Le, A.T.; Langlet, T. Effect of moisture and temperature on thermal properties of three bio-based materials. *Constr. Build. Mater.* **2016**, *111*, 119–127. [CrossRef]
3. Boumhaout, M.; Boukhattem, L.; Hamdi, H.; Benhamou, B.; Nouh, F.A. Thermomechanical characterization of a bio-composite building material: Mortar reinforced with date palm fibres mesh. *Constr. Build. Mater.* **2017**, *135*, 241–250. [CrossRef]
4. Saravanan, N.; Buvaneshwari, M. Experimental Investigation on Behaviour of Natural Fibre Concrete (Sisal Fibre). *Cellulose* **2018**, *54*, 66.

5. Chandel, A.; Shah, T.; Shah, T.; Varde, D. A comparative strength study of coir fiber-reinforced concrete (CFRC) over plain cement concrete (PCC). *J. Mech. Civ. Eng. IOSR-JMCE* **2016**, *13*, 95–97.
6. Andiç-Çakir, Ö.; Sarikanat. M.; Tüfekçi, H.B.; Demirci, C.; Erdoğan, Ü.H. Physical and mechanical properties of randomly oriented coir fiber–cementitious composites. *Compos. Part B* **2014**, *61*, 49–54. [CrossRef]
7. Chikhi, M.; Agoudjil, B.; Boudenne, A.; Gherabli, A. Experimental investigation of new biocomposite with low cost for thermal insulation. *Energy Build.* **2013**, *66*, 267–273. [CrossRef]
8. Bui, H.; Sebaibi, N.; Boutouil, M.; Levacher, D. Determination and review of physical and mechanical properties of raw and treated coconut fibers for their recycling in construction materials. *Fibers* **2020**, *8*, 37. [CrossRef]
9. de Azevedo, A.R.; Marvila, M.T.; Tayeh, B.A.; Cecchi'n, D.; Pereira, A.C.; Monteiro, S.N. Technological performance of açaí natural fiber-reinforced cement-based mortars. *J. Build. Eng.* **2020**, *33*, 101675. [CrossRef]
10. Petrounias, P.; Rogkala, A.; Giannakopoulou, P.P.; Lampropoulou, P.; Golfinopoulos, A. An innovative experimental petrographic study of concrete produced by animal bones and human hair fibers. *Sustainability* **2021**, *13*, 8107. [CrossRef]
11. Oladele, I.O.; Omotoyinbo, J.A.; Adewara, J.O.T. Investigating the effect of chemical treatment on the constituents and tensile properties of sisal fiber. *J. Miner. Mater. Charact. Eng.* **2010**, *9*, 569.
12. Al-Oraimi, S.K.; Seibi, A.C. Mechanical characterisation and impact behaviour of concrete reinforced with natural fibres. *Compos. Struct.* **1995**, *32*, 165–171. [CrossRef]
13. Islam, S.M.; Hussain, R.R.; Morshed, M.A.Z. Fibre-reinforced concrete incorporating locally available natural fibres in normal-and high-strength concrete and a performance analysis with steel fiber-reinforced composite concrete. *J. Compos. Mater.* **2012**, *46*, 111–122. [CrossRef]
14. Claramunt, J.; Ardanuy, M.; García-Hortal, J.A.; Tolêdo Filho, R.D. The hornification of vegetable fibres to improve the durability of cement mortar composites. *Cem. Concr. Compos.* **2011**, *33*, 586–595. [CrossRef]
15. Le Troëdec, M.; Dalmay, P.; Patapy, C.; Peyratout, C.; Smith, A.; Chotard, T. Mechanical properties of hemp-lime reinforced mortars: Influence of the chemical treatment of fibres. *J. Compos. Mater.* **2011**, *45*, 2347–2357. [CrossRef]
16. Raghavendra Rao, H.; Varada Rajulu, A.; Ramachandra Reddy, G.; Hemachandra Reddy, K. Flexural and compressive properties of bamboo and glass fiber-reinforced epoxy hybrid composites. *J. Reinf. Plast. Compos.* **2010**, *29*, 1446–1450. [CrossRef]
17. Zhou, X.; Saini, H.; Kastiukas, G. Engineering properties of treated natural hemp fiber-reinforced concrete. *Front. Built Environ.* **2017**, *3*, 33. [CrossRef]
18. Wei, J.; Meyer, C. Improving degradation resistance of sisal fiber in concrete through fiber surface treatment. *Appl. Surf. Sci.* **2014**, *289*, 511–523. [CrossRef]
19. Long, W.; Wang, Y. Effect of pine needle fiber reinforcement on the mechanical properties of concrete. *Constr. Build. Mater.* **2021**, *278*, 122333. [CrossRef]
20. Wang, Y.; Long, W. Complete stress–strain curves for pine needle fiber-reinforced concrete under compression—sciencedirect. *Constr. Build. Mater.* **2021**, *302*, 124134. [CrossRef]
21. Del Enrique, R.; Tom, A.; Richard, H. Digital image correlation (dic) for measurement of strains and displacements in coarse, low volume-fraction frp composites used in civil infrastructure. *Compos. Struct.* **2019**, *212*, 43–47.
22. Guo, Q.; Wang, H.; Gao, Y.; Jiao, Y.; Liu, F.; Dong, Z. Investigation of the low-temperature properties and cracking resistance of fiber-reinforced asphalt concrete using the dic technique. *Eng. Fract. Mech.* **2020**, *229*, 106951. [CrossRef]
23. Niu, Y.; Huang, H.; Zhang, J.; Jin, W.; Yu, Q. Development of the strain field along the crack in ultra-high-performance fiber-reinforced concrete (uhpfrc) under bending by digital image correlation technique. *Cem. Concr. Res.* **2019**, *125*, 105821. [CrossRef]
24. Shen, X.; Brühwiler, E. Influence of local fiber distribution on tensile behavior of strain hardening uhpfrc using ndt and dic. *Cem. Concr. Res.* **2020**, *132*, 106042. [CrossRef]
25. Donnini, J.; Lancioni, G.; Chiappini, G.; Corinaldesi, V. Uniaxial tensile behavior of ultra-high performance fiber-reinforced concrete (uhpfrc): Experiments and modeling. *Compos. Struct.* **2020**, *258*, 113433. [CrossRef]
26. D'Anna, J.; Amato, G.; Chen, J.F.; Minafò, G.; La Mendola, L. Experimental application of digital image correlation for the tensile characterization of basalt FRCM composites. *Constr. Build. Mater.* **2021**, *271*, 121770. [CrossRef]
27. Hu, J.H.; Sun, M.Q.; Li, J.; Wei, Y.P.; Chen, J.Z. Tensile characteristics of strain-hardening cement-based composites with different curing ages. *Constr. Build. Mater.* **2019**, *221*, 709–719. [CrossRef]
28. Mobasher, B.; Li, A.; Yao, Y.; Arora, A.; Neithalath, N. Characterization of toughening mechanisms in uhpc through image correlation and inverse analysis of flexural results. *Cem. Concr. Compos.* **2021**, *122*, 104157. [CrossRef]
29. Corr, D.; Accardi, M.; Graham-Brady, L.; Shah, S. Digital image correlation analysis of interfacial debonding properties and fracture behavior in concrete. *Eng. Fract. Mech.* **2006**, *74*, 109–121. [CrossRef]
30. Domenico D., D.; Quattrocchi, A.; Alizzio, D.; Montanini, R.; Urso, S.; Ricciardi, G.; Recupero, A. Experimental Characterization of the FRCM-Concrete Interface Bond Behavior Assisted by Digital Image Correlation. *Sensors* **2021**, *21*, 1154. [CrossRef]
31. Golewski, G.L. Validation of the favorable quantity of fly ash in concrete and analysis of crack propagation and its length-using the crack tip tracking (ctt) method—In the fracture toughness examinations under mode II, through digital image correlation. *Constr. Build. Mater.* **2021**, *296*, 122362. [CrossRef]
32. Al-Lami, K.; Colombi, P.; D'Antino, T. Influence of hygrothermal ageing on the fracture energy and mechanical properties of CFRP-concrete joints and their components. *Compos. Struct.* **2020**, *238*, 111947. [CrossRef]

33. Francioso, V.; Moro, C.; Castillo, A.; Velay-Lizancos, M. Effect of elevated temperature on flexural behavior and fibers-matrix bonding of recycled pp fiber-reinforced cementitious composite—sciencedirect. *Constr. Build. Mater.* **2020**, *269*, 121243. [CrossRef]
34. Del Zoppo, M.; Menna, C.; Di Ludovico, M.; Asprone, D.; Prota, A. Opportunities of light jacketing with fiber-reinforced cementitious composites for seismic retrofitting of existing rc columns. *Compos. Struct.* **2021**, *263*, 113717. [CrossRef]
35. Zhao, Y.R.; Wang, L.; Lei, Z.K.; Han, X.F.; Shi, J.N. Study on bending damage and failure of basalt fiber-reinforced concrete under freeze–thaw cycles. *Constr. Build. Mater.* **2018**, *163*, 460–470. [CrossRef]
36. Liu, B.; Guo, J.H.; Wen, X.Y.; Zhou, J.K.; Deng, Z.H. Study on flexural behavior of carbon fibers reinforced coral concrete using digital image correlation. *Constr. Build. Mater.* **2020**, *242*, 117968. [CrossRef]
37. Ma, S.P.; Zhou, H. Evolution characteristics of strain field on specimen surface during rock failure. *J. Rock Mech. Eng.* **2008**, *27*, 1667–1673.
38. Zhang, H.; Huang, G.Y.; Song, H.P.; King, Y.L. Experimental characterization of strain localization in rock. *Geophys. J. Int.* **2013**, *194*, 1554–1558. [CrossRef]
39. *GB5749-2006*; Code for Drinking Water. Chinese Academy of Building Sciences: Beijing, China, 2007.
40. *GB/T 50081-2019*; Standard for Test Methods of Concrete Physical and Mechanical Properties. Chinese Academy of Building Sciences: Beijing, China, 2007.

materials

MDPI

*Article*

# Analytical Hysteretic Behavior of Square Concrete-Filled Steel Tube Pier Columns under Alternate Sulfate Corrosion and Freeze-Thaw Cycles

**Tong Zhang [1,2], Qianxin Wen [1], Lei Gao [3], Qian Xu [1,*] and Jupeng Tang [2]**

1 School of Civil Engineering, Liaoning Technical University, Fuxin 123000, China; zt_1987_zt@163.com (T.Z.); 18242477225@163.com (Q.W.)
2 School of Mechanics and Engineering, Liaoning Technical University, Fuxin 123000, China; jupengt@126.com
3 Liaoning Hanshi Technology Group Co., Ltd., Fuxin 123099, China; gao33lei33@163.com
* Correspondence: eileenxq2022@163.com; Tel./Fax: +86-0418-2162036

**Abstract:** The hysteretic behavior of square concrete-filled steel tube (CFST) stub columns subjected to sulfate corrosion and freeze-thaw cycle is examined by numerical investigation. The constitutive model of steel considered the Bauschinger effect, and compression (tension) damage coefficient was also adopted for the constitutive model of core concrete. The experimental results are used to verify the finite element (FE) model, which could accurately predict the hysteretic behaviors of the CFST piers. Then, the effects of the yield strength of steel, compressive strength of concrete, steel ratio, axial compression ratio, and alternation time on ultimate horizontal load are evaluated by a parametric study. The results showed that the yield strength of steel and the steel ratio have a positive effect of hysteretic behavior. The compressive strength of concrete and alternation time significantly decreased the unloading stiffness which causes the pinching phenomenon. The yield strength of steel, compressive strength of concrete, and alternation time of environmental factors (corrosion-freeze-thaw cycles) has no obvious effect on the initial stiffness, while the steel ratio has a remarkable effect. The ultimate horizontal load increases with the increasing steel ratio, yield strength of steel and compressive strength of concrete. Meanwhile, the decrement of alternation time led to the increase of ultimate horizontal load. This suggests that the confinement coefficient and alternation time are the two main factors that impact the ultimate horizontal load. A formula which considers the reduction coefficient for the ultimate horizontal load of the CFST columns subjected to sulfate corrosion and freeze-thaw cycles is proposed. The formulae can accurately predict the ultimate horizontal load with mean value of 1.022 and standard deviation of 0.003.

**Keywords:** sulfate corrosion; freeze-thaw cycle; alternation effect; CFST pier column; hysteretic behavior



**Citation:** Zhang, T.; Wen, Q.; Gao, L.; Xu, Q.; Tang, J. Analytical Hysteretic Behavior of Square Concrete-Filled Steel Tube Pier Columns under Alternate Sulfate Corrosion and Freeze-Thaw Cycles. *Materials* **2022**, *15*, 3099. https://doi.org/10.3390/ma15093099

Academic Editors: Shao-Bo Kang and Shan Gao

Received: 25 March 2022
Accepted: 21 April 2022
Published: 25 April 2022



## 1. Introduction

Concrete-filled steel tubes (CFST) are composite structural members with an outer steel tube and an inner core of concrete [1,2]. The utilization of CFST columns has been commonplace in recent decades due to the infilling concrete supporting the steel tube, which prevents its inward buckling and postpones its outward buckling [3]. Moreover, the steel tube restricts the core concrete in transverse, and thus, due to the confinement effect, the concrete is under three-dimensional compression, which leads to the improvement of load-bearing capacity and ductility [4,5]. Considerable research of CFST columns has been conducted on buildings and bridges' rib arches [6–9]. In addition, the feasibility of using CFST columns as bridge piers has been reported [10–13].

Tomii et al. [14] firstly studied the seismic behavior of square CFST through a quasi-static method and illustrated that CFSTs were excellent in energy dissipation. K. Nakanishi et al. [15] investigated the peak horizontal load and ductility of bridge pier columns with different composite forms experimentally. The study theoretically verified the analogous rule between full-scale models and scaled-down specimens and found that the CFST pier columns were better in resistant cyclic loading. Aval et al. [16] adopted an inelastic fiber element model considering the axial elongation and curvature of steel tubes and concrete to analyze CFST columns under cyclic load. The results showed that the CFST with studs or reinforcement ribs has a higher dissipation of energy. M. Bruneau et al. [17] evaluated the limitation of existing design codes for circular CFST bridge piers by establishing an experimental database. Then, based on the simple plastic model, a new equation was proposed and the specifications "Guide for Load and Resistance Factor Desing (LRFD) Criteria for Offshore Structures" were recommended for the seismic design of bridge piers. Li [18] carried out a dynamic time history analysis of CFST piers according to practical engineering at a 7-degree ($M = 0.58I + 1.5$, where $M$ is Richter scale, $I$ denotes seismic fortification intensity, $I = 7$ here) protected earthquake intensity. It illustrated that the vertex displacement of pier columns was below the ultimate displacement for all 18 earthquake excitations, which verified that the CFST pier column has good seismic resistant. Wang [19] studied the effect of concrete fill percentage on the seismic behavior of CFST pier columns with filling ratios of 0%, 23%, 27%, 30%, and 31%. The study also considered the slenderness ratio and width-to-thickness ratio. It showed that the supporting role of concrete has the advantage of seismic resistance but the optional filling rate requires further investigation.

The bridge pier is generally used as a pressure-bearing member and works under complex conditions including load action, vehicle impact, harsh environment, and seismic effect, which takes disadvantage of the lifelong service of the bridge [20]. Chen et al. [21] investigated the effect of corrosion ratio and recycled aggregate replacement ratio on the axial compressive behavior of circular CFST columns by experimental and numerical research. It demonstrated that the type of recycled aggregate has a slight effect on loadbearing capacity of specimens under sulfate corrosion. Chen et al. [22–24] consider the effect of corrosion ratio and recycled aggregate replacement ratio to investigate the axial compressive behavior of 16 square CFST columns under sulfate corrosion experimentally and numerically. They demonstrated that the recycled aggregate type has negligible effect on specimens under sulfate corrosion. However, the corrosion time resulted in the decrement of ultimate bearing capacity, ductility, and composite elastic modulus of the specimen. FE models considering the reduction of thickness of the steel tubes could effectively simulate the effect of sulfate corrosion on steel tubes. Zhang [25] carried out a durability test on 20 thin-walled steel tube columns under $SO_4^{2-}$ corrosion. The study found that ultimate bearing capacity, ductility, and stiffness decreased with the corrosion ratio, especially for the specimens with thinner walls. The finial failure mode of the CFST stub columns under sulfate corrosion was due to shear failure. After corrosion by $SO_4^{2-}$, local bucking of the slender steel tube would occur to a large extent.

Yang et al. [26] experimentally dealt with the effect of freeze-thaw cycles, section type, and compressive strength of axially loaded CFST columns under freeze-thaw conditions. The test results demonstrated that the freeze-thaw cycle significantly influenced the load-bearing capacity, composite elastic modulus, and ductility index but slightly impacted the failure mode. They mentioned that the circular specimens performed better with regards to ductility. Wang [27] carried out an experimental test of axially loaded square CFST columns with an end plate to enlarge the effect of freeze-thaw cycles on compressive behavior. Wang [27] found that the cyclic time has a negative linear relationship on loadbearing capacity and initial stiffness of the specimens. Yan [28–30] investigated the axially loaded CFST stub columns under the freezing temperatures of −30 °C, −60 °C, and −80 °C. The test results presented that square specimens were destroyed in crushing failure, and few specimens appeared tensile fracture. Freezing temperature improved the ultimate compressive load and initial stiffness of CFST columns but decreased the ductility. Gao et al. [31] focused on the mechanical behaviors of circular CFST stub columns under freeze-thaw cycles in experimental tests. The effect of cyclic time and compressive strength of concrete for the axially loaded specimens was analyzed. Gao et al. [31] found that axial compressive strength decreased linearly with cyclic time while it increased with a higher compressive strength of concrete.

The durability of CFST pier columns is significantly influenced by a harsh environment, such as corrosion [32] and freeze-thaw [33]. However, few studies have investigated the compound influence of multiple environmental factors, which is in accordance with practical structure. Studies on dynamic properties were concentrated on CFST columns in ambient conditions. The steel tube suffering alternating sulfate corrosion and freeze-thaw cycles will result in higher brittleness and the concrete will crack at the interfacial transition zone due to the mechanical deterioration of the specimen. In addition, the seismic effect coupling with the alternation of sulfate corrosion and freeze-thaw cycles will rapidly cause material deterioration, and thus will cause corrosion fatigue cracks to occur more easily. This study thus carries out a simulation study of hysteretic behavior on 126 square CFST pier columns. Based on the reasonable constitutive model, nonlinear finite element (FE) models are set to analyze the CFST pier columns with a failure mode of shear failure by ABAQUS software. Then, the effect of strength of materials, steel ratio, axial compression ratio, and alternate time on the horizontal load-deformation hysteretic curve and skeleton curve. Ultimately, the reduction coefficient corresponding to the ultimate horizontal load was analyzed and a design formula to predict the ultimate horizontal load of square CFST pier columns after alternate of sulfate corrosion and freeze-thaw cycles was proposed.

## 2. Empirical Experimental Research

The square CFST pier columns were corroded through electrochemical corrosion. The electrolyte was a mixed solution of $Ca(NO_3)_2$, $Na_2SO_4$, and $NH_4Cl$ and adopted $HNO_3$ to adjust pH value, which was set as 4.5. The specimens were fully immersed into the electrolyte. The corrosion rate was defined as the loss of mass of the steel and corrosion rate $\gamma$ = 10%, 20%, and 30% corresponded to the terms of service of 5, 10, and 15 years. Hemispherical pitting was formed on the surface of the steel tubes. The test results reported in previous studies [34–43] were selected to verify the CFST pier columns modeled in ABAQUS software [44]. Table 1 summarizes their experimental test results. The parameters of each specimen were listed in Table 1, and it can be found that the specimen labels dividing different groups were the same as those tested in [34–43]. As shown in Table 1, the length of the specimens ranged from 530 to 1250 mm, corresponding to a length-to-width ($L/B$) ratio of 2.12–5.00. Thus, the influence of the slenderness effect could be avoided [45]. The compressive strength of the concrete cores ranges from 33.0 to 75.1 MPa, which includes normal-strength concrete and high-strength concrete [45] ($f_{ck} \leq 50$ MPa is defined as normal strength concrete). Specimens with reinforcement bars consist of 8Φ16 and 4Φ11, and the ties used are Φ8@100 mm and Φ8@200 mm, which meet the specification requirements [46]. Then, the yield strength of the steel tubes varies from 242.2 to 389.0 MPa, and the axial compression ratio ranges from 0.2 to 0.60. The material properties of steel and infilled concrete can be obtained from tests or specifications such as GB50010 [46], ACI-318-08 [47], EN-1992-1-1 [48], GB50017 [49], AS4100 [50], etc. Figure 1 presents details of the typical testing device with which the hysteretic behavior of the specimens could be tested.

The method of quasi-static tests was generally adopted to investigate the effect of reciprocating cyclic loading on structural members. This method took advantage of easy operation and observation, controlled loading process, and convenience of checking data, and thus the quasi-static test method was selected to investigate the hysteretic behaviors of specimens. According to JGJ/T101-2015 [51], the loading history of cyclic loading included a force control phase and a displacement control phase, as showed in Figure 2. In the force control phase, one cycle was imposed at the load levels of $0.5N_{vy}$, $0.5N_{vy}$, and $N_{vy}$, respectively, where $N_{vy}$ is the estimated value of specimen yield strength. Afterward, displacement-controlled loading was performed at the incremental levels of $1\delta_y$, $2\delta_y$, $3\delta_y$, etc, where $\delta_y$ represents the estimated displacement corresponding to the yield strength, and the cyclic loading was reported twice for each displacement level. The cyclic loading was repeated twice for each displacement level. The yield strength $N_{vy}$ and yield displacement $\delta_y$ were tested with the GYM (general yield moment) method. Horizontal deformation was measured by linear variable differential transformers (LVDTs) and strain was measured by strain gauges. Final failure was defined as horizontal load decreasing to 85% of its peak value [51]. The test results of the specimens in Refs. [34–43] were listed in Table 1.

**Table 1.** Specifications of the tested CFST columns in Refs. [34–43].

| No. | Reference | Labels | $B \times L \times t$ (mm) | $f_{cu,\ test}$ (MPa) | $f_y$ (MPa) | $f_{by}$ (MPa) | $d$ (mm) | $n$ |
|---|---|---|---|---|---|---|---|---|
| 1 | [34,35] | STRC-60-1.5-4 | 200 × 600 × 1.89 | 60.0 | 309.2 | 357.0 | 16.4 | 0.40 |
| 2 | | STRC-60-1.5-6 | | | | | | 0.60 |
| 3 | | STRC-70-4-1.5 | | | | | | 0.35 |
| 4 | [36,37] | STRC-70-5-1.5 | 200 × 600 × 3.00 | 75.1 | 254.0 | 449.4 | 11.0 | 0.45 |
| 5 | | STRC-70-6-1.5 | | | | | | 0.55 |
| 6 | [38] | FGZ4 | 250 × 1250 × 5.00 | 33.0 | 389.0 | | | 0.24 |
| 7 | | A1 | | | | | | 0.20 |
| 8 | [39–41] | A2 | 200 × 600 × 6.00 | 50.1 | 330.0 | | | 0.40 |
| 9 | | A3 | | | | NA | NA | 0.60 |
| 10 | | RS-4L | 250 × 530 × 4.00 | | 265.0 | | | 0.20 |
| 11 | [42,43] | RS-6M | 250 × 530 × 6.00 | 41.0 | 317.6 | | | 0.40 |
| 12 | | RS-10H | 250 × 530 × 10.00 | | 242.2 | | | 0.54 |

Note: NA denotes the abbreviation of not available.

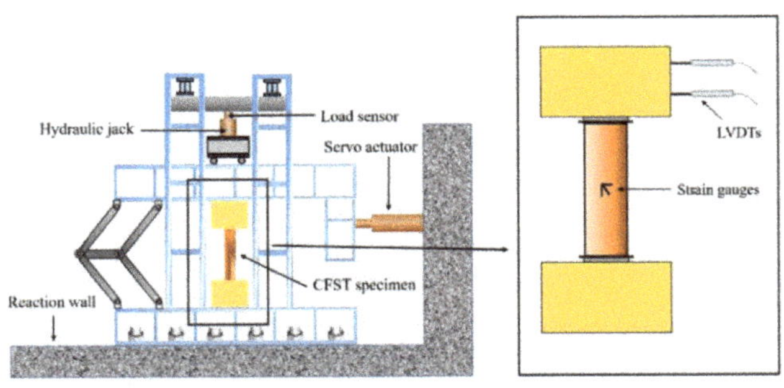


(a) Experimental device (b) Detial of measure device

**Figure 1.** Experimental device and layout of the specimen.

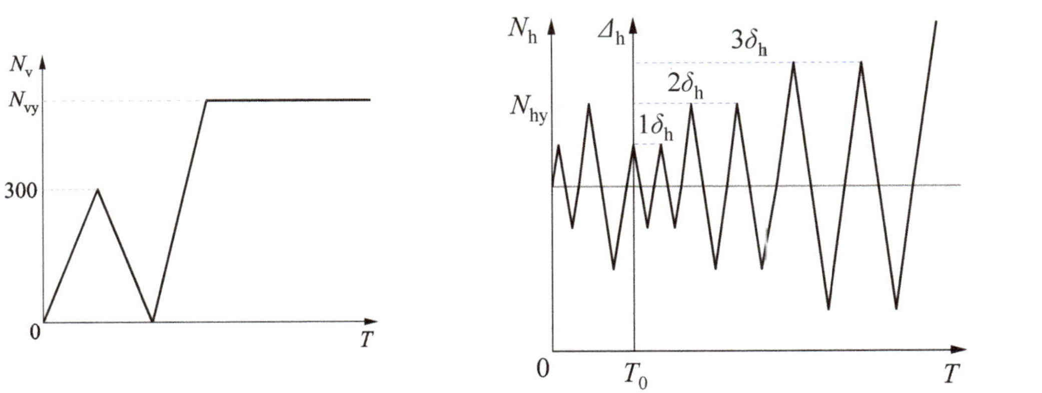


(a) (b)

**Figure 2.** Loading system of specimens. (**a**) Vertical loading system; (**b**) Horizontal loading system.

The test results obtained for the square CFST columns under reciprocating cyclic loading in terms of different main parameters were sketched and compared with the FE results in Section 3.5 in the forms of failure mode, skeleton curve, and horizontal ultimate compressive load in detail.

## 3. Finite Element Modeling

The phenomenological modeling of square CFSTs is typically performed by using differential [52,53] or non-differential models [54,55] and is determined by three main parts, including core concrete, steel tube, and interaction between these two raw materials. In addition, in FE analysis of the specimens, selection of element type and mesh size should balance the accuracy of the simulation results and simulation time. Since square CFST pier columns under cyclic load are not symmetrical, the whole specimen is modelled to investigate the hysteretic behaviors.

### *3.1. Material Properties*

#### 3.1.1. Steel Tube

A secondary plastic flow stress-strain curve [45] was modified according to the effect of corrosion-freeze-thaw cycles on the relationship of stress–strain and then adopted to define the material behavior under sulfate corrosion and freeze-thaw cycles of steel tubes, which include five stages: elastic, elastic-plastic stage, plastic, yield strength, and strengthening and secondary plastic flow stages. The main parameters for defining the stress–strain curve of a steel tube include the limit of proportionality ($f_p$), yield strength of the steel tube ($f_y$), ultimate compressive strength ($f_u$), and the corresponding strain ($\varepsilon_e$, $\varepsilon_{e1}$, $\varepsilon_{e2}$ and $\varepsilon_{e3}$). The five-stage stress–strain model of a steel tube can be defined as Equation (1) shows. Young's modulus and Poisson's ratio in the model adopted were 206 GPa and 0.283 [47,49] respectively.

$$\sigma_s = \begin{cases} E_{se}\varepsilon_s & \varepsilon_s \le \varepsilon_e \\ -A\varepsilon_s^2 + B\varepsilon_s + C & \varepsilon_e < \varepsilon_s \le \varepsilon_{e1} \\ f_{ye} & \varepsilon_{e1} < \varepsilon_s \le \varepsilon_{e2} \\ f_{ye}\left[1 + 0.6\frac{\varepsilon_s - \varepsilon_{e2}}{\varepsilon_{s3} - \varepsilon_{e2}}\right] & \varepsilon_{e2} < \varepsilon_s \le \varepsilon_{e3} \\ 1.6 f_{ye} & \varepsilon_s > \varepsilon_{e3} \end{cases} \quad (1)$$

where $\varepsilon_e = 0.8f_{ye}/E_{se}$, $\varepsilon_{e1} = 1.5\varepsilon_e$, $\varepsilon_{e2} = 10\varepsilon_{e1}$, $\varepsilon_{e3} = 100\varepsilon_{e1}$, $A = 0.2f_y(\varepsilon_{e1} - \varepsilon_e)^2$, $B = 2A\varepsilon_{e1}$, $C = 0.8f_{ye} + A\varepsilon_{e2} - B\varepsilon_e$, $E_{se} = (1 - 0.525\gamma)E_s$, $f_{ye} = (1 - 0.908\gamma)f_y$, and $\beta = \Delta t/t$, $\Delta t = t - t_e$.

When the CFST columns were subjected to the alternation effect of sulfate corrosion and freeze-thaw cycles, not only the thickness, but also the mechanical properties of the steel tube changed. A series of typical stress–strain curves of steel tubes of 345 MPa with corrosion rates of 0%, 10%, 30%, and 50% was presented in Figure 3. The von Mises criterion is adopted to estimate steel yield stress.

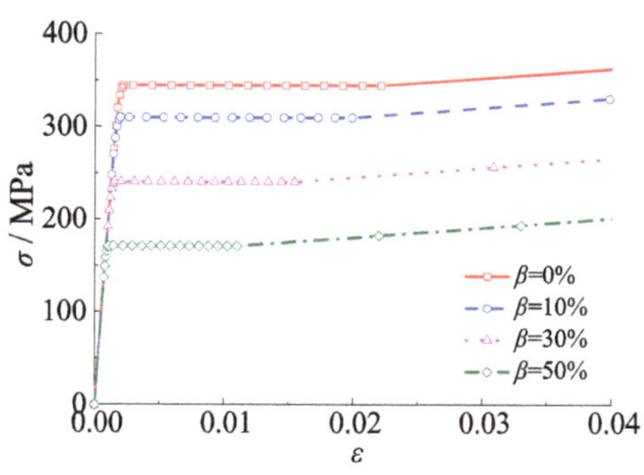


**Figure 3.** Stress–strain curve of steel tube under alternation of sulfate corrosion and freeze-thaw cycles.

The Bouc–Wen modified model [56] is not particularly sensitive to the selection of its parameters. Vaiana et al. [57] proposed a hysteretic mechanical system by combining a novel rate-independent model and an explicit time integration method which avoid the convergence problem for its small time step. Since the hysteretic behaviors of steel tubes under cyclic loading are complex and the effects of corrosion-freeze-thaw cycles have strong randomness, using a bilinear model is more easily describes the complex environment effects. The stress–strain skeleton curve of a steel tube corresponding to its hysteretic behavior is composed of two parts, which are the elastic stage and the strain hardening stage, as shown in Figure 4. It can be seen that unloading occurred in the strain hardening stage, which caused non-negligible residual strain. The residual stress would decrease when reverse loading, which caused dimensional stability issues. Due to the effect of residual strain on the loading and unloading process, the Bausinger effect should be taken into account. The stress–strain model of the steel tube was presented in Equation (2), where the elastic modulus in the strain hardening stage was defined as $0.01E_{se}$.

$$\sigma = \begin{cases} E_{se} & \varepsilon(0 \leq \varepsilon \leq \varepsilon_y) \\ f_{ye} + 0.01E_{se}(\varepsilon - \varepsilon_y) & (\varepsilon \geq \varepsilon_y) \end{cases} \tag{2}$$

where $\sigma$ is the stress of the steel tube, $E_{se}$ is the elastic modulus of steel, $\varepsilon$ is the strain of the steel tube below yield strain, $\varepsilon_y$ is the yield strain corresponding to the yield strength of the steel tube.

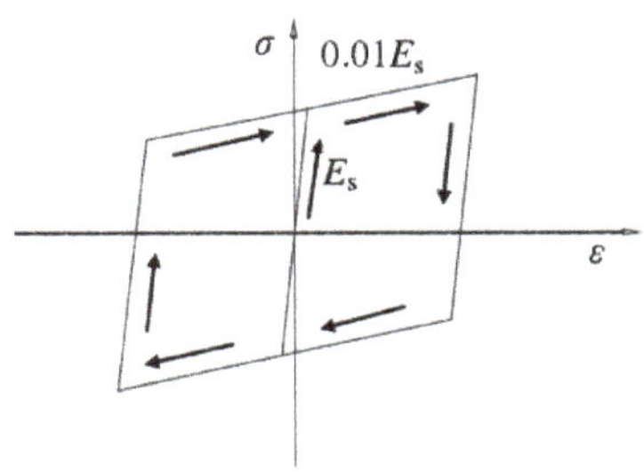


**Figure 4.** Bilinear kinematic hardening model of steel tube.

3.1.2. Confined Core Concrete

The stress–strain model proposed by Han et al. [45] was adopted to simulate the properties of core concrete. The material law for the core concrete under the alternation effect of sulfate corrosion and freeze-thaw cycles was illustrated in Figure 5.

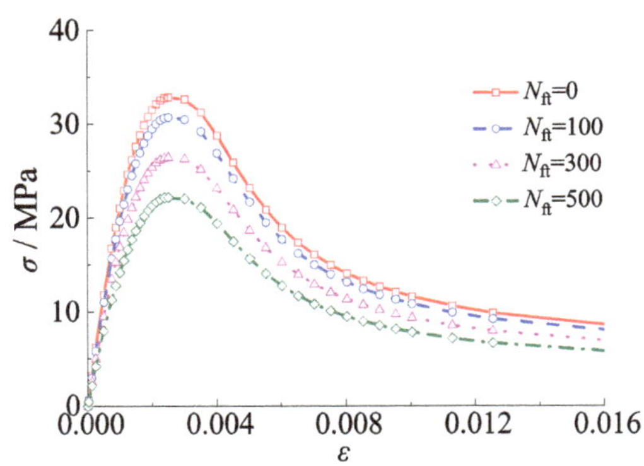


**Figure 5.** Stress–strain model for core concrete in square CFST stub columns subjected to alternating sulfate corrosion and freeze-thaw cycles.

The core concrete is modelled through the concrete damage plasticity (CDP) model. A stress–strain model considering the confinement effect in improving the behavior of core concrete by Han et al. [45] was applied to define the concrete material in the FE model.

$$y = \begin{cases} 2x - x^2 & (x \le 1) \\ \frac{x}{\beta(x-1)^{\eta}+x} & (x > 1) \end{cases} \tag{3}$$

where $x = \varepsilon/\varepsilon_0$, $\varepsilon$ is the strain of the concrete and $\varepsilon_0$ is the peak strain corresponding to peak load, $\varepsilon_0 = (1300 + 12.5f_c' + 800\xi_e{}^2) \times 10^{-6}$, $y = \sigma/\sigma_0$, $\sigma$ is the stress of concrete and $\sigma_0$ is the stress corresponding to the peak load, $\sigma_0 = f_c{}'(1 - 0.035N_{ft}/100)$, $f_c' = [0.76 + 0.2\log_{10}(f_{cu}/19.6)]f_{cu}$, $f_c'$ is the cylinder strength of concrete, $f_{cu}$ is the cube compressive strength of concrete, $\eta = 1.6 + 1.5/x$, $\eta$ is defined as shape coefficient to reflect the properties of shape cross section, $\beta = (f_c')0.1/[1.2(1 + \xi_e)0.5]$, $\xi_e = \alpha_e \times f_{ye}/f_{ck}$, $\xi_e$ is the effective confinement factor, $f_{ye}$ and $f_{ck}$ are the effective yield strength of steel tube and characteristic compressive strength of concrete, respectively; $f_{ck} = 0.88 \times 0.76 \times f_{cu}$, $\alpha_e = A_{se}/A_c$; $\alpha_e$ is the steel ratio, and $A_{se}$ and $A_c$ are the effective cross-sectional areas of the steel tube and core concrete, respectively.

The elastic modulus of core concrete was calculated by Equation (4), which was proposed by ACI 318 [47] as follows:

$$E_c = 4700\sqrt{f_c'} \tag{4}$$

where $E_c$ is the elastic modulus of core concrete and $f_c'$ denotes the cylinder strength of concrete.

The parameters of the CDP model were defined in the numerical model using the formula provided by Tao et al. [58]. The value of the dilation angle ($\psi$), $K_c$, $f_{b0}/f_{c0}$ was determined by Equations (5)–(7). The eccentricity value was selected as 0.1 and the viscosity parameter was chosen as 0.0001 to improve the convergency of the model.

$$\psi = \begin{cases} 56.3(1-\xi_e) & \xi_e \le 0.5 \\ 6.672e^{\frac{7.4}{4.64+\xi_e}} & \xi_e > 0.5 \end{cases} \tag{5}$$

$$K_c = \frac{5.5}{5+2(f_c')^{0.075}} \tag{6}$$

$$f_{b0}/f_{c0} = 1.5(f_c')^{0.075} \tag{7}$$

where $\xi_e$ is the effective confinement factor, $f_c'$ is the cylinder strength of concrete, $f_{b0}$ is the initial biaxial strength of concrete, and $f_{c0}$ is the initial axial strength of concrete.

Since the stress–strain relationship of concrete under hysteretic load could be replaced by that under uniaxial load [45], the stress–strain model of uniaxial load was adopted herein to be the skeleton curve of the hysteretic relationship. However, the concrete of the CFSTs under hysteretic load experienced the phenomenon of strain softening and stiffness degradation, which were caused by the development of cracks. Based on the characteristics of energy absorption and energy dissipation in the process of plastic deformation, the concrete damage model proposed by Britel and Mark [59] was introduced to the constitutive model of core concrete

$$D = \frac{W_0 - W_\varepsilon}{W_0} = \frac{W_0 - \sum_{j=1}^{a} W_{\varepsilon j}}{W_0} \tag{8}$$

where $D$ is the damage variable of core concrete, $W_0$ presents the strain energy in non-destructive state, $W_\varepsilon$ presents the strain energy under damage state, and $a$ denotes the number of calculation steps.

$$d_{c(t)} = 1 - \frac{\sigma_{c(t)}E_c^{-1}}{\varepsilon_{c(t)}^{pl}\left(1/b_{c(t)} - 1\right) + \sigma_{c(t)}E_c^{-1}} \tag{9}$$

where $\sigma_{c(t)}$ is the compression (tensile) stress under a damaged state and $E_c$ denotes the elastic modulus of concrete. In order to ensure the convergence and rationality, coefficient $b_{c(t)}$ is selected as 0.5 and 0.25 corresponding to compression and tension condition, respectively.

The tension recovery of core concrete should be considered in the CDP model under low cyclic load. This characteristic was determined by compression stiffness recovery parameter $w_c$ and tension stiffness recovery parameter $w_t$. According to a former study, micro cracks in concrete could convey compression under compressive load, while concrete with micro cracks could not convey tensile strength [60]. In this model, $w_c = 1.0$ and $w_t = 0$ were employed.

### 3.2. Element Type and Mesh

According to a former study [61], core concrete was modeled using 3D solid element (C3D8R) with reduced integration. The steel tube was modelled using a 4-node 3D shell element (S4R), where 9 integrations were set along the thickness of shell elements using the Simpson integration rule [62]. The steel bar selected was a three-dimensional two-node element (T3D2). All of the element types are available in the ABAQUS program library. The whole structural meshing technic in ABAQUS was adopted to obtain an evenly distributing mesh style. A form of the model in simulation was presented in Figure 6. The element size in longitudinal and lateral directions was selected to be $B/10$ for both the core concrete and the steel tube. The 3D model consisted of a mesh which was deemed fine enough to accurately calculate the hysteretic behaviors of CFSTs with a reasonable computational time.

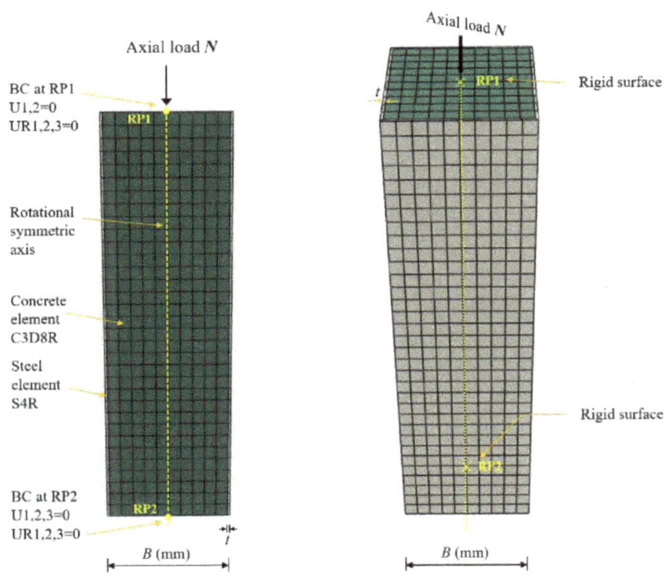


**Figure 6.** Schematic views of FE meshing and boundary conditions.

### 3.3. Boundary Condition and Load Application

The centers of the top and bottom surfaces were set as reference point 1 (RP1) and reference point 2 (RP2), respectively. The top and bottom surfaces of the specimen were constrained with RP1 and RP2 through a rigid body constraint. The top surface of the CFST column is free in all directions except in the loading direction. The axial load was exerted on the RP1 in the Z direction and the horizontal load was also exerted on RP1 in the X direction in the displacement control mode as shown in Figures 6 and 7. The bottom surface of the specimen is restrained in all degrees of freedom. The boundary condition in the model can simulate the CFST column effectively, which corresponds to one end hinged and one end fixed in reality as shown in Figure 8, where $P$ presents the lateral load, $\Delta_m$ denotes lateral displacement, and $L_0$ defined as calculation length.

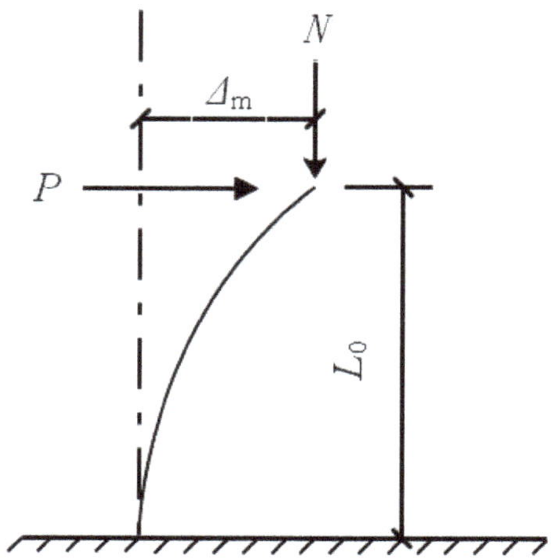


**Figure 7.** Schematic views of CFST stub columns under cyclic load.

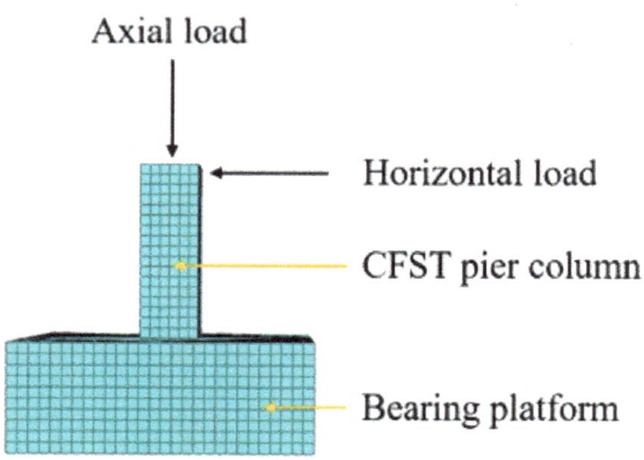


**Figure 8.** Schematic views of FE model under cyclic load.

The model includes three steps: (1) an initial analysis step, where boundary conditions were set up; (2) step 2, where axial load is applied in the load-control mode; (3) step 3, where horizontal load exerted to RP1 with desired displacement in the displacement-control mode. The time of each analysis step was separated according to the displacement loading amplitude curve. The calculated result was obtained by Gauss integral iteration and output at the end of each analysis step.

### 3.4. Concrete-Steel Tube Interface Modeling

The concrete core and steel tube employed surface-to-surface contact to ensure the concrete core and the steel tube maintained an adequate bond, which created the CFST column as a whole. Since the stiffness of the concrete core was higher than that of the steel tube, the concrete core was selected as the master surface, while the steel tube was set as the slave surface. Moreover, the formula of small sliding was adopted in this section since the relative slip of the concrete core and the steel tube was small. Mechanical properties of tangential behavior were simulated by a penalty and friction coefficient, which is 0.6 [62].

The value of the shear stress limit was selected as 0.15 MPa [63]. Normal behavior adopted hard contact, which resulted in surface contact with no limitation.

*3.5. Verification of Finite Element Model*

3.5.1. Failure Mode

The typical failure mode of specimens under cyclic loading at ambient temperature (20 ± 2 °C) was depicted in Figure 9. As shown in the figure, the FE model presents a shear failure mode with outward buckling at the end of the steel tube. The results of the FE models and test specimens were consistent, which suggests that the FE model was effective.

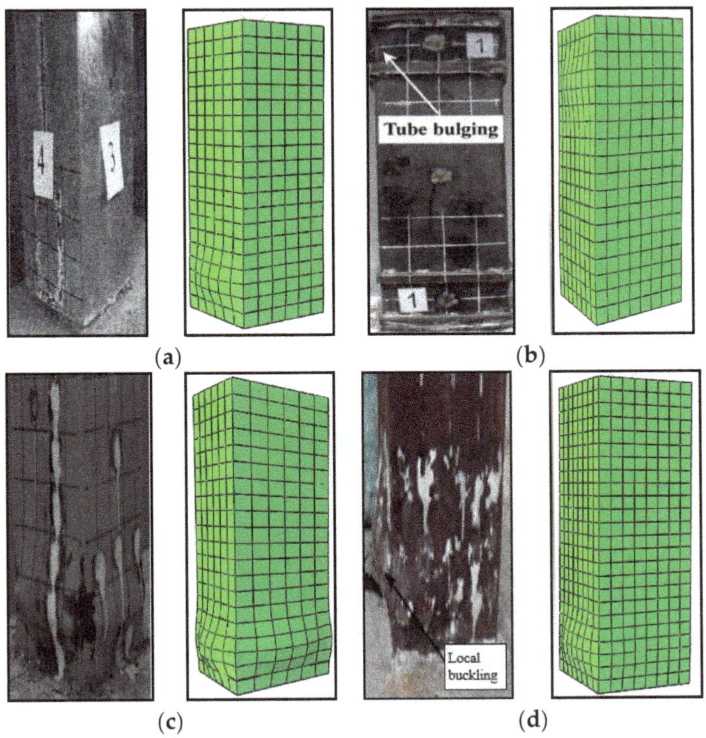


**Figure 9.** Comparison of failure modes between test specimens and FE models under cyclic loading under ambient temperature. (**a**) STRC-60-1.5-4; (**b**) STRC-70-1.5-4: (**c**) FGZ4; (**d**) A1.

3.5.2. Skeleton Curve

The experimental results were compared with those obtained via FE modelling in terms of skeleton curve, as depicted in Figure 10. It can be seen in the figure that in the elastic stage, the skeleton curves of specimens in accordance with the FE models.

However, in the elastic plastic stage, the slope of FE results was higher than test results. For the twelve verified specimens, the skeleton curves match the FE curves with a subtle difference in slope, which is likely to be associated with a geometric and loading defect generated through the experiment. These results were then employed in investigating hysteretic behavior.

3.5.3. Horizontal Ultimate Compressive Strength

The horizontal ultimate compressive strength was obtained from the FE model and compared with the test specimens. Table 2 compares the peak load of the specimens with the capacity derived from the FE modelling. The mean and standard deviation values highlight a favorable efficiency of the FE model in predicting the desired measures.

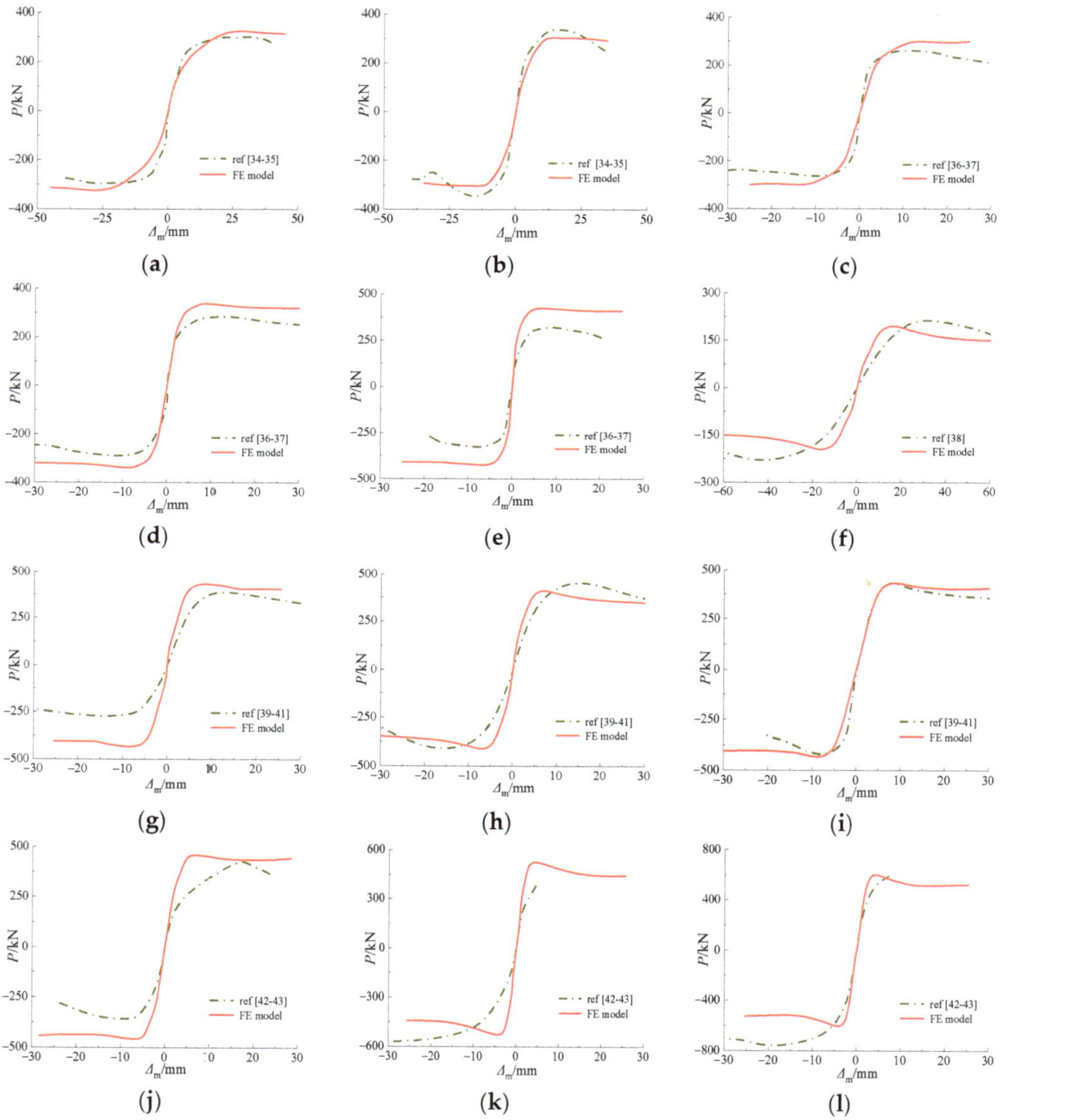


**Figure 10.** Comparison of skeleton curve between test specimen and FE model under cyclic loading at ambient temperature. (**a**) STRC-60-1.5-4; (**b**) STRC-60-1.5-6; (**c**) STRC-70-1.5-4; (**d**) STRC-70-1.5-5; (**e**) STRC-70-1.5-6; (**f**) FGZ4; (**g**) A1; (**h**) A2; (**i**) A3; (**j**) RS-4L; (**k**) RS-6M; (**l**) RS-10H.

### 3.5.4. Hysteretic Behavior of CFSTs after Sulfate Corrosion

The experimental results were compared with the results of the FE model in terms of hysteretic behavior, as depicted in Figure 11. The rationality of the FE model on calculating ultimate horizontal load was verified by the existing test results [64–66], as shown in Figure 12.

**Table 2.** Comparison of ultimate horizontal load between specimens and FE model.

| No. | Ref. | Specimen Labels | $P_{ue1}$ (kN) | $P_{ue2}$ (kN) | $P_{fe1}$ (kN) | $P_{fe2}$ (kN) | $P_{ue1}/P_{fe1}$ | $P_{ue2}/P_{fe2}$ |
|---|---|---|---|---|---|---|---|---|
| 1 | [34,35] | STRC-60-1.5-4 | 304 | 296 | 335 | 335 | 0.907 | 0.884 |
| 2 | | STRC-60-1.5-6 | 337 | 348 | 307 | 307 | 1.098 | 1.134 |
| 3 | | STRC-70-4-1.5 | 262 | 267 | 260 | 260 | 1.008 | 1.027 |
| 4 | [36,37] | STRC-70-5-1.5 | 285 | 290 | 258 | 258 | 1.105 | 1.124 |
| 5 | | STRC-70-6-1.5 | 324 | 329 | 426 | 426 | 0.761 | 0.772 |
| 6 | [38] | FGZ4 | 216 | 240 | 195 | 195 | 1.108 | 1.231 |
| 7 | | A1 | 397 | 275 | 433 | 433 | 0.917 | 0.635 |
| 8 | [39–41] | A2 | 454 | 424 | 412 | 412 | 1.102 | 1.029 |
| 9 | | A3 | 445 | 449 | 436 | 436 | 1.021 | 1.030 |
| 10 | | RS-4L | 434 | 361 | 458 | 458 | 0.948 | 0.788 |
| 11 | [42,43] | RS-6M | 556 | 572 | 544 | 544 | 1.022 | 1.051 |
| 12 | | RS-10H | 773 | 771 | 614 | 614 | 1.259 | 1.256 |
| | | | Mean | | | | 1.064 | 1.022 |
| | | | SD | | | | 0.046 | 0.032 |

Note: Specimen labels in the table are the same as the test specimen in Refs. [34–43].

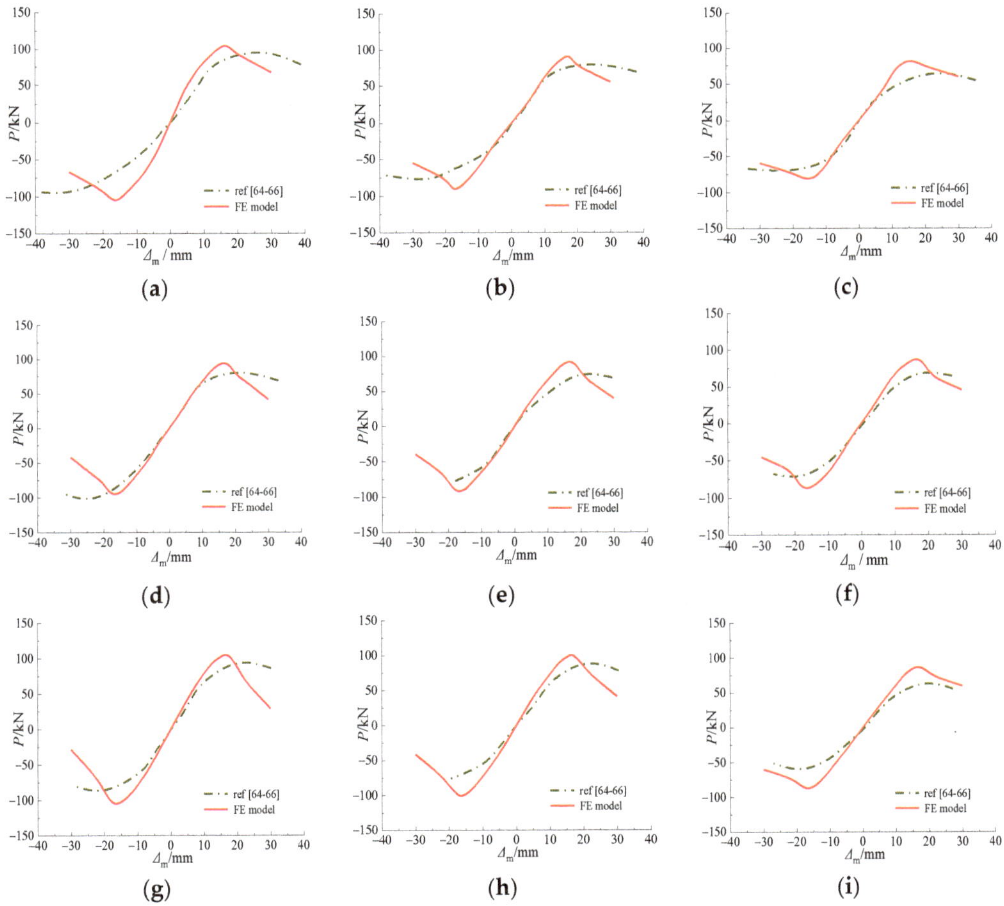


**Figure 11.** Comparison between FE model and test results of specimens after sulfate corrosion. (**a**) RC-0A; (**b**) RC-10A; (**c**) RC-20A; (**d**) RC-0B; (**e**) RC-10B; (**f**) RC-20B; (**g**) RC-0C; (**h**) RC-10C; (**i**) RC-20C.

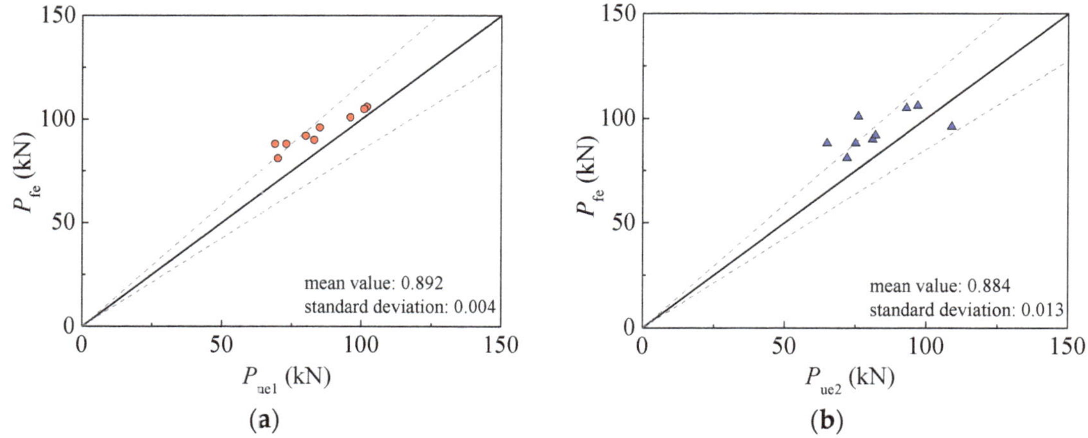


**Figure 12.** Comparison between $P_{fe}$ and $P_{ue1}$, $P_{ue2}$ values of the specimens in Refs. [57–59]. (**a**) $P_{fe}/P_{ue1}$; (**b**) $P_{fe}/P_{ue2}$.

## 4. Analysis and Discussion

In this paper, the typical specimen had a side length ($B$) of 400 mm, shear span ratio ($\lambda$) equal to 1.5, and other parameters as tabulated in Table 3. The number of cycles consisted of 100 of freeze-thaw cycles and 10% sulfate corrosion.

**Table 3.** Main parameters of CFST pier columns in FE modeling.

| Type | Parameter Value | Default Value |
|---|---|---|
| $\alpha$ | 0.06<br>0.10<br>0.18<br>0.20 | 0.10 |
| $f_{cu}$ (MPa) | 30<br>40<br>50<br>60 | 40 |
| $f_y$ (MPa) | 235<br>335<br>390<br>420 | 235 |
| $n$ | 0.05<br>0.10<br>0.15<br>0.20<br>0.25<br>0.30 | 0.15 |
| $i$ | 0<br>1<br>2<br>3<br>4<br>5 | 3 |

### 4.1. Yield Strength of Steel Tube

Figure 13 presents the effects of yield strength of a steel tube on hysteretic curves and skeleton curves of CFST stub columns under horizontal load after an alternation of sulfate corrosion and freeze-thaw cycles. It can be seen that the effect of yield strength of the steel tube was significant on the shape of the curves and cyclic displacement corresponding to ultimate bearing capacity. Regarding the hysteretic curves, with the increase of the yield strength of the steel tube, they became fuller, which resulted in a larger surrounded area. When the hysteretic curve enters the second half, the cyclic displacement declined faster with the larger yield strength of the steel tube due to the damage accumulation of the steel tube under cyclic reciprocating load after sulfate corrosion.

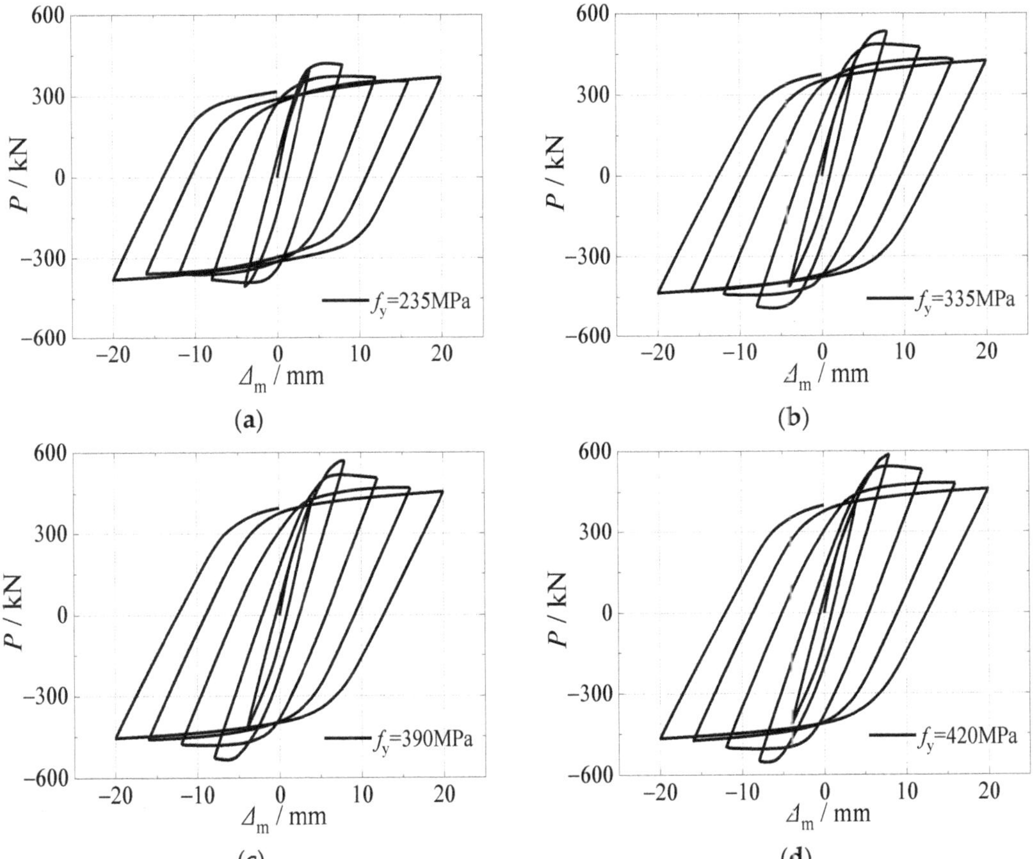


**Figure 13.** Effect of yield strength of steel tube on hysteretic curve. (**a**) $f_y$ = 235 MPa; (**b**) $f_y$ = 335 MPa; (**c**) $f_y$ = 390 MPa; (**d**) $f_y$ = 420 MPa.

Skeleton curves presented in Figure 14 kept a similar trend in the elastic stage, which illustrated the effect of yield strength of the steel tube on initial stiffness. However, the effect of yield strength of the steel tube on horizontal load at peak point was significant, especially when the yield strength of the steel tube increased from 235 to 335 MPa. The ultimate horizontal load increased by 26.3%, 35.0%, and 39.2%, respectively as the yield strength of the steel tube increased from 235 to 335, 390, and 420 MPa. Meanwhile, the effect of yield strength of the steel tube on the slope of the curves in the descending stage and ductility was negligible.

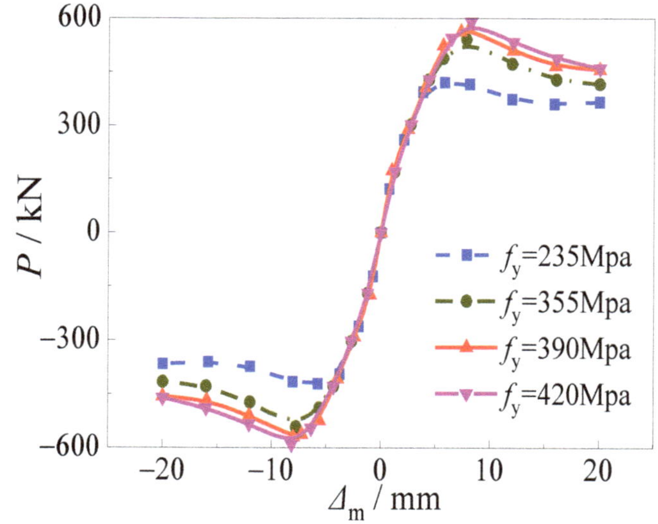


**Figure 14.** Effect of yield strength of steel tube on skeleton curve.

*4.2. Compressive Strength of Core Concrete*

Figures 15 and 16 depict the effects of the compressive strength of concrete on hysteretic curves and skeleton curves of CFST stub columns under horizontal load after an alternation of sulfate corrosion and freeze-thaw cycles. It can be seen that the effect of the compressive strength of concrete was significant on the shape of the curves and cyclic displacement corresponding to ultimate bearing capacity. For the hysteretic curves, when the compressive strength of the concrete increased, the stiffness of the specimen was weaker, which resulted in the "S" shape of the $P$-$\Delta_m$ curve. For skeleton curves, all four curves showed a similar trend in the elastic stage which illustrated that the effect of the compressive strength of the concrete on initial stiffness was slight. However, the effect of the compressive strength of the concrete on horizontal load at peak point was significant. The ultimate horizontal load increased by 16.0%, 27.7%, and 33.7%, respectively as the compressive strength of concrete increased from 30 to 40, 50, and 60 MPa. The compressive strength of concrete had some effect on the slope of the curves in the descending stage. With the increase of the compressive strength of the concrete, the descending stage of the skeleton curve decreased faster, which demonstrated that the ductility of the specimens was weaker due to the brittleness of the high strength of the concrete core. Although high compressive strength of concrete could improve the horizontal ultimate bearing capacity to some degree, the ductility would decrease, which should be focused on.

*4.3. Steel Ratio*

Figures 17 and 18 depicted the variations of hysteretic curves and skeleton curves with different steel ratios under alternating sulfate corrosion and freeze-thaw cycles. The effects of steel ratio on hysteretic curve, skeleton curve, and bearing capacity corresponding to each cycle were significant. With the increase of steel ratio, the hysteretic curves changed from flat to full and the area within the curves became larger, which illustrated that the specimens were sufficient in the plastic deformation capacity.

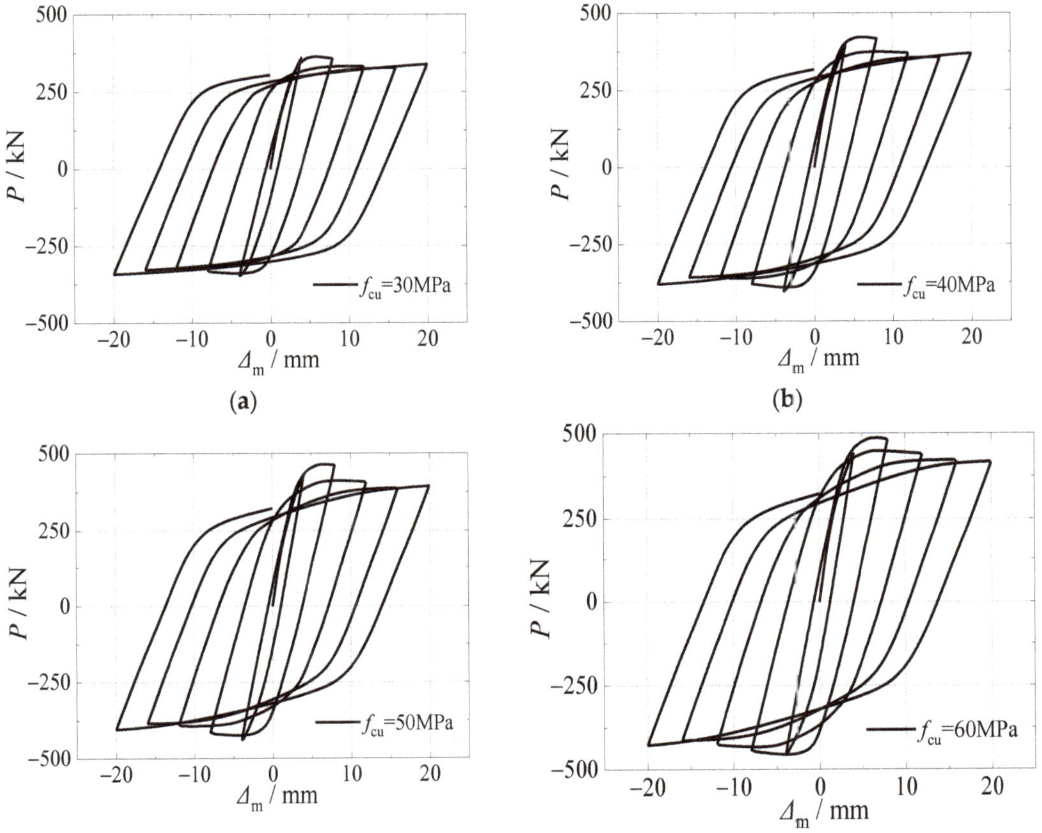


**Figure 15.** Effect of compressive strength of core concrete on hysteretic curve. (**a**) $f_{cu}$ = 30 MPa; (**b**) $f_{cu}$ = 40 MPa; (**c**) $f_{cu}$ = 50 MPa; (**d**) $f_{cu}$ = 60 MPa.

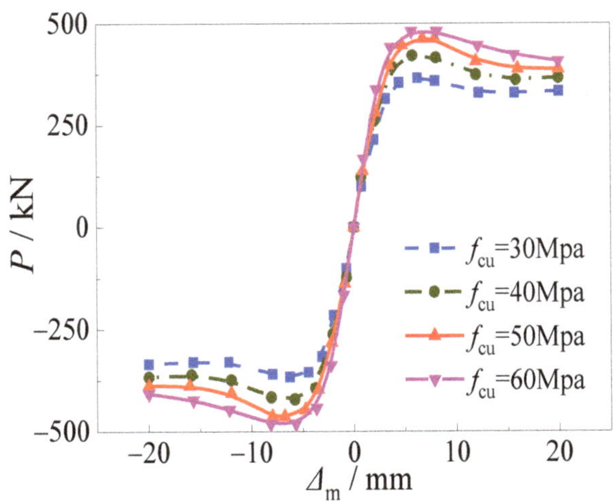


**Figure 16.** Effect of compressive strength of core concrete on skeleton curve.

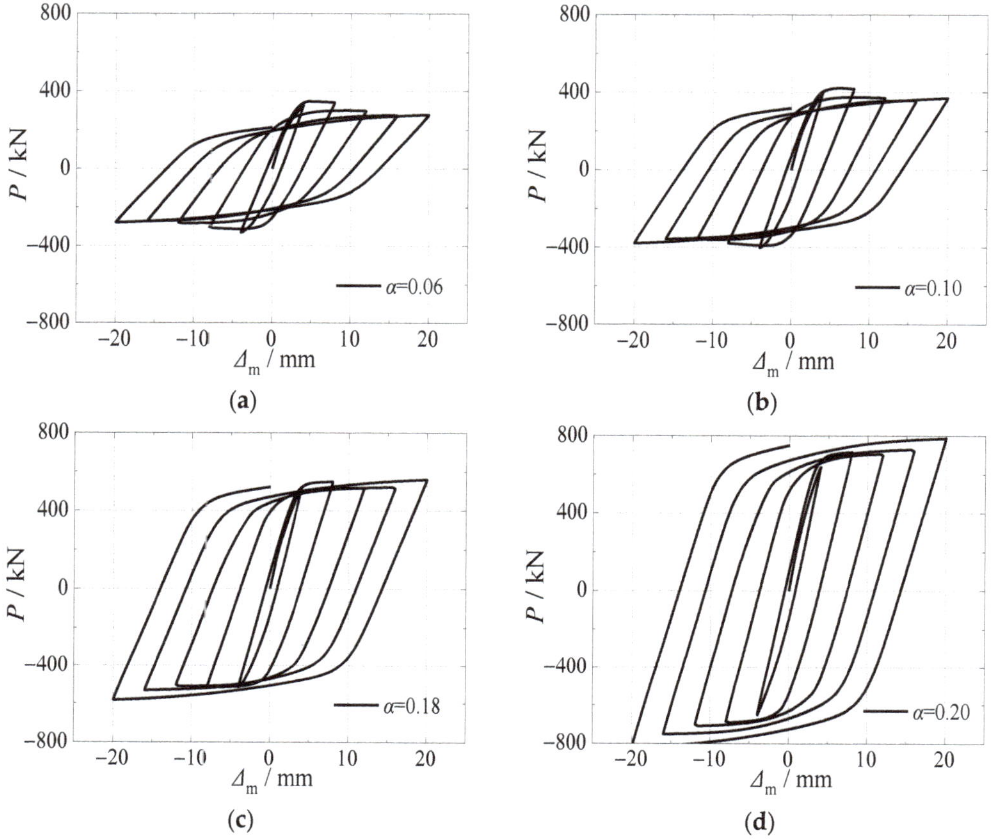


**Figure 17.** Effect of steel ratio on hysteretic curve. (**a**) $\alpha$ = 0.06; (**b**) $\alpha$ = 0.10; (**c**) $\alpha$ = 0.18; (**d**) $\alpha$ = 0.20.

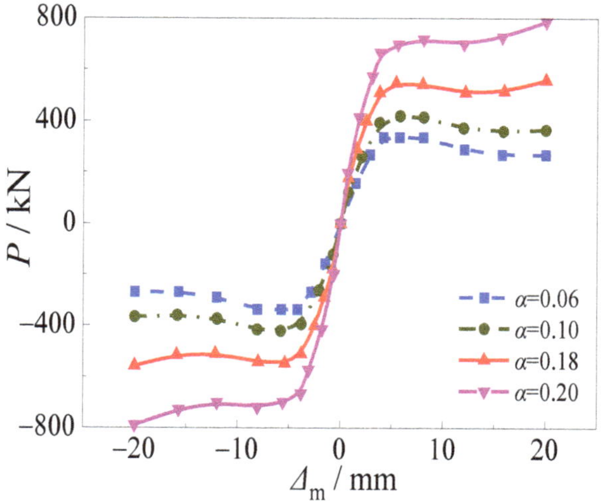


**Figure 18.** Effect of steel ratio on skeleton curve.

It can be seen in Figure 18 that the slope of the four skeleton curves in the elastic stage was obviously different. The higher steel ratio resulted in higher elastic stiffness and ultimate horizontal load. The ultimate horizontal load increased by 21.4%, 61.0%, and 126.4% when the steel ratio increased from 0.06 to 0.10, 0.18, and 0.20, respectively. Moreover, the skeleton curves decreased slowly and even presented strain hardening phenomenon due to the confinement effect of the steel tube, which presented ductile behavior.

### *4.4. Axial Compression Ratio*

Figures 19 and 20 presented the effects of the axial compression ratio on the hysteretic curves and skeleton curves of CFST stub columns under horizontal load after an alternation of sulfate corrosion and freeze-thaw cycles. All the shapes of hysteretic curves demonstrated that the effect of the axial compression ratio was slight. The ultimate horizontal load slightly increased by 4.2%,6.6%, 8.1%, 9.0%, and 9.3%, respectively as the axial compression ratio increased from 0.05 to 0.1, 0.15, 0.20, 0.25, and 0.30 when the alternation time was three. The descending stage of the skeleton curves decreased more with the increase of the axial compression ratio, which demonstrated that the ductility of the specimens was weaker. This was for the reason that the material damage became more serious with the increasing axial compressive load due to the increasing axial compression ratio.

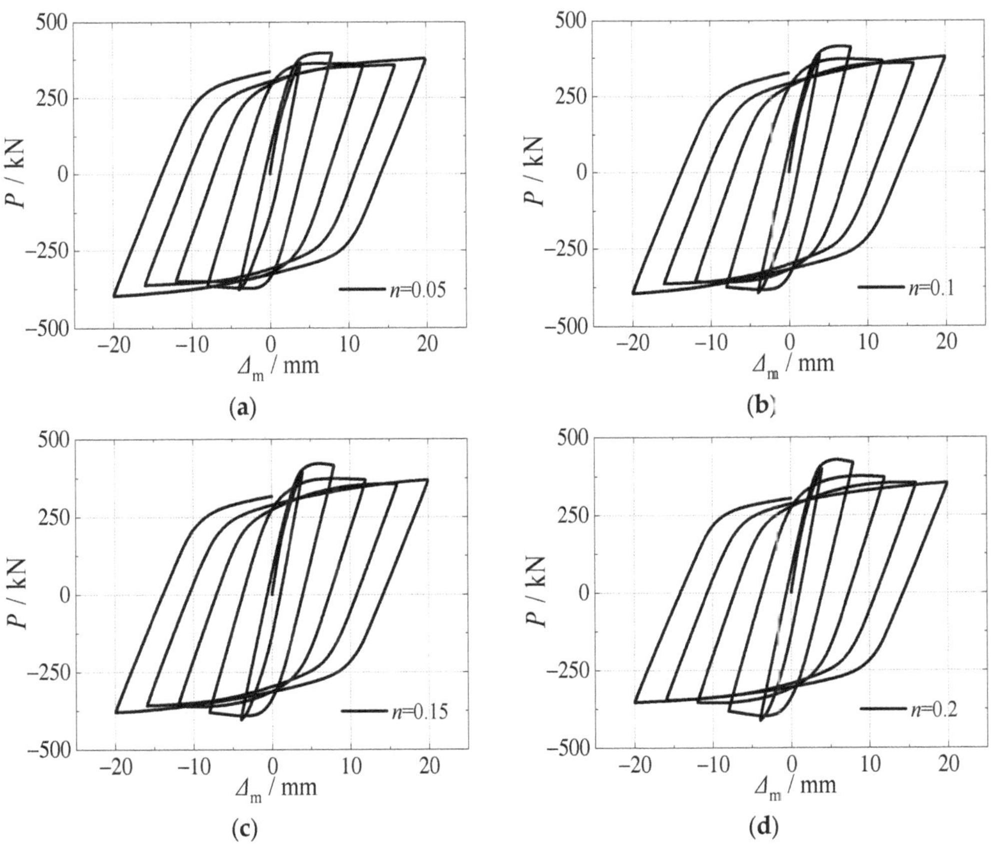


**Figure 19.** *Cont.*

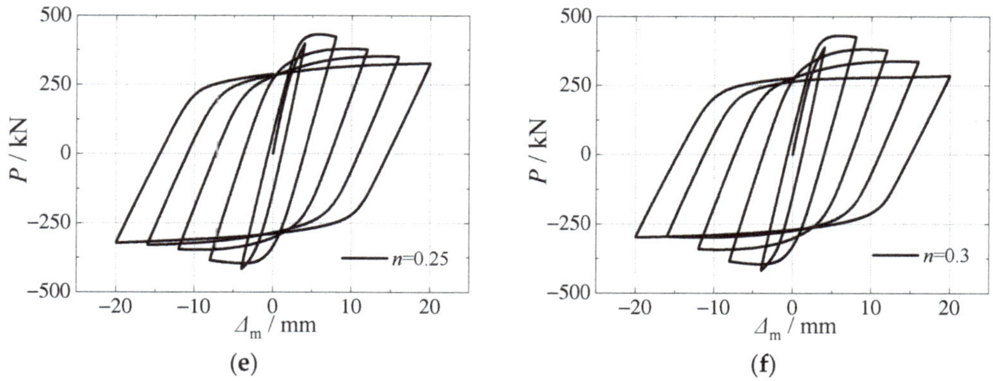


**Figure 19.** Effect of axial load ratio on hysteretic curve. (**a**) $n$ = 0.05; (**b**) $n$ = 0.10; (**c**) $n$ = 0.15; (**d**) $n$ = 0.20; (**e**) $n$ = 0.25; (**f**) $n$ = 0.30.

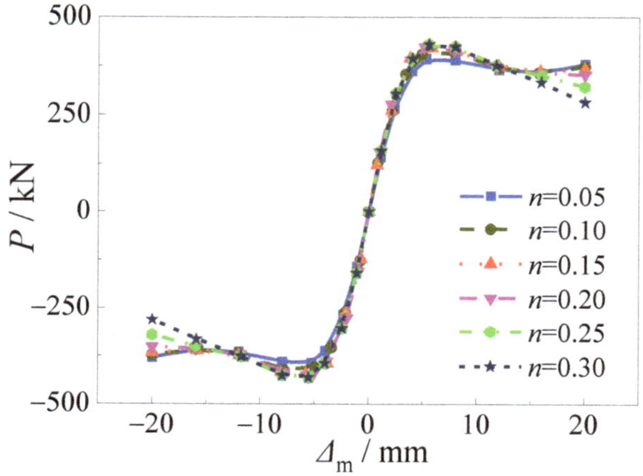


**Figure 20.** Effect of axial load ratio on skeleton curve.

### *4.5. Sulfate Corrosion and Freeze-Thaw Cycles*

Figure 21 depicted the effects of alternation time on hysteretic curves of CFST piers under horizontal load after an alternation of sulfate corrosion and freeze-thaw cycles. It can be seen that the effect of alternation time was significant on the shape of the curves and cyclic displacement corresponding to ultimate bearing capacity. For hysteretic curves, with the increase of alternation time, the phenomenon of pinch occurred, which resulted in the smaller area.

For skeleton curves as showed in Figure 22, five curves kept a similar trend in the elastic stage, which illustrated that the effect of alternate time on initial stiffness could be ignored. However, the effect of alternation time on the horizontal load at peak point was significant. The ultimate horizontal load decreased by 16.2%, 29.8%, 40.5% and 50.2%, respectively as the alternation time increased from 1 to 2, 3, 4, and 5. Meanwhile, alternation time has effects on the slope of the curves in the descending stage. The skeleton curve has a lower ductility, as the skeleton curves decreased faster in the descending stage due to the serious material damage caused by the alternate time on load resistance.

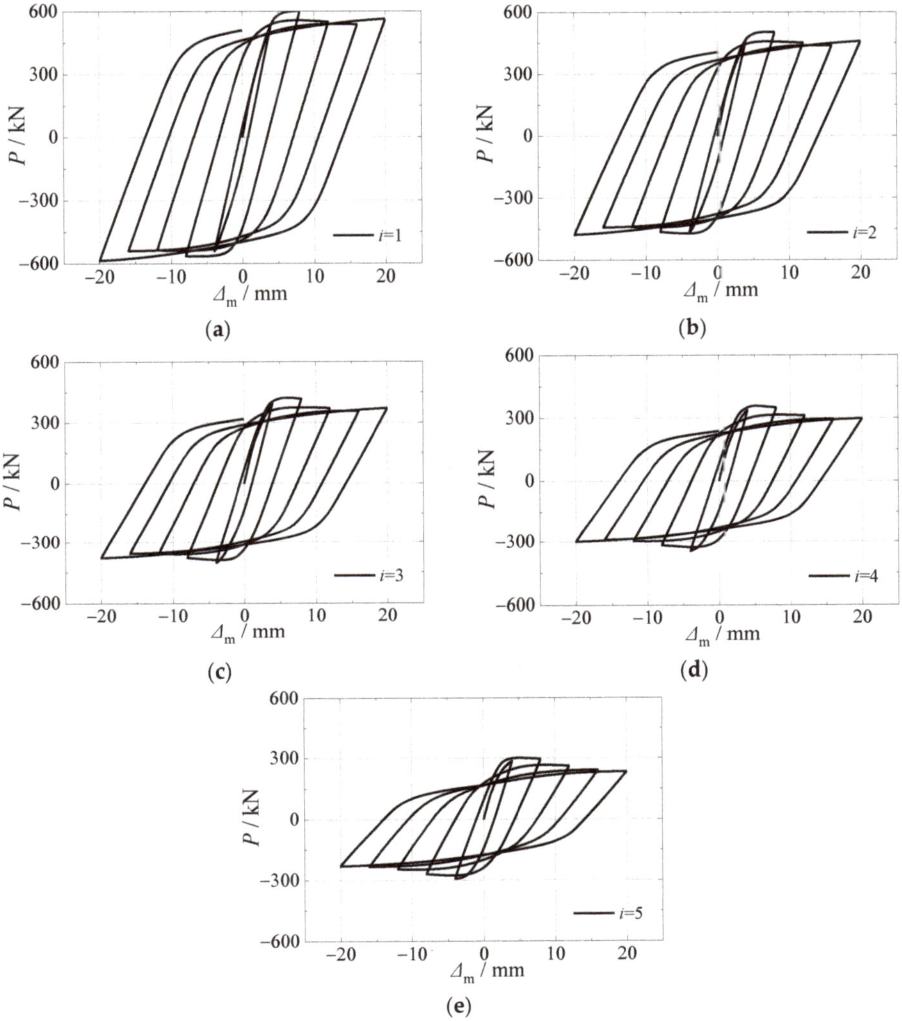


**Figure 21.** Effect of alternation time on hysteretic curve. (**a**) $i$ = 1; (**b**) $i$ = 2; (**c**) $i$ = 3; (**d**) $i$ = 4; (**e**) $i$ = 5.

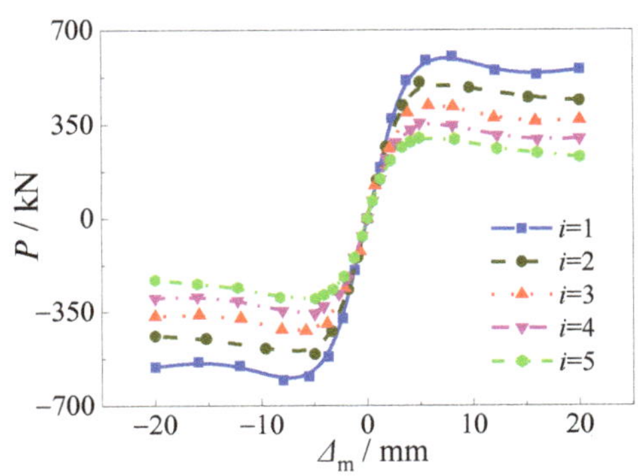


**Figure 22.** Effect of alternate times on skeleton curve.

## 5. Simplified Formula for Horizontal Ultimate Compressive Strength

According to the parameter analysis in Section 4, the reduction factor of the horizontal ultimate compressive load of CFST under alternation between sulfate corrosion and freeze-thaw cycles is mainly determined by confinement coefficient ($\xi$) and alternation time ($i$). In this study, according to practical engineering, the range of parameters was selected as $\xi$ = 0.6–2.1 and $i$ = 0–5.

### 5.1. Design Method

According to Figure 7, the bending moment of CFST stub columns under low cycle reciprocating load was expressed in Equation (10) as follows:

$$M_{\rm u} = P_{\rm u0}L + N\Delta \tag{10}$$

where $M_{\rm u}$ represents the flexure capacity; $P_{\rm u0}$ denotes the horizontal ultimate axial load; $L$ is the length of the CFST column; $N$ stands for the axial compressive load; and $\Delta$ corresponds to lateral deformation.

Based on CJJ166-2011 [67], the axial compression ratio of CFST pier columns with regular section in earthquake fortification zone should be less than 0.3, i.e., $n \leq 0.3$. Meanwhile, lateral deformation was far less than the calculated length of CFST pier column, so the bending moment caused by the axial compressive load could be ignored. Thus, the ultimate horizontal load was expressed in Equation (11) as follows:

$$P_{\rm u0} = M_{\rm u}/L_0 \tag{11}$$

The design formula to calculate the flexure capacity in GB50936-2014 was described in Equations (12)–(14) as follows:

$$M_{\rm u} = \gamma_{\rm m} W_{\rm sc} f_{\rm sc} \tag{12}$$

$$W_{\rm sc} = r_0^3/4 \tag{13}$$

$$\gamma_{\rm m} = -0.483\xi + 1.962\sqrt{\xi} \tag{14}$$

where $\gamma_{\rm m}$ represents the plastic adaption coefficient of cross section; $W_{\rm sc}$ denotes the modulus of section of CFST pier columns (mm$^3$); $r_0$ corresponding to the equivalent circle radius (mm); and $M_{\rm u}$ is the flexure capacity ($N\cdot$mm).

### 5.2. Modified Reduction Coefficient

To evaluate reduction of loading bearing capacity of the square CFST pier columns after alternate sulfate corrosion and freeze-thaw cycles, the reduction coefficient $k_{\rm sq}$ was defined in Equation (15) as follows:

$$k_{\rm sq} = \frac{P_{\rm ue}}{P_{\rm u0}} \tag{15}$$

where, $P_{\rm ue}$ is the ultimate horizontal load under sulfate corrosion and freeze-thaw cycles; $P_{\rm u0}$ is the ultimate horizontal load under ambient condition. The major effect was presented in Figure 23.

The reduction coefficient varied with alternation time ($i$) linearly. With the increase of alternation time, the reduction coefficient decreased rapidly and decreased up to 75.0% compared to the specimens under ambient conditions, which illustrated that $i$ was the major factor that affects the ultimate horizontal load. In Figure 23a, the reduction coefficient decreased with the steel ratio significantly. The reduction coefficient decreased up to 64.0%, 61.0%, 59.1%, 59.0%, 53.1%, and 50.7%, respectively as the steel ratio increased from 0.06 to 0.08, 0.10, 0.15, 0.18, and 0.20, with the alternation time of 3. Thus, the steel ratio was selected to be a major factor of the reduction coefficient.

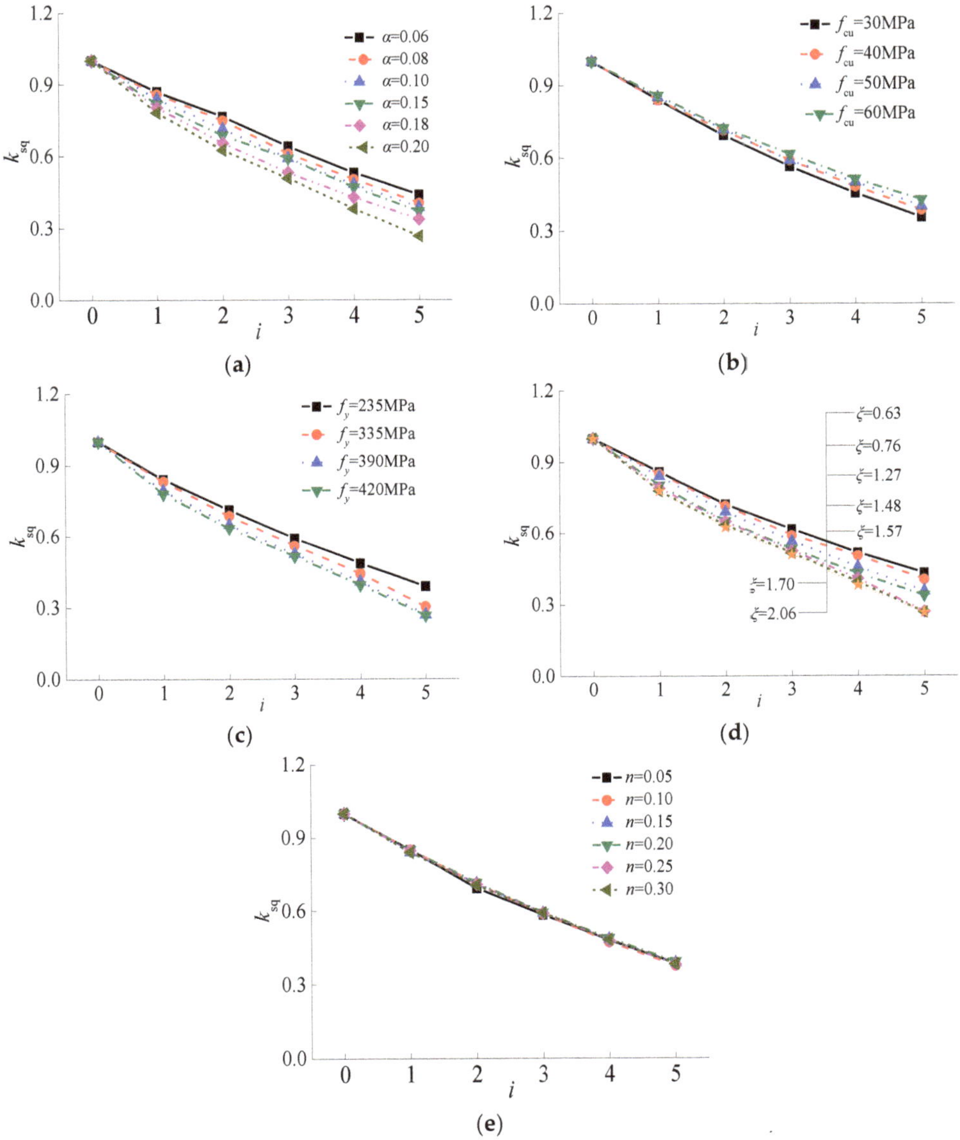


**Figure 23.** Effects of main parameters on reduction factor of horizontal bearing capacity. (**a**) Steel ratio; (**b**) Compressive strength of concrete; (**c**) Yield strength of steel tube; (**d**) Confinement coefficient; (**e**) Axial compression ratio.

It can be seen in Figure 23b,c the reduction coefficient decreased greatly with different compressive strengths of core concrete and yield strengths of steel tube with increasing alternation time. The specimens with lower compressive strength of core concrete or higher yield strength of steel tube corresponded to smaller reduction coefficients. The reduction coefficient decreased up to 61.6%, 59.1%, 59.0% and 56.4%, respectively corresponding to compressive strengths of core concrete of 50, 40, and 30 MPa, compared to the specimens with a compressive strength of core concrete of 60 MPa. Meanwhile, $k_{sq}$ decreased up to

59.1%, 56.0%, and 52.3%, respectively as the yield strength of steel tube increased from 235 to 335, 390, and 420 MPa.

Then, the reduction coefficient decreased obviously with the increasing confinement coefficient, as shown in Figure 23d. Since the confinement coefficient ($\xi = \alpha f_y/f_c$) was determined by the steel ratio and the relative ratio of yield strength of the steel tube to compressive strength of the core concrete, the reduction coefficient considering $\xi$ could reflect three above parameters. As shown in Figure 23e, the curves almost overlapped with the axial compression ratio ($n$) when it ranged from 0.05 to 0.30. This illustrated that the effect of axial compression ratio was negligible. Therefore, alternation time and confinement coefficient were adopted to be the major factors for evaluating the reduction coefficient of square CFST pier columns after alternate sulfate corrosion and freeze-thaw cycles. The reduction coefficient was described in Equation (16) as follows:

$$k_{\mathrm{sq}} = 1 - \left(0.008\xi^2 + 0.004\xi + 0.119\right)i \tag{16}$$

where $\xi$ is the confinement coefficient; $i$ presents the alternate time.

### 5.3. Verified Design Model

The reduction coefficient of the horizontal ultimate compressive load of square CFST pier columns after sulfate corrosion and freeze-thaw cycles was calculated based on Equations (15) and (16). All the reduction coefficients corresponding to FE models were compared with the calculated results, as shown in Figure 24. It can be seen that the formula for predicting reduction coefficient has high accuracy with a mean value of 1.028 and a standard deviation of 0.001.

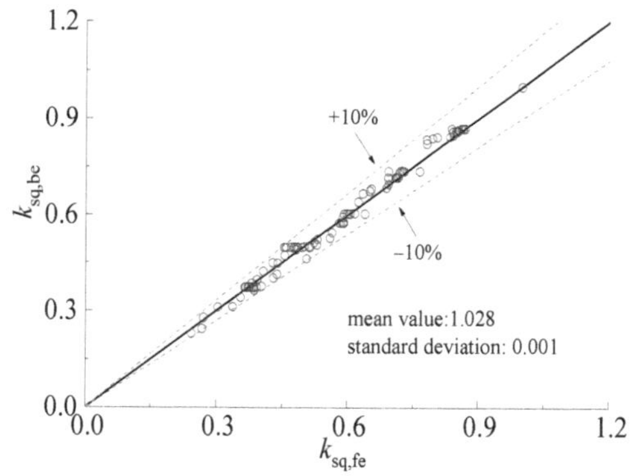


**Figure 24.** Comparison between calculated and simulated values of horizontal reduction coefficient.

For the square CFST pier column after alternate of sulfate corrosion and freeze-thaw cycles, ultimate horizontal load can be calculated by modifying the formula under ambient condition, which is expressed as follows:

$$P_{\mathrm{be}} = k_{\mathrm{sq}} P_{\mathrm{u0}} = \left[1 - \left(0.008\xi^2 + 0.004\xi + 0.119\right)(0.00375N_{\mathrm{ft}} + 6.25\gamma)\right] P_{\mathrm{u0}} \tag{17}$$

where $k_{\mathrm{sq}}$ is the reduction coefficient; $P_{\mathrm{u0}}$ presents the ultimate horizontal load under ambient conditions; the relationship between $N_{\mathrm{ft}}$ and $\gamma$ was defined as $i$ in Ref. [68].

The comparison between the calculated values and numberical values of the ultimate horizontal load of square CFST pier columns after alternate sulfate corrosion and freeze-thaw cycles was presented in Figure 25. The error of the modified simplified formula in GB50936-2014 [62] was within 15%, with a mean value of 1.022 and standard deviation of 0.003, which illustrated that the proposed model was accurate in predicting the ultimate horizontal load. Therefore, the modified formula for GB50936-2014 [69] could be adopted to evaluate the ultimate horizontal load of square CFST pier columns after alternate sulfate corrosion and freeze-thaw cycles.

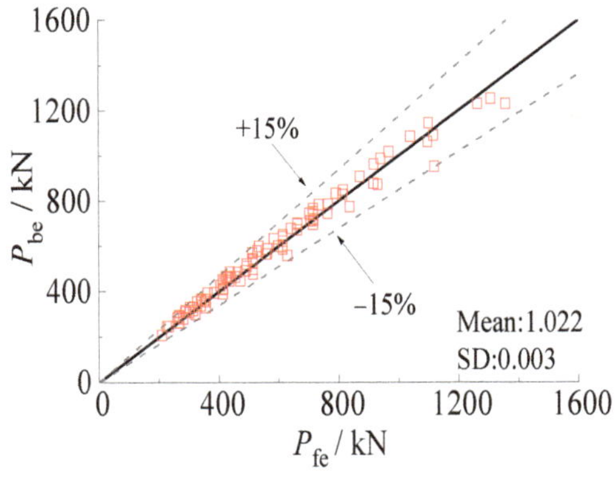


**Figure 25.** Comparison of ultimate horizontal load between calculated values and numerical values.

It should be mentioned that the modified simplified formula could be adopted to calculate the ultimate horizontal load of square CFST pier columns after alternate sulfate corrosion and freeze-thaw cycles with the limited parameters: (1) the section side length ranges from 200–1200 mm; (2) compressive strength ranges from 30–60 MPa; (3) yield strength of steel tubes ranges from 235–420 MPa, (4) alternation time ranges from 0–500; and (5) corrosion ratio ranges from 0%–50%.

## 6. Conclusions

The study presented a nonlinear finite element (FE) model to analyze the hysteretic behavior of concrete-filled steel tube (CFST) pier columns subjected to alternate of sulfate corrosion and freeze-thaw cyclic. Quantitative and qualitative verifications performed in this paper indicate that the FE model can accurately predict the hysteretic behavior of CFST pier columns. According to the numerical analysis, the following conclusions can be drawn:

(1) The nonlinear finite element (FE) model considering the dual environmental impact of sulfate corrosion and freeze-thaw cycles was accurate in simulating the degradation of loading stiffness and unloading stiffness, bearing load, and ductility on the basis of reasonable damage factors.

(2) The steel ratio, yield strength of steel tube, compressive strength of core concrete and alternation time have a significant effect on hysteretic behaviors of square concrete-filled steel tube (CFST) pier columns after alternate sulfate corrosion and freeze-thaw cycles. The cyclic curves of hysteresis loops increased with increasing steel ratio and yield strength of steel tube caused the increasing initial stiffness of the specimens. The skeleton curves corresponding to the hysteresis loops presented ductile behavior with increasing steel ratio and yield strength of steel tube. The effect of yield strength of the steel tube, compressive strength of core concrete, and alternation time on initial stiffness was negligible. The ultimate horizontal load increased by 21.4–126.4%, 26.3–39.2%, and 16.0–33.7% as the steel ratio increased from 0.06 to 0.20, the yield strength of steel tube increased from 235 MPa to 420 MPa, and compressive strength of core concrete increased from 30 MPa to 60 MPa, respectively while the ultimate horizontal load decreased by 16.2–50.2% as alternation time increased from 1 time to 5 times.

(3) The steel ratio, compressive strength of core concrete, axial compression ratio, and alternation time have a significant effect on ductility behavior. The ductility of the specimens increased with the increasing steel ratio since the post stage of the skeleton

curve varied from the descending stage to the ascending stage, while the ductility of the specimens decreased with the increasing compressive strength, axial compression ratio, and alternation time for the increasing brittleness caused by the damage of raw materials.

(4) The confinement coefficient and alternation time were the two major factors that affect the reduction coefficient of the ultimate horizontal load of square CFST pier columns after alternate sulfate corrosion and freeze-thaw cycles. The horizontal bearing capacity decreased by 16.2–50.2% with the alternation time increasing from 1 time to 5 times. The horizontal bearing capacity decreased up to 17.7% as the confinement coefficient increased from 0.63 to 2.06.

(5) A simplified formula for predicting the reduction coefficient of the horizontal bearing load with high accuracy with a mean value of 1.028 and standard deviation of 0.001 was developed.

(6) A modified equation for predicting the ultimate horizontal load of square CFST pier columns after alternate sulfate corrosion and freeze-thaw cycles was proposed within the error of 15%.

**Author Contributions:** Conceptualization, Q.X.; methodology, Q.X. and T.Z.; software, T.Z. and Q.W.; investigation, Q.X., T.Z. and L.G.; writing—original draft preparation Q.X. and T.Z.; writing—review and editing, Q.X. and J.T. All authors have read and agreed to the published version of the manuscript.

**Funding:** This research received no external funding.

**Institutional Review Board Statement:** Not applicable.

**Informed Consent Statement:** Not applicable.

**Data Availability Statement:** The study did not report any data.

**Conflicts of Interest:** The authors declare no conflict of interest.

## References


1. Han, L.H.; Li, W.; Bjorhovde, R. Developements and advanced applications of concrete-filled steel tubular (CFST) structures: Members. *J. Constr. Steel Res.* **2014**, *100*, 211–228. [CrossRef]
2. Sakino, K.; Nakahara, H.; Morino, S.; Nishiyama, I. Behavior of centrally loaded concrete-filled steel tube short columns. *J. Struct. Eng.* **2004**, *130*, 180–188. [CrossRef]
3. Li, W.; Wang, T.; Han, L.H. Seismic performance of concrete-filled double skin steel tubes after exposure to fire: Experiments. *J. Constr. Steel Res.* **2019**, *154*, 209–223. [CrossRef]
4. Tao, Z.; Han, L.H.; Wang, D.Y. Strength and ductility of stiffened thin-walled hollow steel structural stub columns filled with concrete. *Thin-Walled Struct.* **2008**, *45*, 1113–1128. [CrossRef]
5. Uy, B. Local and post-local buckling of concrete filled steel welded box columns. *J. Constr. Steel Res.* **1998**, *47*, 47–72. [CrossRef]
6. Zheng, Y.Q.; Zeng, S.X. Design of L-shaped and T-shaped concrete-filled steel tubular stub columns under axial compression. *Eng. Struct.* **2020**, *207*, 110262. [CrossRef]
7. Zhang, Y.B.; Han, L.H.; Zhou, K.; Yang, S.T. Mechanical performance of hexagonal multi-cell concrete-filled steel tubular (CFST) stub columns under axial compression. *Thin-Walled Struct.* **2019**, *134*, 71–83. [CrossRef]
8. Zheng, J.L.; Wang, J.J. Concrete-filled steel tube arch bridges in China. *Engineering* **2018**, *4*, 143–155. [CrossRef]
9. Gao, Q.F.; Ma, Q.L.; Cui, K.M.; Li, J.; Xu, C. Numerical investigation of the characteristics of the dynamic load allowance in a concrete-filled steel tube arch bridge subjected to moving vehicles. *Shock. Vib.* **2020**, *2020*, 8819137. [CrossRef]
10. Yue, C.; Xiong, G.; Li, L.A.; Wang, S.B.; Li, H.Q.; Tong, J.W.; Mu, X.X. Reset monitoring for the cross—bridge of much across continuous beam. *J. Exp. Mech.* **2004**, *19*, 29–33.
11. Fan, X.J. The application of concrete filled steel tube in columns piers. *Shanxi Archit.* **2007**, *33*, 321–322.
12. Huang, Y.F.; Wu, Q.X.; Yuan, H.H. Vehicle vibration analysis of lightweight bridge with CFST composite truss girder and lattice pier. *J. Guangxi Univ. (Nat. Sci. Ed.)* **2018**, *43*, 1640–1650.
13. Ou, Z.J. The practice of concrete filled steel tube piers to bridges: A review. *Appl. Mech. Mater.* **2013**, *2685*, 1602–1605. [CrossRef]
14. Tomii, M.; Sakino, K. Experimental studies on the ultimate moment of concrete filled square steel tubular beam-columns. *Trans. Archit. Inst. Jpn.* **1979**, *275*, 55–65. [CrossRef]
15. Nakanishi, K.; Kitada, T.; Nakai, H. Experimental study on ultimate strength and ductility of concrete-filled steel columns under strong earthquake. *J. Constr. Steel Res.* **1999**, *51*, 297–319. [CrossRef]
16. Aval SB, B.; Saadeghvaziri, M.A.; Golafshani, A.A. Comprehensive composite inelastic fiber element for cyclic analysis of concrete-filled steel tube columns. *J. Eng. Mech.* **2002**, *128*, 428–437. [CrossRef]

17. Bruneau, M.; Marson, J. Seismic design of concrete-filled circular steel bridge piers. *J. Bridge Eng.* **2004**, *9*, 24–34. [CrossRef]
18. Li, J. *Dynamic Time-History Analysis and Cumulative Damage Assessment of Concrete-Filled Steel Tube Columns*; Hunan University: Changsha, China, 2016.
19. Wang, Z.F.; Zhang, H.; Zhang, M.J.; Sui, W.N.; Toshitaka, Y. Analytical study on seismic performance partially concrete-filled steel bridge piers. *J. Shenyang Jianzhu Univ. (Nat. Sci.)* **2008**, *24*, 949–953.
20. Chen, Z.Y.; Fang, H.; Zhu, L.; Mao, Y.F.; Liu, W.Q. Experimental tests and numerical simulations of circular reinforced concrete piers under ship impact. *Adv. Bridge Eng.* **2020**, *1*, 9–16. [CrossRef]
21. Chen, M.C.; Fang, W.; Huang, H.; Wang, C.; Liu, J.J. Axial compressive behavior of recycled concrete filled corroded circular steel tubular columns. *J. Build. Struct.* **2019**, *40*, 138–146.
22. Chen, M.C.; Wang, C.; Huang, H.; Liu, J.J. Study on mechanical properties of recycled concrete filled steel square tube under acid rain environment. *Concrete* **2016**, *12*, 1–12.
23. Chen, M.C.; Lin, B.Y.; Huang, H. Research on the bearing capacity of corroded square concrete filled steel tubular short column. *Steel Constr.* **2017**, *32*, 110–116.
24. Chen, M.C.; Wang, C.; Huang, H.; Liu, J.J. Experimental study on mechanical properties of the recycled concrete filled steel square tube short column under acid rain environment. *Build Struct.* **2017**, *47*, 35–40.
25. Zhang, F.J.; Xia, J.W.; Guo, L.; Guo, Z.; Chang, H.F.; Wang, K.J. Degradation of axial ultimate loading-bearing capacity of circular thin-walled concrete filled steel tubular stub columns after corrosion. *Materials* **2020**, *13*, 795. [CrossRef] [PubMed]
26. Yang, Y.F.; Cao, K.; Wang, T.Z. Experimental behavior of CFST stub columns after being exposed to freezing and thaw. *Cold Reg. Technol.* **2013**, *89*, 7–21. [CrossRef]
27. Wang, K.; Liu, W.H.; Chen, Y.; Wang, J.; Xie, Y.; Fang, Y.; Liu, C. Investigation on mechanical behavior of square CFST stub columns after being exposed to freezing and thawing. *J. Guangxi Univ. (Nat. Sci. Ed.)* **2017**, *42*, 1250–1255.
28. Yan, J.B.; Dong, X.; Wang, T. Axial compressive behaviours of square CFST stub columns at low temperatures. *J. Constr. Steel Res.* **2020**, *164*, 105812. [CrossRef]
29. Yan, J.B.; Dong, X.; Zhu, J.S. Behaviurs of stub steel tubular columns subjected to axial compression at low temperatures. *Constr. Build. Mater.* **2019**, *228*, 116788. [CrossRef]
30. Yan, J.B.; Dong, X.; Zhu, J.S. Compressive behaviours of CFST stub columns at low temperatures relevant to the Arctic environment. *Constr. Build. Mater.* **2019**, *223*, 503–519. [CrossRef]
31. Gao, S.; Peng, Z.; Guo, L.H.; Fu, F.; Wang, Y.G. Compressive behavior of circular concrete-filled steel tubular columns under freeze-thaw cycles. *J. Constr. Steel Res.* **2020**, *166*, 1–9. [CrossRef]
32. Li, G.; Hou, C.; Shen, L.M.; Yao, G.H. Performance and strength calculation of CFST columns with localized pitting corrosion damage. *J. Constr. Steel Res.* **2022**, *188*, 107011. [CrossRef]
33. Zhang, T.; Lyu, X.T.; Liu, H.Q.; Zhang, L.Q.; Wang, J.F.; Gao, S. Axial performance degradation of squared CFST stubs in severe cold and acid rain area. *Constr. Build. Mater.* **2020**, *282*, 120612. [CrossRef]
34. Gan, D. *Static and Seismic Behavior of Steel Tube Confined Concrete Short Columns*; Lanzhou University: Lanzhou, China, 2012.
35. Wu, B. *Experimental Research on Seismic Behavior of Tubed RC Short Columns with High Longitudinal Reinforcement Ratio*; Harbin Institute of Technology: Harbin, China, 2008.
36. Ye, J.X. *Experimental Research on Seismic Behavior of Steel Tube Confined Reinforced High-Strength Concrete Short Columns*; Harbin Institute of Technology: Harbin, China, 2010.
37. Zhou, X.H.; Liu, J.P. Seismic behavior and strength of tubed steel reinforced concrete (SRC) short columns. *J. Constr. Steel Res.* **2010**, *66*, 885–896. [CrossRef]
38. Wu, C.; Xu, S.C.; Zhao, Q.S.; Yang, Y.K. Experimental research on seismic behavior of ultrahigh-performance concrete filled square steel tubular short columns. *Ind. Constr.* **2019**, *49*, 188–194.
39. Wang, Y.T.; Cai, J.; Long, Y.L.; Yang, C.; Chen, Q.J.; Zuo, Z.L. Experimental study on seismic behavior of square concrete-filled steel tubular stub columns with binding bars. *J. Build. Struct.* **2015**, *36*, 18–25.
40. Wang, Y.T. *Seismic Behavior of Rectangular Concrete-Filled Steel Tubular Stub Columns with Binding Bars*; South China University of Technology: Guangzhou, China, 2015.
41. Liao, X.S. *Research on Seismic Behavior of Square CFST Columns with Binding Bars*; South China University of Technology: Guangzhou, China, 2014.
42. Wang, T.C.; Zhang, L.; Zhao, H.L.; Du, C. Analysis of parameter influence on seismic behavior of concrete filled square steel tubular columns. *J. Build. Struct.* **2013**, *34*, 339–344.
43. Li, L.M.; Li, N.; Chen, Z.H.; Jiang, X.L. Anti-seismic test on concrete-filled square steel tube column. *J. Jilin Univ. (Eng. Technol. Ed.)* **2008**, *38*, 817–822.
44. ABAQUS. *ABAQUS Standard User's Manual*; Version 6.13; Dassault Systemems Crop.: Providence, RI, USA, 2013.
45. Han, L.H. *Concrete Filled Steel Tubular Structures: Theory and Practice*, 3rd ed.; Science Press: Beijing, China, 2016.
46. Ministry of Housing and Urban-Rural Development of the People's Republic of China (MOHURD). *GB50010-2010: Code for Design of Concrete Structures*; China Architecture & Building Press: Beijing, China, 2011. (In Chinese)
47. American Concrete Institute (ACI). *ACI318-08: Building Code Requirements for Structural Concrete and Commentary*; ACI: Farmington Hills, MI, USA, 2008.

48. *EN-1992-1-1*; Eurocode 2-Design of Concrete Structures, Part 1-1: General Rules and Rules for Buildings. European Committee for Standardization: Brussels, Belgium, 2004.
49. Ministry of Housing and Urban-Rural Development of the People's Republic of China (MOHURD). *GB50017-2017: Standard for Design of Steel Structures*; China Architecture & Building Press: Beijing, China, 2018. (In Chinese)
50. Australian Standard (AS). *AS4100-2020: Steel Structures*; Australian Standard: Sydney, Australia, 2020.
51. Ministry of Housing and Urban-Rural Development of the People's Republic of China (MOHURD). *JGJ/T 101-2015: Specification for Seismic Test of Buildings*; China Architecture & Building Press: Beijing, China, 2015. (In Chinese)
52. Wen, Y.K. Method for random vibration of hysteretic systems. *J. Eng. Mech. Div.* **1976**, *102*, 249–263. [CrossRef]
53. Wen, Y.K. Equivalent linearization for hysteretic systems under random excitation. *J. Appl. Mech.* **1980**, *47*, 150–154. [CrossRef]
54. Nicolo, V.; Salvatore, S.; Francesco, M.; Luciano, R. A class of uniaxial phenomenological models for simulating hysteretic phenomena in rate-independent mechanical systems and materials. *Nonlinear Dyn.* **2018**, *93*, 1647–1669.
55. Nicolo, V.; Salvatore, S.; Luciano, R. A generalized class of uniaxial rate-independent models for simulating asymmetric mechanical hysteresis phenomena. *Mech. Syst. Signal Processing* **2021**, *146*, 106984.
56. Charalampakis, A.E.; Koumousis, V.K. Identification of Bouc-Wen hysteretic systems by a hybrid evolutionary algorithm. *J. Sound Vib.* **2008**, *314*, 571–585. [CrossRef]
57. Nicolo, V.; Salvotore, S.; Francesco, M.; Luciano, R. Nonlinear dynamic analysis of hysteretic mechanical systems by combining a novel rate-independent model and an explicit time integration method. *Nonlinear Dyn.* **2019**, *98*, 2879–2901.
58. Tao, Z.; Wang, Z.B.; Yu, Q. Finite element modeling of concrete-filled steel tube columns under axial compression. *J. Constr. Steel Res.* **2013**, *89*, 121–131. [CrossRef]
59. Wilfried, B.; Krazig, R.P. An elasto-plastic damage model for reinforced concrete with minimum number of material paramters. *Comput. Struct.* **2004**, *82*, 1201–1215.
60. Wang, Z.Q.; Yu, Z.W. Concrete damage model based on energy loss. *J. Build. Mater.* **2004**, *7*, 365–369.
61. Lyu, W.Q.; Han, L.H.; Hou, C. Axial compressive behaviour and design calculations on recycled aggregate concrete-filled steel tubular (RAC-FST) stub columns. *Eng. Struct.* **2021**, *241*, 112452. [CrossRef]
62. Han, L.H.; Yao, G.H.; Tao, Z. Performance of concrete-filled thin-walled steel tubes under pure torsion. *Thin-Walled Struct.* **2007**, *45*, 24–36. [CrossRef]
63. Yan, R. *Post-Fire Behaviors of Axially Loaded Reinforced Concrete Stub Columns Confined by Square Steel Tubes*; Harbin Institute of Technology: Harbin, China, 2017.
64. Zhang, F.M. *An Seismic Performance Research on Components of Square Concrete Filled Steel Tubes after the Simulated Acid Rain Corrosion*; East China Jiaotong University: Nanchang, China, 2016.
65. Chen, M.C.; Zhang, M.F.; Huang, H.; Wang, C. Study on seismic performance of concrete filled square steel tubes subjected to simulated acid rain attack. *J. China Railw. Soc.* **2018**, *44*, 106–114.
66. Yuan, F.; Chen, M.C.; Huang, H. Square CFST columns under cyclic load and acid rain attack: Experiments. *Steel Compos. Struct.* **2019**, *30*, 171–183.
67. Ministry of Housing and Urban-Rural Development of the People's Republic of China (MOHURD). *CJJ166-2011: Code for Seismic Design of Urban Bridges*; China Architecture & Building Press: Beijing, China, 2011. (In Chinese)
68. Zhang, T.; Lyu, X.T.; Yang, Y. Prediction and analysis of the residual capacity of concrete-filled steel tube stub columns under axial compression subjected to combined freeze-thaw cycles and acid rain corrosion. *Materials* **2019**, *12*, 3070. [CrossRef]
69. Ministry of Housing and Urban-Rural Development of the People's Republic of China (MOHURD). *GB50936-2014: Code for Design of Concrete Filled Steel Tubular Structures*; China Architecture & Building Press: Beijing, China, 2014. (In Chinese)

materials

Article

# A Mesoscale Study on the Dilation of Actively Confined Concrete under Axial Compression

Peng Chen [1,2,*], Xiaomeng Cui [2], Huijun Zheng [3] and Shengpu Si [2]

[1] Shaanxi Key Laboratory of Safety and Durability of Concrete Structures, Xijing University, Xi'an 710123, China
[2] Department of Engineering Mechanics, Hebei Key Laboratory of Mechanics of Intelligent Materials and Structures, Shijiazhuang Tiedao University, Shijiazhuang 050043, China
[3] Department of Construction Engineering, Hebei Petroleum University of Technology, Chengde 067000, China
* Correspondence: chenpeng@stdu.edu.cn

**Abstract:** The confinement of concrete enhances its strength and ductility by restraining lateral dilation. The accuracy of a confinement model depends on how well it captures the dilation of concrete. In the current paper, a mesoscale model is established to study the dilation properties of concrete in active confinement, where the heterogeneity of concrete is considered. The stress–strain and lateral–axial strain curves of concrete in active confinement were used to demonstrate the validity of the mesoscale model. Subsequently, the distribution of lateral strain and the influences of the strength grade and confinement ratio on the dilation of concrete were investigated in a simulation. The results show that the distribution of the lateral strain along the radial or longitudinal directions is not uniform on the specimen when compressive failure occurs. The confinement ratio has a more significant influence on the concrete's transverse dilation than the strength grade. Finally, an expression of the lateral–axial strain relationship of concrete in active confinement is proposed. The proposed formula can reflect the simulation results of the mesoscale model and is in good agreement with the prediction of existing formulas.

**Keywords:** dilation property of concrete in active confinement; secant strain ratio; lateral–axial strain relationship; mesoscale model of concrete

check for updates

**Citation:** Chen, P.; Cui, X.; Zheng, H.; Si, S. A Mesoscale Study on the Dilation of Actively Confined Concrete under Axial Compression. *Materials* **2022**, *15*, 6490. https://doi.org/10.3390/ma15186490

Academic Editor: Alessandro P. Fantilli

Received: 17 August 2022
Accepted: 17 September 2022
Published: 19 September 2022



## 1. Introduction

The mechanical properties of confined concrete have been extensively studied in recent years, illustrating that confinement can enhance concrete's strength and ductility by inhibiting transverse dilation [1–4]. Many practical design situations, such as the anchorage of pre-stressing reinforcement, bridge piles, containment vessels, and columns of high-rise buildings, require the prediction of concrete behavior under multiaxial stress states [5]. According to the loading path of confining pressure, the confinement can be divided into passive and active confinement [4,6]. In passively confined concrete, the confining pressure increases with the development of transverse dilation as well as restraining the transverse dilation of concrete. Thus, the key to predicting concrete properties in passive confinement lies in constructing a relationship between transverse dilation and confining pressure. Nevertheless, the development trend of confining pressure is greatly influenced by the type of confining material, which makes the study more complicated. In comparison to passively confined concrete, the force in actively confined concrete is relatively simple, in which the confining pressure remains constant. Therefore, it is more convenient to study the relationship between transverse dilation and confining pressure. Moreover, the relevant study can be applied to study the dilation of passively confined concrete by changing the confining pressure. The researchers have also adopted the same method to study the mechanical behavior of passively confined concrete [1,2,4,6–8].

The properties of concrete in active confinement have been experimentally studied recently [5,9–11]. However, the study focused more on the mechanical properties, while the study of dilation properties in the lateral direction is limited. Candappa et al. [12] tested the stress–lateral strain curves of concrete in active confinement. The analysis showed that the stress ratio versus the secant strain ratio was independent of the concrete strength and confinement ratio, and then a simplified formula was proposed. Binici [3] proposed a dilation model for concrete in active confinement, in which, the secant strain ratio remains constant in the elastic stage and tends to the limit value with the increase in the axial strain. Subsequently, another two lateral–axial strain models [13,14] were proposed by analyzing the experimental results of confined and unconfined concrete, in which the influence of confining pressure was explicitly introduced. The comparative analysis showed that the existing dilation models presented differences in form and the prediction results, which is caused by the selection of different test data. In addition, the measurement of strain presents a considerable discreteness, which is easily affected by the equipment, personnel, and testing method. Therefore, it is far from adequate to investigate the dilation property of confined concrete from the limited test data. The development of high-performance computers has facilitated mesoscopic simulations, which can consider the heterogeneous properties of concrete and the interaction among phases. The mesoscopic simulation can overcome the shortcomings of tests, such as the systematic error and the sensitive problem of strain measurement, which provide a new way to study the dilation behavior of concrete. To the best of the authors' knowledge, no relevant studies have been conducted from this perspective. As an important supplement to the experimental study, the mesoscopic simulation conducted in this paper expands our understanding of concrete dilation.

This paper aims to develop a new method for studying the dilation properties of concrete in active confinement. To accomplish this objective, the concrete is regarded as a combination of the mortar phase, aggregate phase, and interfacial transition zone (ITZ), and then, a numerical concrete model is developed according to the theory of random distribution. The validity of the mesoscale model is first verified, and then a parameter analysis is conducted to study the influences of strength grade and confinement ratio on the dilation properties. Based on the simulation results, the lateral–axial strain curve of the concrete column in active confinement is proposed, and then it is compared to the existing models presented in the literature.

## 2. Mesoscale Model of Concrete

### 2.1. Random Aggregate Structure

Considering the heterogeneity of concrete at the mesoscale, existing studies [15–20] have shown that mesoscopic simulation is an effective method for investigating concrete's mechanical performance. Thus, the dilation properties of concrete in active confinement were also studied using the same method.

The "take and place" method was used to generate aggregate particles, which was similar to the previous studies [21,22]. The volume content of coarse aggregate within each grading segment was determined according to the particle distribution curve proposed by Fuller [23]. Then, the coarse aggregate was randomly placed into the specimen's size range under the premise that it did not coincide with the previous aggregate. In reality, the content of coarse aggregate is about 40–45% [24]. However, according to the studies conducted by Donza et al. [25] and Meddah et al. [26], relatively small coarse aggregate particles (less than 5 mm) were assumed to be contained in mortar to reduce the computational cost. The same treatment and proportional content are also used in the literature [27,28], and good prediction results have been achieved. Therefore, the volume of coarse aggregate was set as 30% of the concrete volume in the current study. The maximum aggregate size was 25 mm. The coarse aggregate with a particle size less than 5 mm was considered as the mortar, together with hardened cement, fine aggregate, voids, and micro-cracks. To improve the computation efficiency, 2 mm mesh was utilized to describe the ITZ around the particles.

Figure 1 reflects the random aggregate structure of concrete established through the method described above, and the height-diameter ratio is two. In this figure, the gray zone represents the mortar phase, the red zone denotes the ITZ phase, and the green area represents the aggregate phase. During the simulation, the interaction between the concrete specimen and loading plate was set as surface-to-surface contact, and the movement of the lower loading plate was fixed. To investigate the dilation properties of concrete in active confinement, the radial confining pressure $p$ and vertical displacement $u$ were applied to the specimen and upper loading plate, respectively.

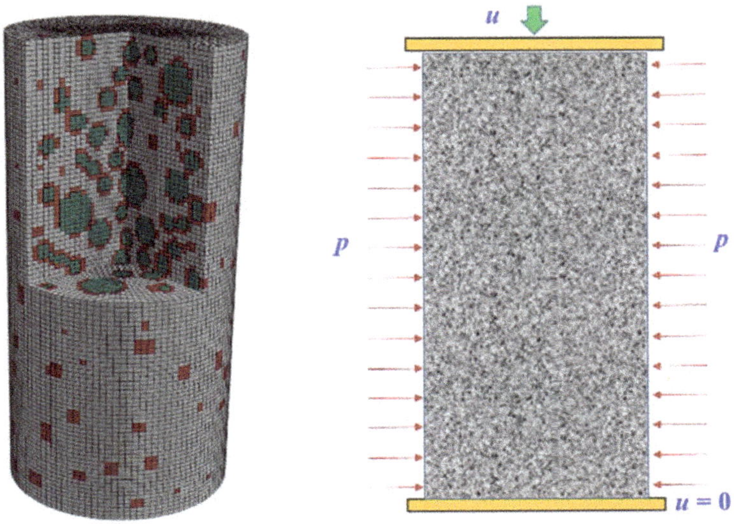


**Figure 1.** Random aggregate structure and loading method of concrete specimen.

### 2.2. Material Model

The damage plasticity model developed by Lee and Fenves [29] is extensively used in concrete definition, which has been incorporated into the software of ABAQUS. In this model, the failure of concrete can be divided into tension cracking and compression crushing, and two isotropic damage coefficients are introduced to describe the stiffness degradation of concrete. Figure 2 presents the damage plasticity model of concrete in uniaxial tension and uniaxial compression, respectively. In conditions of uniaxial tension, the stress–strain curve grows linear before reaching the tensile strength. Then, strain-softening occurs, and the tensile stiffness degrades. In conditions of uniaxial compression, the stress–strain curve is linear before reaching the proportional limit, then the plastic strain starts to develop, and the compressive stiffness begins to degrade. The relationship between stress and strain under uniaxial loading is defined by Equations (1) and (2) [29], in which, $d_t$, $\varepsilon_t$ and $\widetilde{\varepsilon}_t^{pl}$ represent the damage coefficient, the total strain, and plastic strain under axial tension, respectively. Similarly, the subscript c in the symbol represents the variables under axial compression.

$$\sigma_t = (1 - d_t)E_0\left(\varepsilon_t - \widetilde{\varepsilon}_t^{pl}\right) \tag{1}$$

$$\sigma_c = (1 - d_c)E_0\left(\varepsilon_c - \widetilde{\varepsilon}_c^{pl}\right) \tag{2}$$

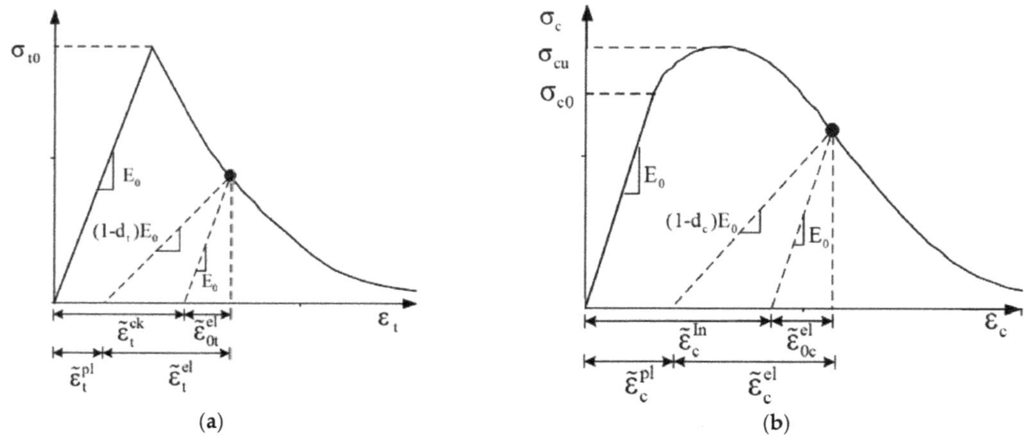

**Figure 2.** Concrete damage plasticity model. (**a**) Tensile stress–strain curve; (**b**) compressive stress–strain curve.

Grote [30] highlighted that concrete and mortar have similar mechanical properties, thus the plastic damage model is also suitable for the mortar phase. Due to the high porosity of the ITZ phase, it can be assumed to be the mortar phase with reduced strength. Similar to the previous studies [31–33], the coefficient of 0.8 was used to describe the strength relationship between the ITZ and mortar phases. The coarse aggregate was usually regarded as the elastic body in the mesoscopic simulation accounting for its high strength and modulus. In general, the key of the material model lies in the definition of the mortar phase.

Experimental results show that the elastic modulus, peak strain, and decreasing index of the mortar phase differ from those of concrete with the same strength [34]. A parameter analysis was conducted in the authors' previous work [35], and three correction constants were proposed to account for the differences. Finally, the constitutive relationship of the mortar phase is presented by Equations (3) and (4). It should be noted that the tensile failure of concrete was defined in terms of fracture energy, and the stress–cracking relation instead of the stress–strain relation was adopted to avoid mesh-size dependency [19,36].

$$y = \begin{cases} 2x - x^2 & x \leq 1 \\ \frac{xr}{r-1+x^r} & x > 1 \end{cases} \tag{3}$$

where $y = \sigma_m / f_m$, $x = \varepsilon_m / \varepsilon_0$, $\varepsilon_0 = \begin{cases} \varepsilon_{m0} + x(\varepsilon_{mc} - \varepsilon_0) & x \leq 1 \\ \varepsilon_{mc} & x > 1 \end{cases}$, $\varepsilon_{m0} = 1260 + 310\sqrt{f_m}$, $\varepsilon_{mc} = (1 + 15p/f_m)\varepsilon_{m0}$; $r = E_m/(E_m - E_p)$, $E_m = 4000\sqrt{f_m}$, $E_p = 1.44 f_m / \varepsilon_{m0}$.

$$\sigma_t = \begin{cases} f_t\left(1 - \frac{w}{w_0}\right) & w \leq w_0 \\ 0 & w > w_0 \end{cases} \tag{4}$$

where $w_0 = 2G_F/f_t$, $f_t = 1.4(f_m/10)^{2/3}$, $G_F = G_{F0}(f_m/10)^{0.7}$ [37].

In a word, the elastic and plastic damage models were used for the aggregate and mortar phases in the random aggregate structure. Meanwhile, the ITZ phase can be regarded as a mortar phase with reduced strength. The main parameters of the constitutive model include the compressive strength of mortar $f_m$, and the confining pressure $p$. Table 1 presents the mechanical properties of each phase.

**Table 1.** Parameters used in the mesoscale model.

| | Aggregate | Mortar Matrix | ITZ |
|---|---|---|---|
| Mass density (kg/m$^3$) | 2400 | 2400 | 2400 |
| Compressive strength | - | $f_m$ | $0.8f_m$ |
| Tensile strength | - | $1.4(f_m/10)^{2/3}$ | $1.12(f_m/10)^{2/3}$ |
| Elastic modulus (GPa) | 60 | $4000\sqrt{f_m}$ | $3200\sqrt{f_m}$ |
| Fracture energy (N/mm) | - | $G_{F0}(f_m/10)^{0.7}$ | $0.8G_{F0}(f_m/10)^{0.7}$ |
| Poisson's ratio | 0.20 | 0.20 | 0.20 |

*2.3. Strength Relationship between Concrete and Mortar*

To illustrate the accuracy of the mesoscale model, the stress–strain curves of concrete columns were compared between the simulation results and specification results, as presented in Figure 3a. The comparison indicates that the simulation results coincide well with the specification results, and the mesoscopic model is an efficient tool for investigating the properties of plain concrete.

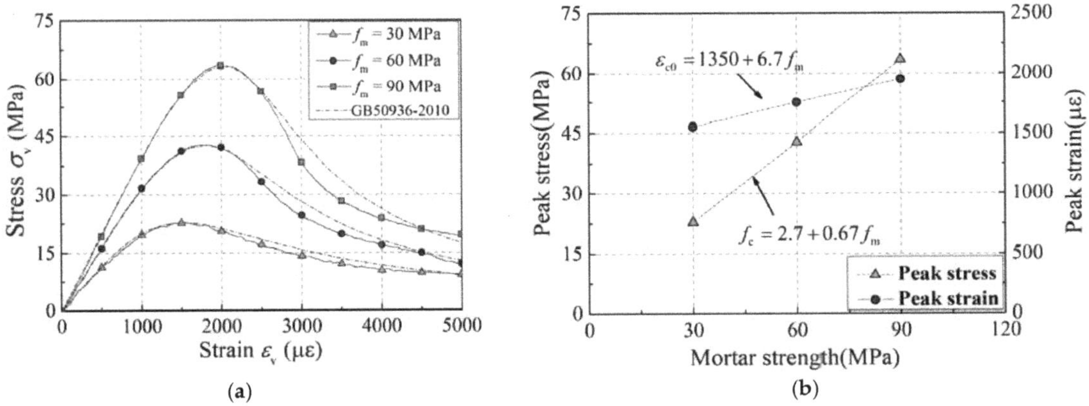


**Figure 3.** Influence of the mortar strength on performance of unconfined concrete. (**a**) Stress–strain curve; (**b**) strength relationship.

Figure 3b presents the strength relationship between concrete and mortar by analyzing the peak values of these curves. This figure indicates that both peak strain and peak stress of concrete are linearly related to the strength of the mortar phase, and the strength relationship is presented in Equation (5).

$$f_c = 2.7 + 0.67f_m \quad (5)$$

In this section, the numerical model of concrete with different strengths was established. The proposed model was efficient in describing the compressive stress–strain curve of unconfined concrete. Based on this model, it is valuable to study the dilation property of concrete in active confinement by applying the confining pressure.

## 3. Validation of the Mesoscale Model

*3.1. Stress–Strain Curve*

In order to verify the prediction accuracy of the mesoscale model on the performance of actively confined concrete, concrete specimens with four strengths, such as 35.8, 40, 50, and 60 MPa, were collected from the literature [5,9,11] to create a comparison with the simulation results, as presented in Figure 4. The dotted lines represent the experimental stress–strain curves of concrete under different confinement levels, while the solid lines represent the simulation results of the mesoscale model. One can notice that the peak strain, peak stress, and the ductility of concrete are enhanced with the increase in the confinement level. The development of axial strain in the test is slightly faster than that in the simulation, and this may be connected with the sensitivity and discreteness of the strain measurement. Fortunately, the maximum stress on the stress–strain curve in the simulation was consistent with the test result. Moreover, the numerical model can also reflect the development trend of the stress–strain curve in the test. Figure 5 presents the influence of the confinement ratio on the enhancement of peak stress and peak strain. In the figure, the scatter points represent the peak stress or peak strain of the numerical simulation, while the solid lines represent the calculation results of theoretical formulas presented in the literature. With the increase in the confinement ratio, the growth rate of the stress ratio decreases and presents a nonlinear development trend. However, the strain ratio increases linearly with the increase in the confinement ratio. It can be observed that the simulation results agree well with the theoretical formulas describing the stress and strain enhancements, presented by Equations (6) and (7), which were introduced by Mander et al. [38] and Jiang et al. [4], respectively. The above comparison illustrates the validity of the numerical model in forecasting the mechanical behavior of concrete in active confinement.

$$\frac{f_{cc}}{f_{co}} = 2.254\sqrt{1 + 7.94\frac{p}{f_{co}}} - 2\frac{p}{f_{co}} - 1.254 \tag{6}$$

$$\frac{\varepsilon_{cc}}{\varepsilon_{co}} = 1 + 17.5\frac{p}{f_{co}} \tag{7}$$

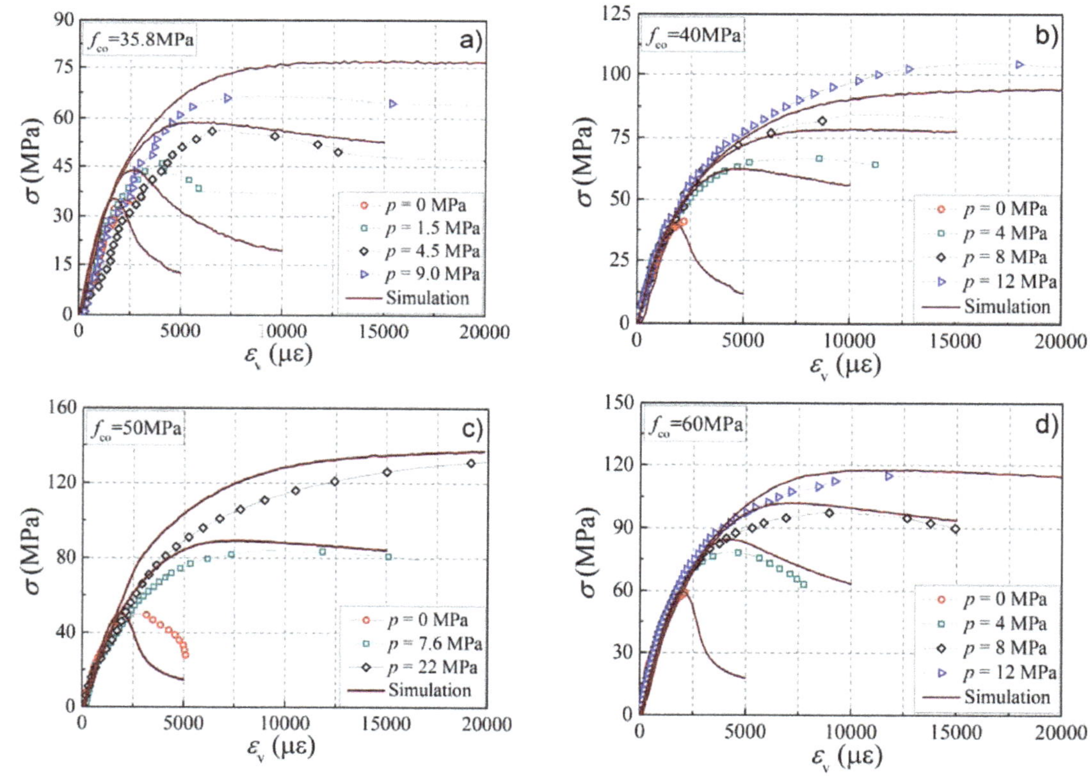


**Figure 4.** Comparison of the stress–strain curves between simulation and test results. (**a**) $f_{co}$ = 35.8 MPa; (**b**) $f_{co}$ = 40 MPa; (**c**) $f_{co}$ = 50 MPa; (**d**) $f_{co}$ = 60 MPa.

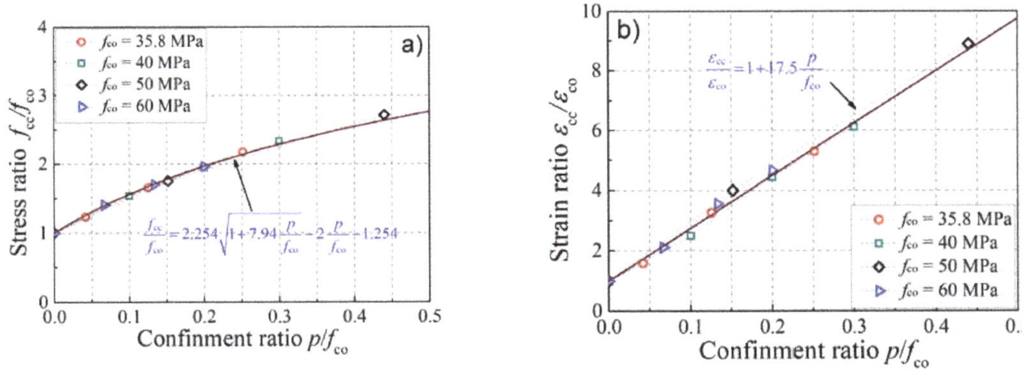


**Figure 5.** The enhancement of stress and strain in the simulation. (**a**) Stress enhancement; (**b**) strain enhancement.

### 3.2. Lateral–Axial Strain Curve

The lateral–axial strain relationship is an important curve reflecting the dilation properties of concrete. Nevertheless, the strain measurement presents considerable discreteness due to the influence of the tester, equipment, and strain gauge displacement. The numerical model previously established above can overcome the shortcomings of tests and provide an effective means for studying concrete's lateral–axial strain curve. Figure 6 compares the lateral–axial strain curves in the test and simulation. In the simulation, a two-step loading method was used. The confining pressure in the radial direction was first applied, and then the displacement in the longitudinal direction was applied. Therefore, the lateral strain was less than zero when the axial strain was zero. The figure presents that the lateral–axial strain curve tends to be flat with the increase in the confining pressure. Although some differences exist between the simulation and test results, the numerical model can reflect the development trend of the lateral–axial strain curve well. Thus, the mesoscale simulation can be regarded as an effective tool for studying the dilation properties of concrete in active confinement accounting for the sensitivity of the strain measurement.

The analysis described above shows that the proposed mesoscale model can make a good estimation on the stress–strain and lateral–axial strain curves of concrete in active confinement, illustrating that it is an efficient tool for investigating the mechanical and deformational behavior of concrete.

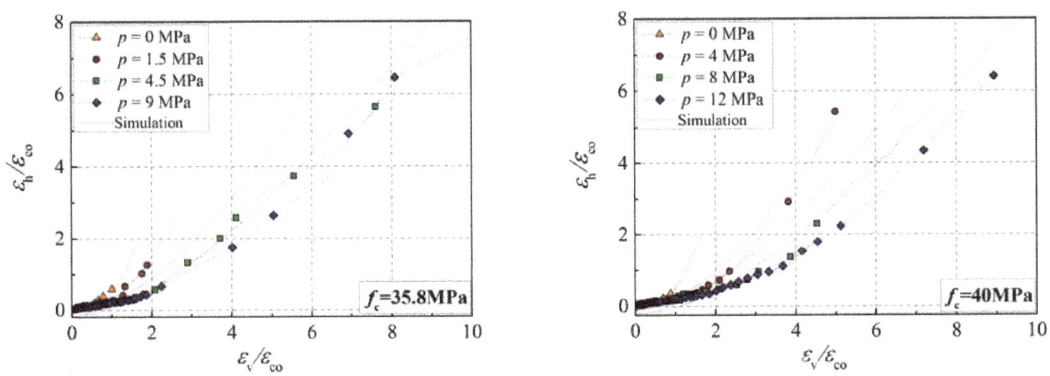


(**a**) (**b**)

**Figure 6.** *Cont.*

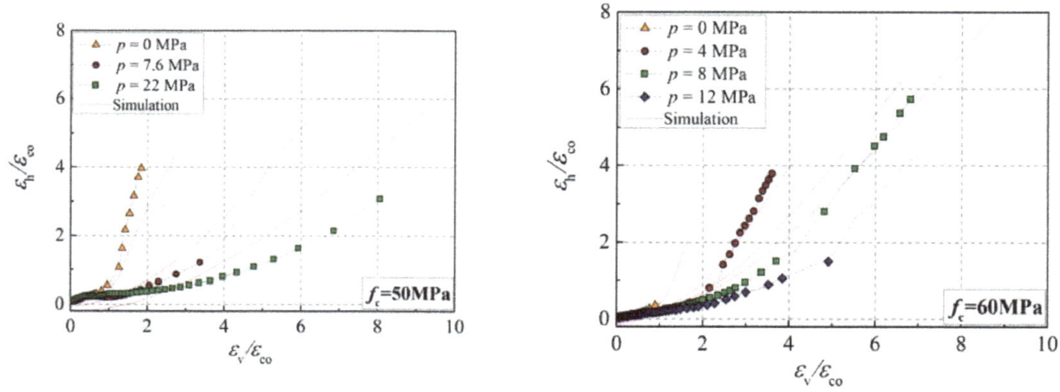


(c) (d)

**Figure 6.** Comparison of the lateral–axial strain curves between simulation and test results. (**a**) $f_{co}$ = 35.8 MPa; (**b**) $f_{co}$ = 40 MPa; (**c**) $f_{co}$ = 50 MPa; (**d**) $f_{co}$ = 60 MPa.

## 4. Parameter Analysis

To study the dilation properties of concrete in active confinement, the simulations of a concrete cylinder with different strengths and confinement ratios were conducted based on the mesoscale model. The strengths of the concrete cylinder were $f_c$ = 30 MPa, 40 MPa, 50 MPa, and 60 MPa. Moreover, the confinement ratios were $p/f_c$ = 0, 0.1 0.2, 0.3, and 0.4.

### *4.1. Failure Mode and Strain Distribution*

The failure modes of the concrete cylinder having different confinement ratios are presented in Figure 7, and the concrete's strength was 40 MPa. It can be observed from Figure 6 that shear failure occurs in the concrete cylinder with no confining pressure. With the increase in the confinement ratio, the specimen's failure mode changes from shear failure to expansion failure, illustrating that the ductility of concrete increases with the development of confinement. The difference in failure modes indicates that the dilation of plain concrete is localized, while the dilation of concrete with a high confinement ratio is more homogeneous.

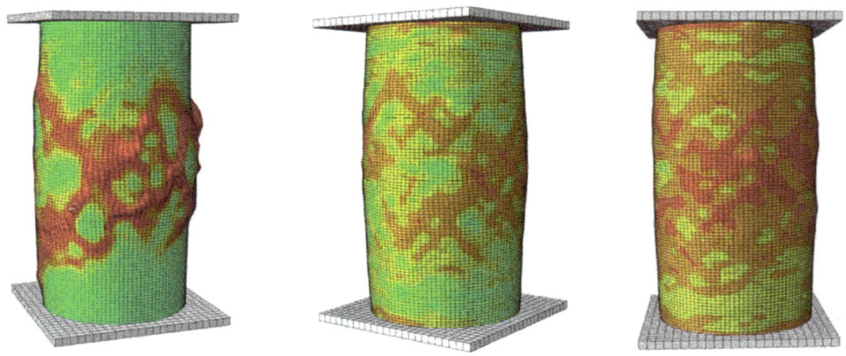

(a) (b) (c)

**Figure 7.** Influence of confinement ratio on the failure mode of a concrete cylinder. (**a**) $p/f_{co}$ = 0; (**b**) $p/f_{co}$ = 0.2; (**c**) $p/f_{co}$ = 0.4.

The failure of the concrete cylinder with no or low confinement ratios was mainly concentrated in the middle of the specimen, where the dilation was the most significant; thus, the measurement of the lateral strain in the test was focused here. To investigate the strain distribution along the longitudinal direction, nine sections were selected along the specimen's height in the simulation. The total height of the concrete cylinder was 200 mm, and the spacing between sections was 25 mm. Figure 8 presents the schematic diagram of the section location and strain calculation method.

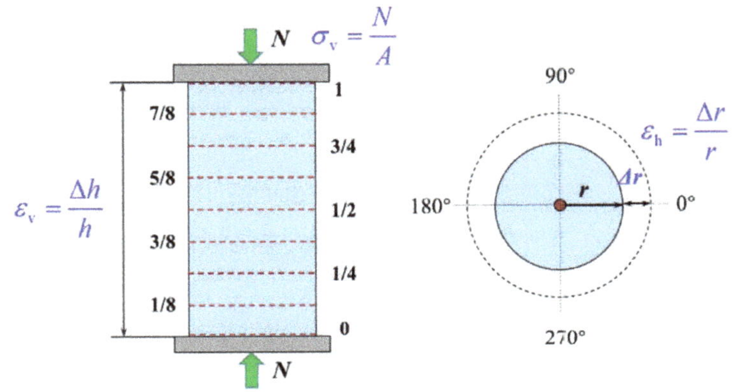


**Figure 8.** Schematic diagram of section location and strain calculation method.

Figure 9a presents the distribution of the lateral strain in the middle section (1/2 height). It can be observed that the lateral strain is uniformly distributed in a circumferential direction as the axial strain ratio, $\varepsilon_v/\varepsilon_{co}$, is set as 1.75 or 3.5. Nevertheless, the distribution of the lateral strain begins to become non-uniform as the axial strain ratio, $\varepsilon_v/\varepsilon_{co}$, exceeds 5.25. The peak strain of the confined concrete is about 4.5 times that of plain concrete when the confinement ratio is set as 0.2 according to Equation (5), namely, $\varepsilon_{cc}/\varepsilon_{co} = 4.5$. Thus, it can be concluded that the transverse dilation of concrete is not uniformly distributed as the axial strain exceeds the peak strain of confined concrete. A similar phenomenon was also observed in plain concrete in our previous work [39]. Figure 9b describes the development of lateral strain along the longitudinal direction; the concrete's strength is 40 MPa and the confinement ratio is set as 0.2. It can be observed that the lateral strain is not uniformly distributed along the specimen's height, which is almost in a sinusoidal form.

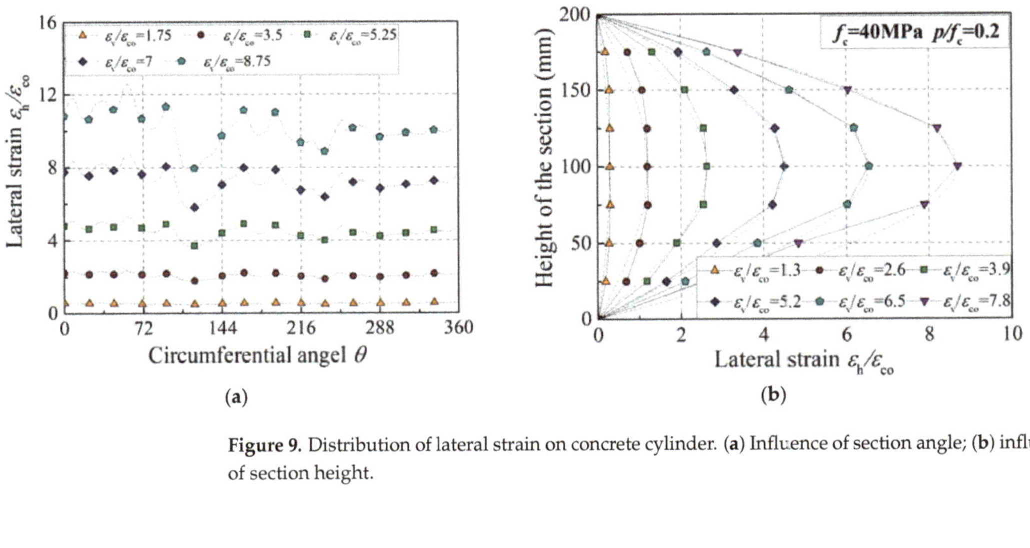


**Figure 9.** Distribution of lateral strain on concrete cylinder. (**a**) Influence of section angle; (**b**) influence of section height.

Generally, the distribution of the lateral strain along the circumference and the height of the specimen is not uniform when compressive failure occurs in concrete. Therefore, great discreteness exists in the strain measurement in the tests. In order to reduce the influence of discreteness, the average strain was used in the analysis of the simulation results. Specifically, the lateral strain at each section is the average of 100 points along the circumferential direction, and the average strain at several sections can be regarded as the specimen's lateral strain. Considering that the failure of the concrete cylinder with no or low confinement ratio was prone to local damage, while the failure of the specimen with a high confinement ratio was relatively uniform, the difference was considered in the current study when obtaining the average value. By trial and error and comparing with the test data, the authors observed that if 3, 5, and 7 sections were selected for the specimens with low (0 or 0.1), middle (0.2), and high (0.3 or 0.4) confinement ratios, respectively, presented in Figure 10, the simulation would produce a satisfactory estimation. The zone proportions were 1/4, 1/2, and 3/4 of the specimen height, and the average values were approximately 0.97, 0.90, and 0.78 of the maximum value in the middle section.

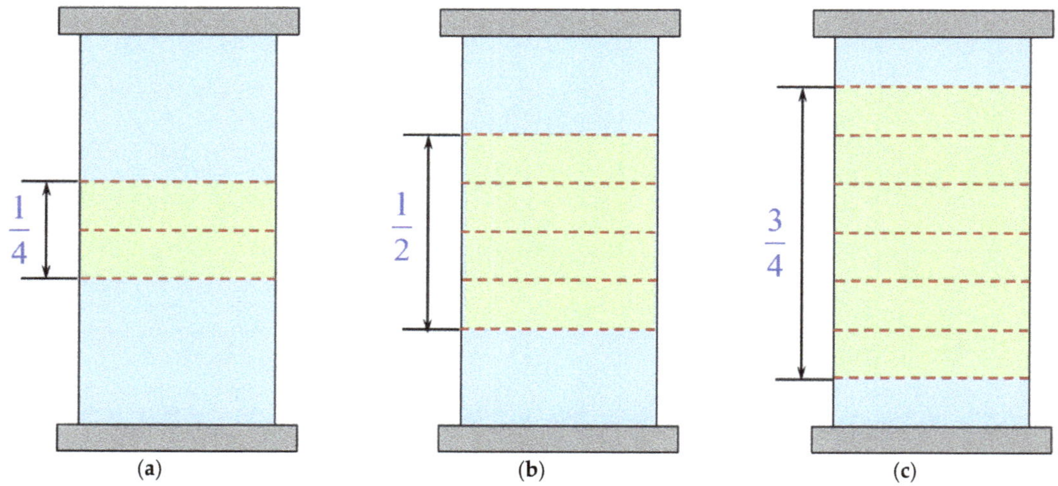


**Figure 10.** Average zone of the specimen with different confinement ratios. (**a**) $p/f_{co}$ = 0 or 0.1; (**b**) $p/f_{co}$ = 0.2; (**c**) $p/f_{co}$ = 0.3 or 0.4.

## 4.2. Secant Strain Ratio of The Specimen

Figure 11 presents the development of the lateral strain of the specimen with a confinement ratio of 0.2, and the concrete's strength is 40 MPa. Deformations in the lateral direction are usually described by the secant strain ratio, $\mu_s$ ($\mu_s = \varepsilon_h/\varepsilon_v$), which is also plotted in this figure. In the elastic stage, the secant strain ratio of unconfined concrete equaled the Poisson's ratio, ranging from 0.15 to 0.20. The initial secant strain ratio of the specimen mentioned above was about 0.12, which was slightly less than the Poisson's ratio of plain concrete, because the secant strain ratio is influenced by confining pressure. It should be noticed that the growth rate gradually decreased with the development of the axial strain. The strength grade and confinement ratio were important parameters affecting the properties of the concrete, thus, the influences of these two parameters on the dilation properties of the concrete in active confinement were analyzed in the following sections.

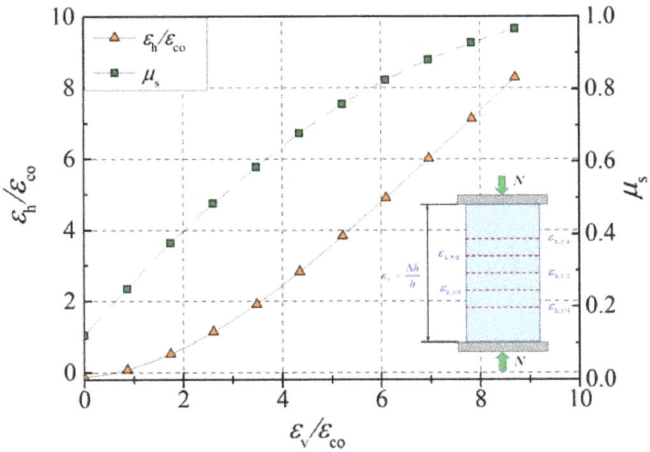


**Figure 11.** Transverse dilation of concrete cylinder with a confinement ratio of 0.2.

*4.3. Influence of the Compressive Strength*

Figure 12 presents the influence of strength grade on the secant strain ratio of plain concrete. It can be observed that the development of the secant strain ratio is generally divided into three stages. The secant strain ratio remains constant in the first stage and equals the Poisson's ratio of the concrete, as the micro-cracks slowly develop in this stage. It should be noticed that the platform segment was affected by the strength of the concrete; the higher the strength, the wider the platform segment. This was because the high-strength concrete had a greater range of elastic segments. In the second stage, the value and the growth rate continue to increase as the axial strain increases due to the accumulation of internal cracks. As the cracks develop slowly in the high-strength concrete specimen, the secant strain ratio is relatively smaller. In the third stage, the growth rate gradually decreases, although the secant strain ratio continues to increase. Due to the brittleness of the high-strength concrete, its secant strain ratio becomes larger than that of the low-strength concrete in this stage. The secant strain ratio is about 0.5 as the axial strain approaches the peak strain, which was demonstrated by the studies in the literature [3,7,40].

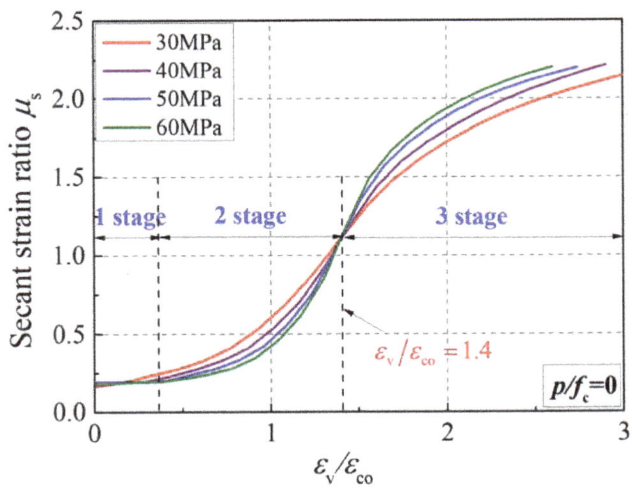


**Figure 12.** Influence of strength grade on secant strain of plain concrete.

Figure 13 reflects the influence of strength grade on the secant strain ratio of concrete in active confinement. One can observe that its value continues to increase as the axial strain increases. Similar to plain concrete, the growth rate of the secant strain ratio of actively confined concrete increases during the early stage, and slows down later on. The influence of strength grade on the secant strain ratio of concrete in active confinement is slightly different from that of plain concrete. Specifically, the influence of strength on the secant strain ratio of confined concrete is monotonic. The greater the strength, the smaller the secant strain ratio. This is because the greater the concrete's strength, the higher the confining pressure as the confinement ratio is constant, inhibiting the brittle failure of the high-strength concrete.

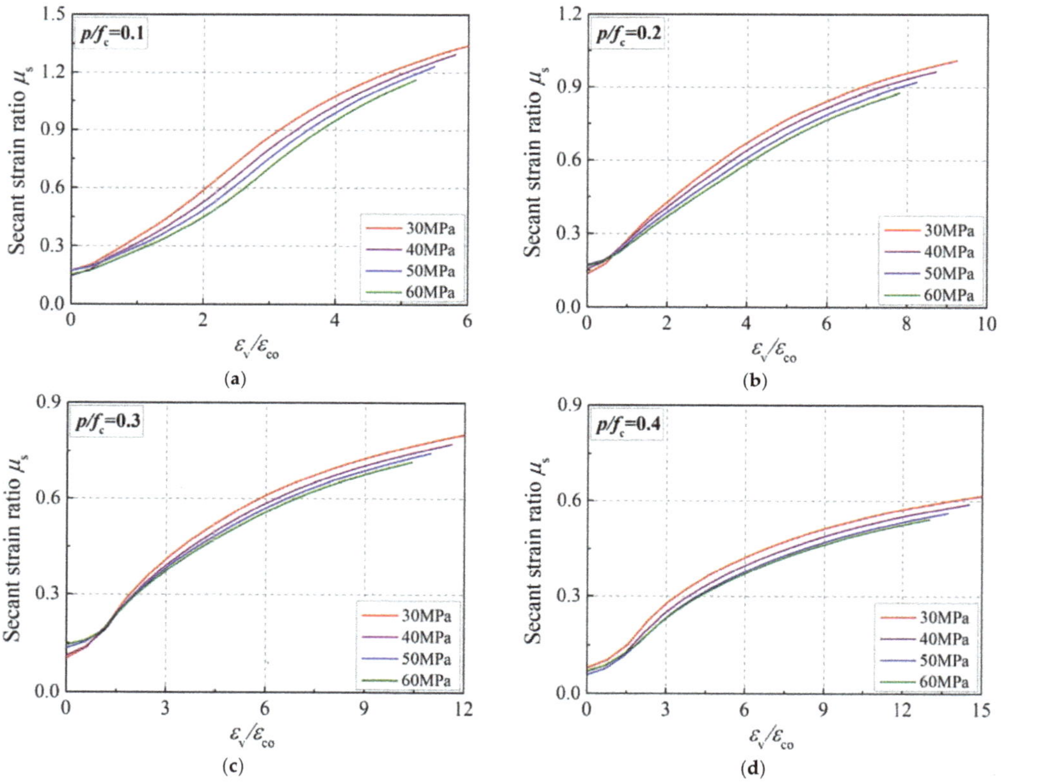


**Figure 13.** Influence of strength grade on secant strain ratio of concrete in active confinement. (**a**) $p/f_c = 0.1$; (**b**) $p/f_c = 0.2$; (**c**) $p/f_c = 0.3$; (**d**) $p/f_c = 0.4$.

Overall, the strength grade influenced the development of concrete's secant strain ratio, especially for concrete specimens with no or low confinement ratios.

## *4.4. Influence of the Confinement Ratio*

The influence of the confinement ratio on the dilation of the specimen with different strengths is similar. Figure 14a presents the simulation results of the concrete specimen with a strength of 40 MPa. With the increase in the confinement ratio, the transverse dilation is inhibited, and then the secant strain ratio slowly develops. Figure 14b presents the influence of the confinement ratio on the secant strain ratio at zero strain and $3\varepsilon_{co}$. When the axial strain is three times the peak strain of plain concrete, the secant strain ratio is 2.25, 0.81, 0.54, 0.40, and 0.25 as the confinement ratio is set as 0, 0.1, 0.2, 0.3, and 0.4, respectively. Moreover, the initial secant strain ratio is 0.177, 0.156, 0.130, 0.113, and 0.070 when the confinement ratio increases from 0 to 0.4, illustrating that the confining pressure can inhibit the development of the secant strain ratio. The analysis indicates that the confinement ratio is an important parameter affecting the transverse dilation of concrete, which should be considered in the formula.

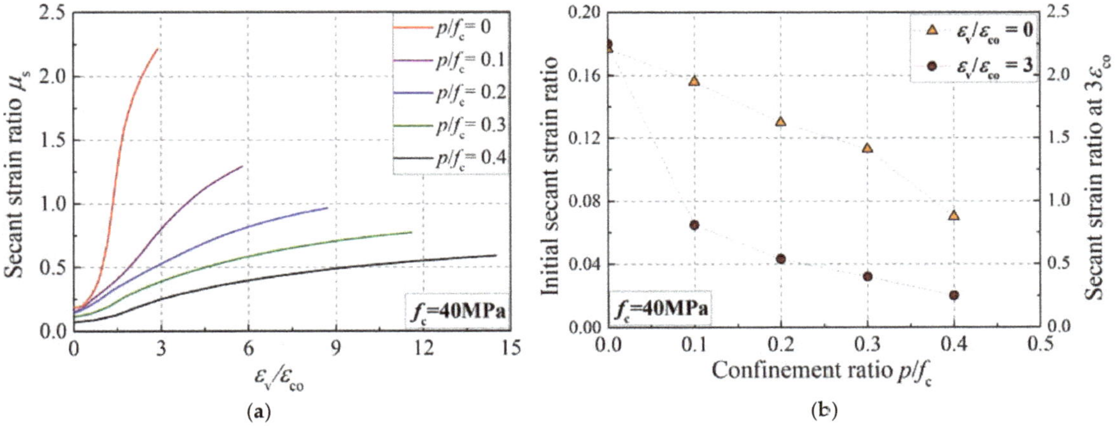


**Figure 14.** Influence of confinement ratio on secant strain ratio of concrete cylinder. (**a**) Development of secant strain ratio; (**b**) secant strain ratio at zero strain and $3\varepsilon_{co}$.

*4.5. Dilation Model for Actively Confined Concrete*

The simulation results show that the confinement ratio has a more significant influence on transverse dilation than that of concrete's strength, as presented in Figure 15a. Considering the weak effect of concrete's strength, the average curves were adopted to establish the dilation formula of actively confined concrete, which is presented in Figure 15b. The dotted lines with colors of red, purple, blue, green and black in Figure 15b represent the equation results for the specimens with confinement ratios of 0, 0.1, 0.2, 0.3 and 0.4, respectively. The key to predicting the secant strain ratio lies in the establishment of the lateral–axial strain relationship of concrete. Moreover, it is crucial to express the axial strain as an explicit function of the lateral strain, because it is more convenient to calculate the confining pressure and axial strain. Otherwise, it is not feasible.

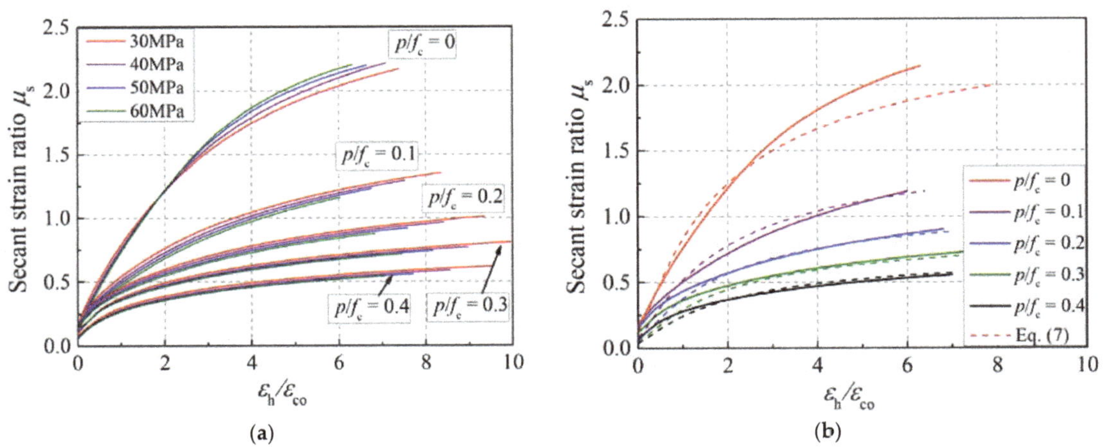


**Figure 15.** Influence of strength grade and confinement ratio on secant strain ratio. (**a**) Strength grade; (**b**) confinement ratio.

Figure 16a presents the lateral–axial strain curves of concrete cylinders with varying confinement ratios. The dotted lines with colors of red, purple, blue, green and black in Figure 16a represent the equation results for the specimens with confinement ratios of 0, 0.1, 0.2, 0.3 and 0.4, respectively. According to the study in the literature [13], the following expression was adopted to describe the lateral–axial strain relationship of concrete in active confinement, as presented by Equation (8).

$$\frac{\varepsilon_{\mathrm{v}}}{\varepsilon_{\mathrm{co}}} = \left(\alpha + \beta\frac{p}{f_{\mathrm{c}}}\right) A\left\{\left[1 + B\left(\frac{\varepsilon_{\mathrm{h}}}{\varepsilon_{\mathrm{co}}}\right)\right]^{C} - \exp\left[D\left(\frac{\varepsilon_{\mathrm{h}}}{\varepsilon_{\mathrm{co}}}\right)\right]\right\} \tag{8}$$

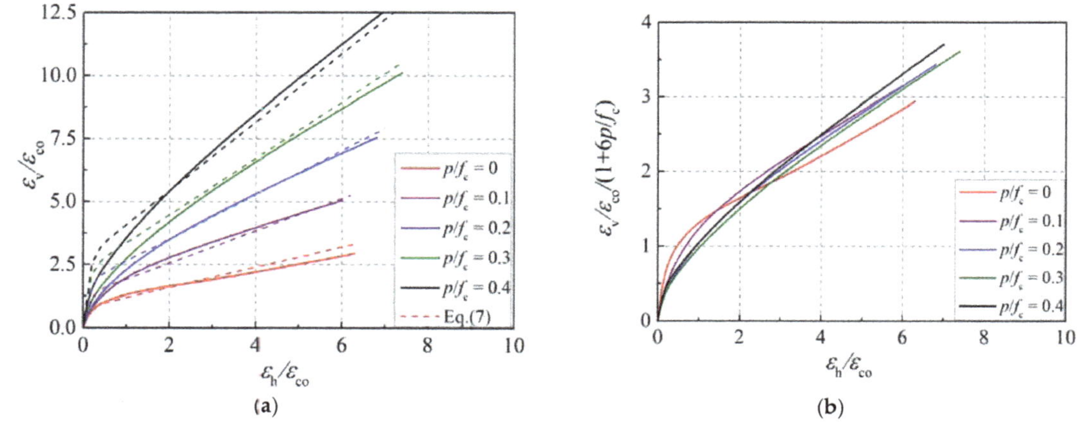


**Figure 16.** Lateral–axial strain curves of concrete with varying confinement ratios. (**a**) Lateral–axial strain curves; (**b**) equivalent lateral–axial strain curves.

As the formula can apply to unconfined concrete, the value of parameter $\alpha$ equals 1.0. The confinement ratio $p/f_c$ can be set as zero for unconfined concrete. Previous studies [3,7,39] have shown that the secant strain ratio is approximately 0.5 as the axial strain approaches the peak strain of unconfined concrete, which needs to be considered in the formula. The coefficients A, B, C, and D can be determined according to the simulation results of unconfined concrete. By analyzing the lateral–axial strain curve in the simulation, the coefficients A, B, C, and D equal to 0.8, 0.5, 1.0, and −9, respectively. Coefficient $\beta$ reflects the influence of the confinement ratio, and a satisfactory prediction can be achieved when it is set as 6.0. Figure 16b presents the equivalent lateral–axial strain relationship of the concrete cylinder. The curves present a slight difference after considering the equivalent reduction in the confinement ratio. Therefore, the final expression of the lateral–axial strain relationship of the concrete in active confinement is presented by Equation (9).

$$\frac{\varepsilon_{\mathrm{v}}}{\varepsilon_{\mathrm{co}}} = 0.8\left(1 + 6\frac{p}{f_{\mathrm{c}}}\right)\left\{1 + 0.5\left(\frac{\varepsilon_{\mathrm{h}}}{\varepsilon_{\mathrm{co}}}\right) - \exp\left[-9\left(\frac{\varepsilon_{\mathrm{h}}}{\varepsilon_{\mathrm{co}}}\right)\right]\right\} \tag{9}$$

To illustrate the performance of the proposed lateral strain equation, it was compared with the simulation results presented in Figures 15b and 16a. Moreover, the lateral–axial strain curves proposed by Binici et al. [3], Teng et al. [13], and Lim et al. [14] were utilized to have a comparison, as shown in Figure 17. It can be observed that Lim's model can produce a good estimation of the dilation of confined concrete, but it overestimates the dilation of plain concrete. Although Binici's model can reflect the dilation behavior of the concrete cylinder, the influence of the confinement ratio on the initial secant strain ratio was not considered. Teng's model can produce a good estimation of the development of the secant strain ration in the simulation.

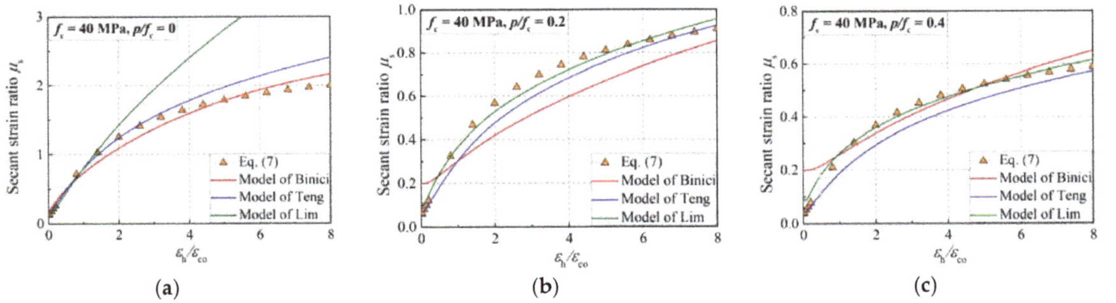


**Figure 17.** Secant strain ratio prediction of the proposed and existing formulas. (**a**) $p/f_c$ = 0; (**b**) $p/f_c$ = 0.2; (**c**) $p/f_c$ = 0.4.

In general, the proposed lateral strain formula can reflect the simulation results of the mesoscale model, and it is consistent with the prediction of the existing formulas. Therefore, it is an effective formula for predicting the dilation properties of actively confined concrete.

## 5. Conclusions

A mesoscale concrete model was established to investigate the dilation properties of actively confined concrete in the current study. The following conclusions were formulated through the analysis:

1. The simulation results of the mesoscale model were in good agreement with the stress–strain curves of the plain concrete from codes, based on which the strength relationship between the concrete and mortar was established.
2. The mesoscale model made a good estimation of the stress–strain and lateral–axial strain curves of the concrete in active confinement, illustrating that it was an efficient tool for investigating the concrete's mechanical and deformational behavior.
3. The distribution of the lateral strain along the circumference and the height of the specimen was not uniform when compressive failure occurred in the concrete.
4. The confinement ratio had a more significant influence on the transverse dilation than that of the concrete's strength. With the increase in the confinement ratio, the transverse dilation was inhibited, and then the concrete's secant strain ratio slowly developed.
5. An expression of the lateral–axial strain curve of the concrete in active confinement was proposed, which can reflect the simulation results of the mesoscale model, and it was consistent with the prediction of the existing formulas.

Mesoscopic analysis can reflect the heterogeneity of concrete and the interaction among different phases, which is an efficient method to study the mechanical and deformation behavior of concrete. At present, the content of our research is limited, and we anticipate conducting further research using this approach in the future, such as the influence of nano silica or fiber on the performance of concrete, and the interaction mechanisms of the steel tube and concrete core.

**Author Contributions:** Conceptualization, P.C.; methodology, P.C.; software, X.C.; validation, X.C.; formal analysis, P.C. and H.Z.; investigation, H.Z.; writing—original draft preparation, S.S.; writing—review and editing, H.Z.; visualization, S.S.; supervision, H.Z.; funding acquisition, P.C. All authors have read and agreed to the published version of the manuscript.

**Funding:** This research was funded by the National Natural Science Foundation of China (grant number 51908381), the Natural Science Foundation of Hebei Province (E2022210080), and the program from Shaanxi Key Laboratory of Safety and Durability of Concrete Structures (XJKFJJ201905).

**Institutional Review Board Statement:** Not applicable.

**Informed Consent Statement:** Not applicable.

**Data Availability Statement:** Not applicable.

**Conflicts of Interest:** The authors declare no conflict of interest.

## References


1. Mirmiran, A.; Shahawy, M. Dilation characteristics of confined concrete. *Mech. Cohesive-Frict. Mater.* **1997**, *2*, 237–249. [CrossRef]
2. Mirmiran, A.; Shahawy, M. A new concrete-filled hollow FRP composite column. *Compos. B. Eng.* **1996**, *27*, 263–268. [CrossRef]
3. Binici, B. An analytical model for stress-strain behavior of confined concrete. *Eng. Struct.* **2005**, *27*, 1040–1051. [CrossRef]
4. Jiang, T.; Teng, J.G. Analysis-oriented stress-strain models for FRP-confined concrete. *Eng. Struct.* **2007**, *29*, 2968–2986. [CrossRef]
5. Sfer, D.; Carol, I.; Gettu, R.; Etse, G. Study of the behavior of concrete under triaxial compression. *J. Eng. Mech.* **2002**, *128*, 156–163. [CrossRef]
6. Harries, K.A.; Kharel, G. Experimental investigation of the behavior of variably confined concrete. *Cem. Concr. Res.* **2003**, *33*, 873–880. [CrossRef]
7. Spoelstra, M.R.; Monti, G. FRP-confined concrete model. *J. Compos. Constr.* **1999**, *3*, 143–150. [CrossRef]
8. Fam, A.Z.; Rizkalla, S.H. Confinement model for axially loaded concrete confined by circular fiber-reinforced tubes. *Struct. J.* **2001**, *98*, 451–461.
9. Lahlou, K.; Aitcin, P.C.; Chaallal, O. Behavior of high-strength concrete under confined stresses. *Cem. Concr. Compos.* **1992**, *14*, 185–193. [CrossRef]
10. Imran, I.; Pantazopoulou, S.J. Experimental study of plain concrete under triaxial stress. *ACI Mater. J.* **1996**, *93*, 589–601. [CrossRef]
11. Lu, X.; Hsu, C.T. Stress-strain relations of high-strength concrete under triaxial compression. *J. Mater. Civ. Eng.* **2007**, *19*, 261–268. [CrossRef]
12. Candappa, D.C.; Sanjayan, J.G.; Setunge, S. Complete triaxial stress-strain curves of high-strength concrete. *J. Mater. Civil Eng.* **2001**, *13*, 209–215. [CrossRef]
13. Teng, J.; Huang, Y.L.; Lam, L.; Ye, L.P. Theoretical model for fiber-reinforced polymer-confined concrete. *J. Compos. Constr.* **2007**, *11*, 201–210. [CrossRef]
14. Lim, J.C.; Ozbakkaloglu, T. Lateral strain-to-axial strain relationship of confined concrete. *J. Struct. Eng.* **2015**, *141*, 04014141. [CrossRef]
15. Nguyen, D.; Lawrence, C.; Borderie, C.; Matallah, M.; Nahas, G. A mesoscopic model for a better understanding of the transition from diffuse damage to localized damage. *Eur. J. Environ. Civ. Eng.* **2010**, *14*, 751–776. [CrossRef]
16. Caballero, A.; Carol, I.; Lopez, C.M. 3D meso-mechanical analysis of concrete specimens under biaxial loading. *Fatigue Fract. Eng. Mater. Struct.* **2007**, *30*, 877–886. [CrossRef]
17. Dupray, F.; Malecot, Y.; Daudeville, L.; Buzaud, E. A mesoscopic model for the behavior of concrete under high confinement. *Int. J. Num. Methods Geomech.* **2009**, *33*, 1407–1423. [CrossRef]
18. Zhang, Y.; Chen, Q.; Wang, Z.; Zhang, J.; Wang, Z.; Li, Z. 3D mesoscale fracture analysis of concrete under complex loading. *Eng. Fract. Mech.* **2019**, *220*, 106646. [CrossRef]
19. Jin, L.; Du, M.; Li, D.; Du, X.; Xu, H. Effects of cross section size and transverse rebar on the behavior of short squared RC columns under axial compression. *Eng. Struct.* **2017**, *142*, 223–239. [CrossRef]
20. Jin, L.; Yu, W.; Du, X.; Yang, W. Dynamic size effect of concrete under tension: A numerical study. *Int. J. Impact Eng.* **2019**, *132*, 103318. [CrossRef]
21. Wriggers, P.; Moftah, S.O. Mesoscale models for concrete: Homogenisation and damage behavior. *Finite Elem. Anal. Des.* **2006**, *42*, 623–636. [CrossRef]
22. Du, X.; Jin, L.; Ma, G. Numerical simulation of dynamic tensile-failure of concrete at mesoscale. *Int. J. Impact Eng.* **2014**, *66*, 5–17. [CrossRef]
23. Fuller, W.B.; Thompson, S.E. The laws of proportioning concrete. *Trans. Am. Soc. Civ. Eng.* **1907**, *59*, 67–143. [CrossRef]
24. Jin, L.; Xia, H.; Jiang, X.; Du, X. Size effect on shear failure of CFRP-strengthened concrete beams without web reinforcement: Mesoscale simulation and formulation. *Compos. Struct.* **2020**, *236*, 111895. [CrossRef]
25. Donza, H.; Cabrera, O.; Irassar, E.F. High-strength concrete with different fine aggregate. *Cem. Concr. Res.* **2022**, *32*, 1755–1761. [CrossRef]
26. Meddah, M.S.; Zitouni, S.; Belâabes, S. Effect of content and particle size distribution of coarse aggregate on the compressive strength of concrete. *Constr. Build. Mater.* **2010**, *24*, 505–512. [CrossRef]
27. Jin, L.; Fan, L.; Li, P.; Du, X. Size effect of axial-loaded concrete-filled steel tubular columns with different confinement coefficients. *Eng. Struct.* **2019**, *198*, 109503. [CrossRef]
28. Jin, L.; Chen, H.; Wang, Z.; Du, X. Size effect on axial compression failure of CFRP-wrapped square concrete columns: Tests and simulations. *Compos. Struct.* **2020**, *254*, 112843. [CrossRef]
29. Lee, J.; Fenves, G.L. Plastic-damage model for cyclic loading of concrete structures. *J. Eng. Mech.* **1998**, *124*, 892–900. [CrossRef]
30. Grote, D.L.; Park, S.W.; Zhou, M. Dynamic behavior of concrete at high strain rates and pressures: I. experimental characterization. *Int. J. Impact Eng.* **2001**, *25*, 869–886. [CrossRef]
31. Mondal, P.; Shah, S.P.; Marks, L.D. *Nanomechanical Properties of Interfacial Transition Zone in Concrete*; Nanotechnology in Construction 3; Springer: Berlin/Heidelberg, Germany, 2009; pp. 315–320.

32. Chen, H.; Xu, B.; Mo, Y.L.; Zhou, T. Behavior of mesoscale heterogeneous concrete under uniaxial tensile and compressive loadings. *Constr. Build. Mater.* **2018**, *178*, 418–431. [CrossRef]
33. Prokopski, G.; Halbiniak, J. Interfacial transition zone in cementitious materials. *Cem. Concr. Res.* **2000**, *30*, 579–583. [CrossRef]
34. Su, J. *The Research on the Size Effect of Concrete Behavior in Compression and Tension*; Hunan University: Hunan, China, 2013. (In Chinese)
35. Chen, P.; Liu, J.; Cui, X.; Si, S. Mesoscale analysis of concrete under axial compression. *Constr. Build. Mater.* **2022**, *337*, 127580. [CrossRef]
36. Huang, Y.J.; Yang, Z.J.; Chen, X.W.; Liu, G.H. Monte Carlo simulations mesoscale dynamic compressive behavior of concrete based on X-ray computed tomography images. *Int. J. Impact Eng.* **2016**, *97*, 102–115. [CrossRef]
37. CEB. *CEB-FIP Model Code 1990*; CEB Bulletin D'information: London, UK, 1990; pp. 35–37.
38. Mander, J.B.; Priestley, M.J.; Park, R. Theoretical stress-strain model for confined concrete. *J. Struct. Eng.* **1988**, *114*, 1804–1826. [CrossRef]
39. Chen, P.; Cui, X.; Zheng, H.; Si, S. Mesoscale study on dilation behavior of plain concrete under axial compression. *Buildings* **2022**, *12*, 908. [CrossRef]
40. Marques, S.P.; Marques, D.C.; Lins, S.J.; Cavalcante, M.A. Model for analysis of short columns of concrete confined by fiber-reinforced polymer. *J. Compos. Constr.* **2004**, *8*, 332–340. [CrossRef]

materials

MDPI

*Article*

# Experimental Study on Long-Term Mechanical Properties of Prestressed Glulam Continuous Beams

**Nan Guo [1], Shouting Zhou [1], Yan Zhao [2], Lidan Mei [1,*] and Yunan Zhang [1]**

[1] Department of Civil Engineering, Northeast Forestry University, Harbin 150040, China; snowguonan@163.com (N.G.); zhoushouting0606@163.com (S.Z.); zhangyunan07262021@163.com (Y.Z.)
[2] School of Civil Engineering and Architecture, Wuyi University, Wuyishan 354300, China; zhaoyan_hit@163.com
* Correspondence: mld@nefu.edu.cn

**Abstract:** To study the effect of wood creep on the long-term performance of prestressed glulam continuous beams, a 180-day test was carried out on beams configured with different numbers of steel wires (2, 4, 6) and with different prestress values (0, 7, 14 kN). By investigating the stress loss of the steel wires in the beam and the change in the mid-span deflection over time, the factors influencing the creep of the continuous beam were analyzed. Three models were selected to fit the creep process of the test beams. Moreover, the creep deformation coefficient $\theta$ was introduced to reflect the influence of glulam creep on the deflection change in the test beams and to predict the total deflection of the beam within 50 years. The results showed that with increasing the number of steel wires and the prestress value on the beams, the total stress of the steel wires declined more and faster. Increasing the number of steel wires or decreasing the prestress force value could effectively restrain the change speed of the mid-span long-term deflection of the beam. Three models were compared, and the power-law equation was the most accurate. At familiar steel wire quantities and force levels, the $\theta$ value of the test beams within the design service life of 50 years was determined to be 1.28–2.29.

**Keywords:** glulam continuous beam; long-term; creep; deformation coefficient; prestress

check for updates

**Citation:** Guo, N.; Zhou, S.; Zhao, Y.; Mei, L.; Zhang, Y. Experimental Study on Long-Term Mechanical Properties of Prestressed Glulam Continuous Beams. *Materials* **2022**, *15*, 4182. https://doi.org/10.3390/ma15124182

Academic Editors: Shao-Bo Kang and Shan Gao

Received: 23 May 2022
Accepted: 10 June 2022
Published: 13 June 2022



## 1. Introduction

Timber structure has a long history and is an important element of traditional architectural culture. Various structural systems represented by light-frame timber structures, glued laminated timber structures, and hybrid timber structures have been formed worldwide. In addition, with the signing of the "Paris Agreement" and the policy of "emission peak and carbon neutrality" advocated by China, the research and technology of modern timber structures have gradually expanded. Timber structure has a series of advantages such as green qualities, low energy consumption, livability, and superior seismic properties, and there is a newly developed opportunity in timber-structured buildings [1–4]. Glulam is a main timber structural material, and its cross-sectional dimensions can be freely specified. Furthermore, glulam has fewer defects and higher available strength and stiffness compared to logs. However, although glulam is widely used, it also has some weaknesses. First, because of the limitations of raw material size and processing technology, the traditional glulam beams generally cannot meet the length requirements for continuous beams or long-span members. Second, most of the simply supported beams suffer brittle tensile failure and larger deformation in the bending state. Therefore, the compressive strength is not fully utilized [5–8], and deformation is controlled by section size. Third, the traditional glulam joint connection needs grooving and drilling, and the secondary processing will damage the wood at the joints. Therefore, the location of the joint connection is easily cracked along the damaged area under the action of force, the capacity to transfer bending moment is weak, and the joint stiffness is insufficient.

Based on the above problems, our research group proposed the concept of the non-damaged splint-type prestressed glulam continuous beam joint and then developed the joint connection device [9]. According to this connection device, a type of novel composite prestressed glulam continuous beam was proposed. When the continuous beam was bent, the prestressed steel wires were in tension and the glulam was in compression. It incurred ductile failure, like the prestressed glulam simply supported beam. The advantage was that not only was the stiffness of the joint improved through the connection device and prestress and the semi-rigid connection realized, but also the bending moment distribution was changed, to make further full use of the materials and reduce deformation. The research group carried out related research and tested the short-term flexural performance of continuous beams in the early stage [10–12]. The creep characteristic of timber is the key factor affecting the long-term mechanical performance of glulam continuous beams, and because of the application of prestressing, the force of the beam is more complicated. Therefore, it is important to investigate the long-term mechanical performance of the prestressed glulam continuous beam and to promote the development of this composite member.

At present, the scholarly research in related fields has mainly focused on the reinforcement of timber members, timber connection, and long-term mechanical performance of timber structures. The timber members are reinforced mainly by adding steel bars, steel plates, and fiber-reinforced materials. In addition, because of the impact on building construction, strength, ductility, and the ability to increase structural performance in service, timber connections have been studied widely [13–16]. Eurocode 5 part 1-1 [17] provides simplified equations for calculating the load-carrying capacity of a single-shear and double-shear plane in different connection types, and double-shear timber connections with steel fasteners have been used in heavy wood structures [18]. Fonseca et al. [19] presented methodologies to evaluate the safe design of three types of timber connections in double shear, which was proposed by Eurocode 5, part 1-1, and analyzed the influence of the design parameters. Creep is an important characteristic of wood, so it is of great significance to study its deformation development law. Scholars have discussed the mechanisms of internal and external influences on wood creep from factors such as load size, temperature and humidity, tree species, repeated load, and load type. G. Hunt et al. [20] summarized a set of methods for predicting long-term creep of timber through short-term creep test data, which has been widely considered and applied. Yazdani et al. [21] conducted a full-scale model test on 16 T-section glulam simply supported beams, and obtained parameters closely related to the Burger creep model. Guo et al. [22,23] carried out long-term loading tests and short-term flexural tests on pure glulam beams, internally reinforced glulam beams, and internally reinforced prestressed glulam beams, and systematically analyzed the influence of factors such as loading level, reinforcement rate, and prestress value. In-depth research and discussion have been carried out on glulam simply supported beams with different cross-section and reinforcement forms, and the creep basic theory. It has provided a reference for this study in terms of long-term test methods and considerations, but a reinforcement method for external prestressed steel wires and the long-term performance research on continuous beams are still lacking, which indicates the necessity of carrying out long-term tests on prestressed glulam continuous beams.

In this paper, a total of 10 prestressed glulam continuous beams were subjected to a 180-day long-term loading test. By comparing the changes in steel wire stress and mid-span long-term deflection over time under the influence of creep, the effects of the number of prestressed steel wires and the prestress level on the long-term mechanical performance of prestressed glulam continuous beams were studied. Three models were selected to fit the creep process of the test beams. The creep deformation coefficient that reflected the long-term deflection growth of the test beams was determined, which lay the foundation for establishing the long-term deflection calculation formula for the prestressed glulam continuous beam.

## 2. Materials and Methods

### 2.1. Materials and Specimens

The same batch of spruce materials was used to produce the prestressed glulam continuous beams and the small clear specimens. 1860-grade, 7 mm diameter low-relaxation prestressed steel wires were recommended for all wire. Based on Chinese standards GB/T50329-2012 [24], GB/T1938-2009 [25], and GB/T228.1-2010 [26], material properties experiments were conducted at the laboratory of Northeast Forestry University. The relevant mechanical properties of each material as shown in Table 1.

**Table 1.** Material properties.

| Material | Characteristics | Average (N/mm$^2$) | COV (%) |
|---|---|---|---|
| Spruce | Modulus of elasticity in compression | $8.301 \times 10^3$ | 4.630 |
| | Compression strength parallel to grain | 38.67 | 6.151 |
| | Modulus of elasticity in tensile | $9.750 \times 10^3$ | 6.42 |
| | Tensile strength parallel to grain | 81.32 | 6.184 |
| Steel wire | Modulus of elasticity | $1.92 \times 10^5$ | 1.710 |
| | Tensile strength | $1.640 \times 10^3$ | 0.354 |

The prestressed glulam continuous beam was composed of glulam, high-strength steel wires, and steel components, as shown in Figure 1. The steel components included connection apparatuses, prestressed control devices, steel plates, and anchors. Five pieces of 20 mm-thick spruce sheets bonded together to form glulam with dimensions of 3080 mm × 80 mm × 100 mm (length × width × height). Therefore, the size of the prestressed glulam continuous beam was 6160 mm × 80 mm × 100 mm (Length × width × height). All the specimens were designed by the Chinese standard GB50005-2017 [27] and Canadian standard CSA O86-14 [28].

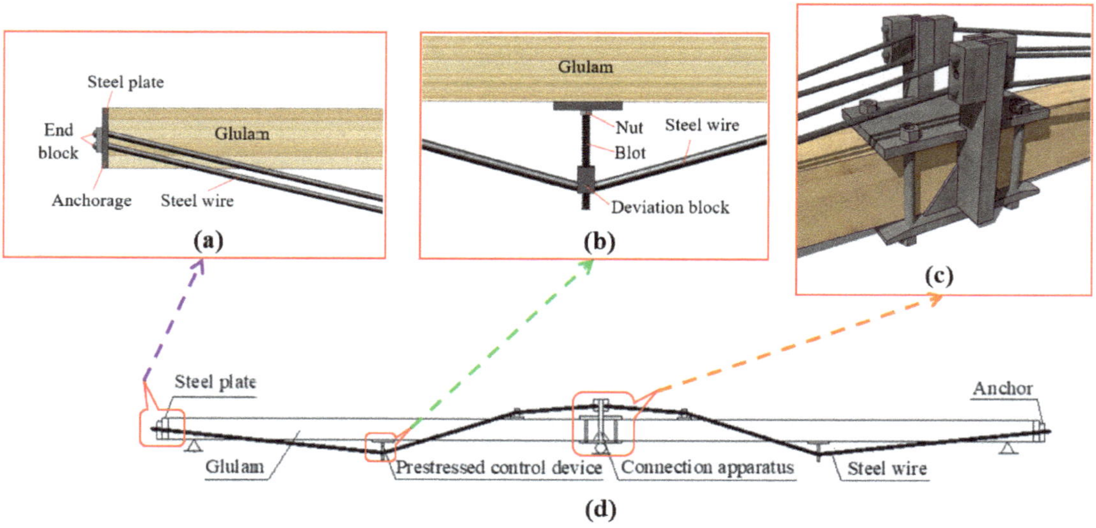


**Figure 1.** *Cont.*

(e) (f)

**Figure 1.** Diagrams of prestressed glulam continuous beam: (**a**) steel plate and anchor, (**b**) prestressed control device, (**c**) connection apparatus, (**d**) details of specimen, (**e**) photo of the connection apparatus, and (**f**) photo of the steel plate and anchor.

To study the effect of the number of steel wires and prestress level on the long-term performance of prestressed glulam continuous beams, the test beams were divided into two groups, A and B. Group A was used to research the effect of the number of steel wires, the first number represents the different number of steel wires, the second number indicates that there are two test beams under the same condition. Group B was used to research the effect of prestress level; the first number represents the different prestress values, and the second number indicates that there are two test beams under the same conditions. More grouping details are shown in Table 2. It is worth noting that the prestress was realized by turning the nut counterclockwise caused the screw to rotate, as shown in Figure 1b. The deviation block moved downward, and thus, the reinforcements were stretched, and the prestress was applied to glulam beams. By adjusting the distance from the deviation block to the bottom of the beam, the value of prestressing can be controlled.

**Table 2.** Groups and information of specimens.

| Group | Number of the Component | Prestress Value (kN) | The Number of Steel Wire | Long-Term Loading Value (kN) | Loading Time (day) |
|---|---|---|---|---|---|
| A | $LA_{1\text{-}1}$ | 7 | 2 | 8.24 | 180 |
| | $LA_{1\text{-}2}$ | 7 | 2 | 8.92 | |
| | $LA_{2\text{-}1}$ * | 7 | 4 | 9.24 | |
| | $LA_{2\text{-}2}$ * | 7 | 4 | 9.24 | |
| | $LA_{3\text{-}1}$ | 7 | 6 | 10.06 | |
| | $LA_{3\text{-}2}$ | 7 | 6 | 9.55 | |
| B | $LB_{1\text{-}1}$ | 0 | 4 | 8.24 | 180 |
| | $LB_{1\text{-}2}$ | 0 | 4 | 8.92 | |
| | $LB_{2\text{-}1}$ * | 7 | 4 | 9.24 | |
| | $LB_{2\text{-}2}$ * | 7 | 4 | 9.24 | |
| | $LB_{3\text{-}1}$ | 14 | 4 | 10.06 | |
| | $LB_{3\text{-}2}$ | 14 | 4 | 9.55 | |

Note: "*" indicates that $LA_{2\text{-}1}$ and $LB_{2\text{-}1}$ in the table are the same beams, but they are in different groups, so different subscript characters are labeled. $LA_{2\text{-}2}$ and $LB_{2\text{-}2}$ are the same situation.

### 2.2. Long-Term Test

#### 2.2.1. Loading Device and Method

A long-term loading test device was designed by our research team, the plane layout and object pictures of the device are shown in Figures 2 and 3. The device was composed of three identical parts, and each part includes four steel columns, two I-type steel beams, and one II-type steel beam. The steel column and I-type steel beams are connected by prefabricated bolt holes to form foundation support, and then the II-type steel beam is placed on top of the two foundation supports. The device can be used to carry out long-term tests on five beams at the same time, with high efficiency. The device has a large bearing capacity and rigidity and has little influence on the test results. In addition to this, it also can adjust the distance between two parts according to the length of the test beams, which has good applicability.

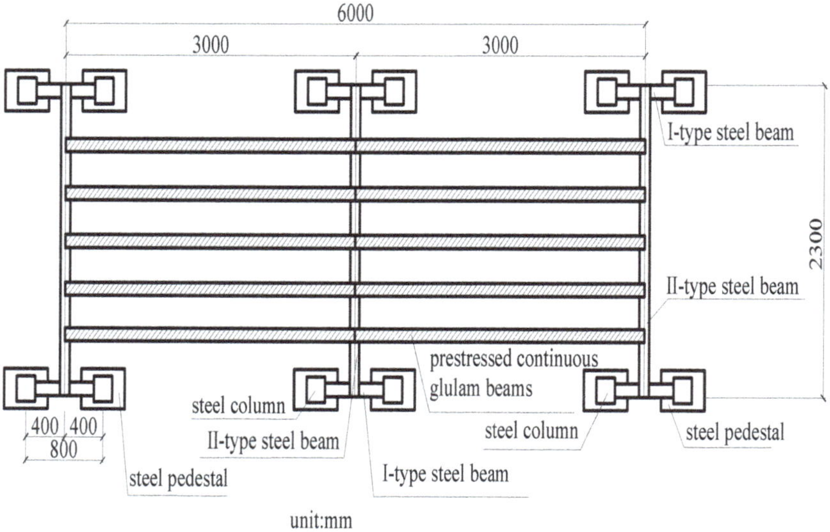


**Figure 2.** The plane layout of the test device.

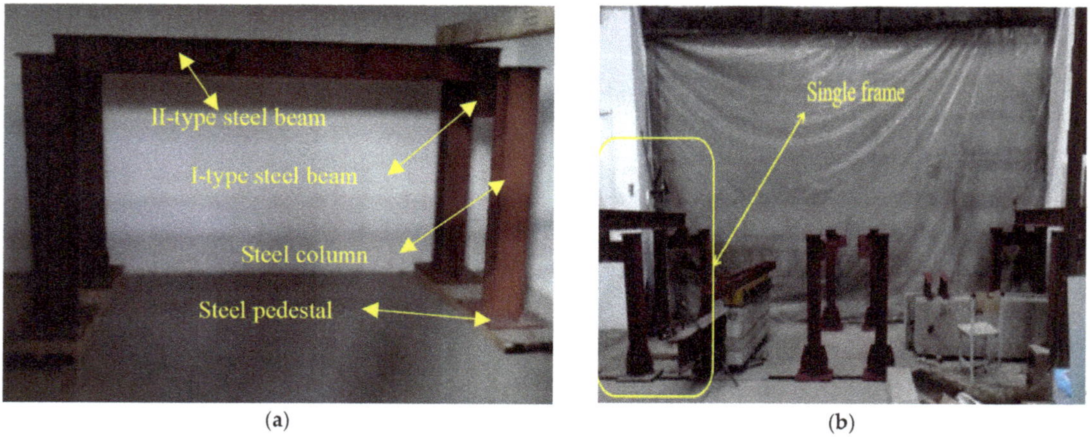


**Figure 3.** Long-term loading test device. (**a**) The composition of the single-loaded frame; (**b**) the whole loading support.

The test adopted the form of four-point bending for long-term loading. To ensure the constant loading value during the loading period, the long-term load was applied by hanging the corresponding weight in the midspan of the left and right spans of the continuous beam, respectively. Two steel plates were placed on the upper surface of the mid-span position of the left and right spans, and a small wooden block was placed on each steel plate, which prevents local compression failure at the loading point. Single-span loading is shown in Figure 4.

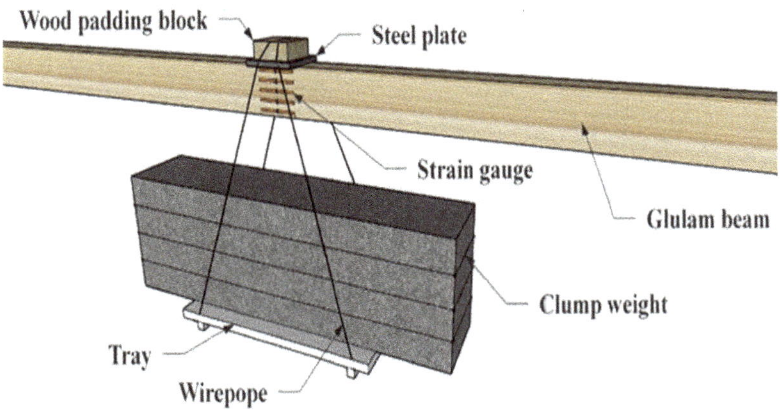


**Figure 4.** Loading method.

2.2.2. Determine Long-Term Loads and Times

According to the Chinese standard GB50009-2012 [29] for the derivation process of the design value of the bearing capacity of the flexural member and the relevant content of the load effect combination. In terms of ultimate bearing capacity, the average value of the ultimate bearing capacity of the beams ($\mu_R$) was obtained based on the short-term bending experiment data of prestressed glulam continuous beams conducted by the research group in the early stage [12]. The standard value of the ultimate bearing capacity ($R_k$) under certain reliability was obtained through the probability distribution, and then the design value ($R$) was obtained through the resistance partial coefficient. In terms of load effects, the load effects ($S_d$) of prestressed members under the serviceability limit state and the ultimate limit state were calculated based on the ratio of dead load and live load effect of floor and office buildings was 2:3. The serviceability limit state shall adopt the standard combination, and the bearing capacity limit state shall adopt the basic combination. The process is shown in Figure 5. It is calculated that the normal service load of the beam was 0.33 times its ultimate bearing capacity. According to the tests performed by previous researchers, the loading level range is generally 30–70% of the ultimate failure load [30], so the long-term load value was determined to be 30% of the ultimate bearing capacity obtained from the short-term loading test. By consulting Chinese standards [31] and referring to previous experience [32–35], the long-term loading time of the beam was determined to be 180 days.

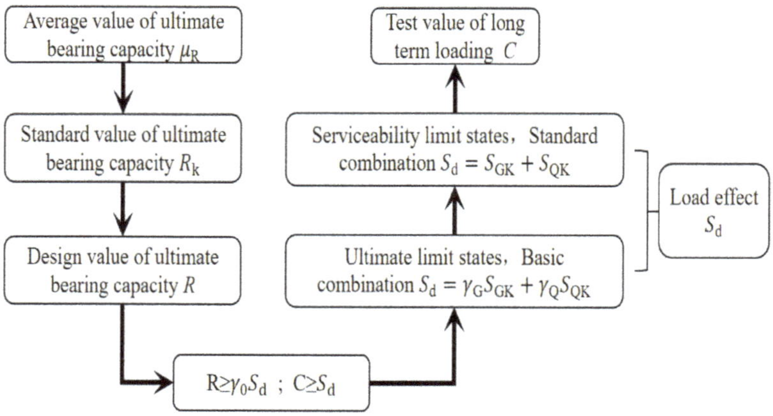


**Figure 5.** Derivation and determination of long-term loading value.

### 2.2.3. Measurement Layout

Test measurements include beam mid-span displacement, glulam beam strain, and steel wire strain. In order to monitor the change in beam mid-span deflection with time, two LVDTs with a range of 150 mm were set in the midspan of the left and right spans. The use range of the LVDT used in the test was between 20% and 80% of the total range, and the LVDT connected to the line was calibrated and placed on the upper surface of the middle span of the test beam through the magnetic seat fixed on the support. To measure the strain of the glulam beams, five 100 mm × 3 mm strain gauges were pasted at equal intervals on the mid-span sides of the left and right spans of the continuous beam, a strain gauge was pasted at the beam top and bottom of the mid-span of the left and right spans, respectively. To measure the strain of the steel wires, the 2 mm × 3 mm strain gauges were pasted to the steel wire surfaces at a distance of 155 mm from the support points at both ends of the beam, the steel wires were polished by an angle grinder in advance. The specific details are shown in Figure 6.

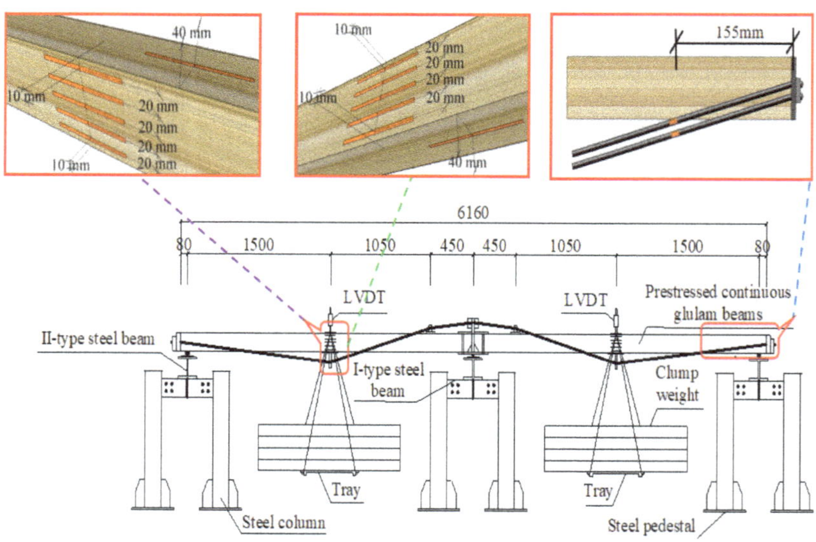


**Figure 6.** Test setup and measuring point layout of the long-term loading test.

First, the beam was applied to prestress, and at the same time, JM3813 static strain acquisition systems (Yangzhou Jing Ming Technology Co., Ltd., Yangzhou, China) were applied to collect the strain and deflection measurements synchronously. The professional Shield wire was used to connect JM3813 static strain acquisition box with LVDT and strain gauge, and in order to make the collected data more accurate, glulam, steel wire and strain acquisition box had connection compensation wire. Secondly, after the beam reached the expected prestress value, the heavyweight was hung to supply the load. According to the development law of the long-term deformation of the beam, the time interval between two data collections gradually increases as the experiment continues. The method of man-machine combination reading data was used, the details of the data collection arrangement are shown in Table 3.

**Table 3.** Information on data collection interval.

| Test Time | Collection Time Interval | Manual Data Collection Time Interval |
|---|---|---|
| Days 1~4 | 0.25 h/time | 0.25 h/time |
| Days 5~12 | 0.25 h/time | 0.5 h/time |
| Day 13~20 | 0.5 h/time | 1 h/time |
| Day 21~28 | 0.5 h/time | 2 h/time |
| Day 29~56 | 1 h/time | 4 h/time |
| Day 57~180 | 2 h/time | 12 h/time |

## 3. Test Results and Analysis

### 3.1. Change in Temperature and Relative Humidity

According to the previous research results [36], the external environment (temperature and moisture content) can obviously affect the creep deformation of wood chipboard under bending. With the increase in the temperature and relative humidity, the creep deformation of wood would increase. In this study, the whole long-term loading process was carried out under normal indoor environment in Harbin, and test beams were divided into two batches and loaded in the same season. During this period, measures such as sprinkling water in summer and heating in winter were taken, and the temperature and humidity of each batch of the beams were regularly monitored so that the relative stability of the indoor environment could be ensured, so the influence of the change in the temperature and humidity on the analysis of the test results could be ignored.

### 3.2. Test Phenomenon

On the first day, the 60th day, the 120th day, and the 180th day of the long-term test, the changes in the test beam were observed, as shown in Figure 7. Compared with the first day of the experiment, cracks at the wood-knot position of the test beam expanded slightly on the 60th day of the test and did not expand on the 120th and 180th days. Since the load applied in the long-term test was equivalent to the normal service load of the beam, and the force state of the beams under normal service condition were simulated, there were no obvious damage phenomenon such as cracks and wrinkles, except that the deflection of the test beam had obvious changes.

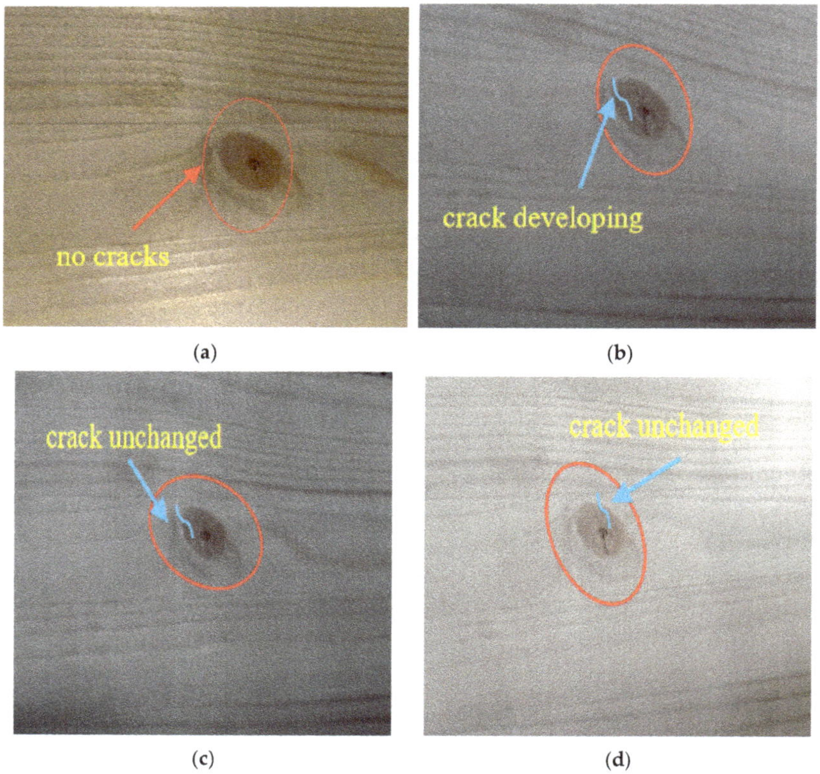


**Figure 7.** Long-term loading phenomenon of prestressed glulam continuous beam. (**a**) Test for 1 day; (**b**) test at 60 days; (**c**) test at 120 days; (**d**) test at 180 days.

### *3.3. Stress Variation Law of Prestressed Steel Wire*

Based on the previous experiments, it has been shown that when the initial stress of the prestressed steel wire was less than 0.5 times the value of the ultimate tensile strength $R_y$, the influence of the relaxation of the steel wire could be ignored [37–42]. The maximum initial prestress applied to the steel wire in this test was $0.3R_y$ ($0.3R_y < 0.5R_y$). Therefore, the influence of the relaxation of the prestressed steel wires($0.3R_y$) could be ignored, which indicates the creep of the glulam results in the main stress loss of steel wires.

For the beams in groups A and B, the time-varying curve of the total stress of the steel wire is shown in Figure 8. The total stress means the stress sum of all steel wires in a test beam. The time process is from the initial loading to the end of loading. For the continuous beams, the stress changes for left and right spans were plotted separately.

It is known from Figure 8 that for the beams in group A, when the prestress value was unchanged and the number of prestressed steel wires was 2, 4, and 6, respectively, from the beginning to the end of the loading, the total stress values of the steel wires decreased by 60.39%, 66.49% and 69.61%, respectively. For the beams in group B, when the number of prestressed steel wires was unchanged and the prestress values were 0, 7, and 14 kN, respectively, from the beginning to the end of the loading, the total stress values of the steel wires decreased by 61.60%, 66.53%, and 68.74%, respectively. It is worth noting that the total stress values were the average values of the left span and the right span of the same continuous beam. Due to the increase in the number of prestressed steel wires or the prestress value, the external load on the continuous beam increased, then the section stress of the glulam also increased, and the wood creep was more obvious, and the total stress of the steel wires decreased more.

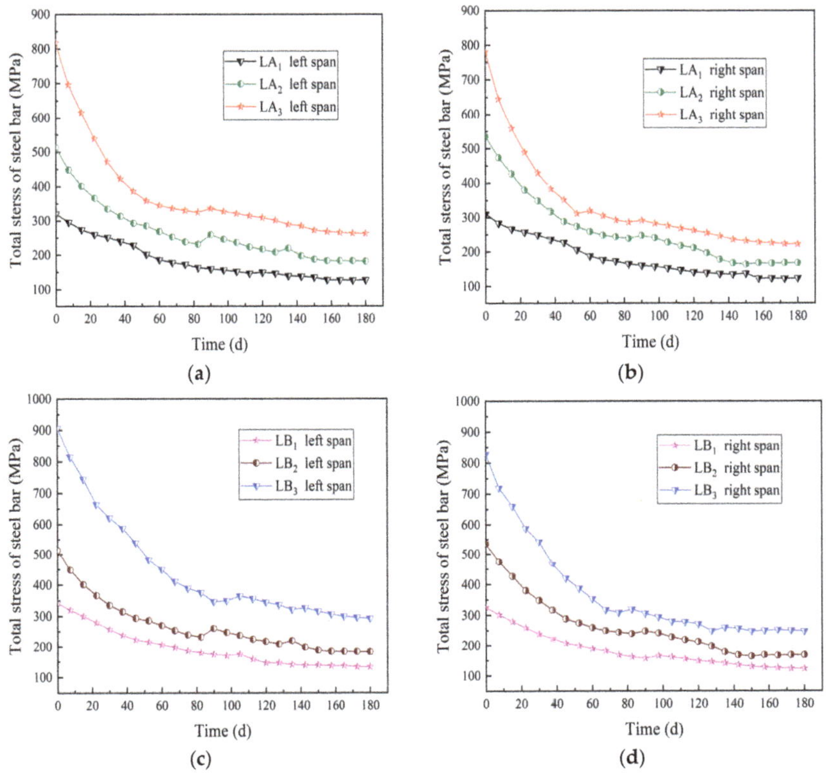


**Figure 8.** Time-varying curve of total stress of the steel wires. (**a**) Left span of group A (different steel wires quantities); (**b**) right span of group A (different steel wires quantities); (**c**) left span of group B (different prestress values); (**d**) right span of group B (different prestress values).

In order to study the influences of the number of prestressed steel wires and the prestress level on the change rate of the steel wire stress value, the initial stress of the steel wire was set to 0, and the variation of steel wire stress value during the experiment was defined as the relative stress value of the steel wire. The time-varying curve of the relative stress of the steel wire is shown in Figure 9.

It can be seen from Figure 9 that for the magnitude of the relative stress change, $LA_2$ is 1.71 times larger than $LA_1$, and $LA_3$ is 2.87 times larger than $LA_1$; in addition, $LB_2$ is 1.58 times larger than $LB_1$, and $LB_3$ is 2.94 times larger than $LB_1$. The relative stress of the left and right spans of continuous beams changes with time in the same law. This shows that by increasing the number of prestressed steel wires and the applied prestress value, the relative stress value increased, the total stress of the steel wires declined faster, and the effect on the creep of glulam was more obvious.

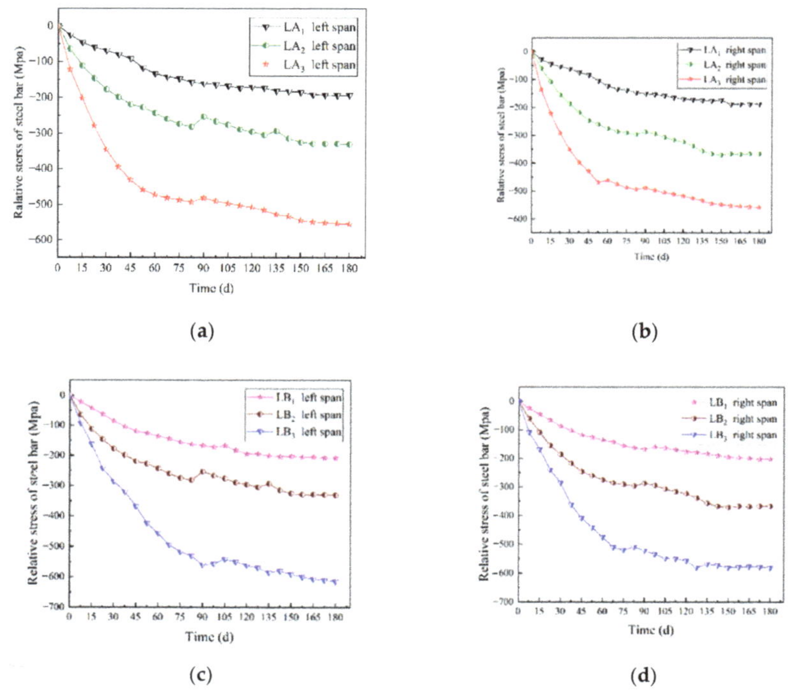


**Figure 9.** Time-varying curve of the relative stress of the steel wires. (**a**) Left span of group A (different steel wires quantities); (**b**) right span of group A (different steel wires quantities); (**c**) left span of group B (different prestress values); (**d**) right span of group B (different prestress values).

### *3.4. Variation Law of Beam Mid-Span Deflection*

The typical creep curve is shown in Figure 10. It can be divided into three stages: the first stage is the transient creep, which means that the wood is applied with a certain load, and produces an instantaneous deformation. The deformation rate of this stage is continuously reduced until it reaches a relatively stable state. The second stage is the steady-state creep, the rate of deformation remains basically unchanged. That is, at the same time, the deformations are approximately equal. This stage lasts the longest and is the most typical during the entire creep development process. The third stage is the accelerated creep, where the deformation rate increases suddenly until the material damages.

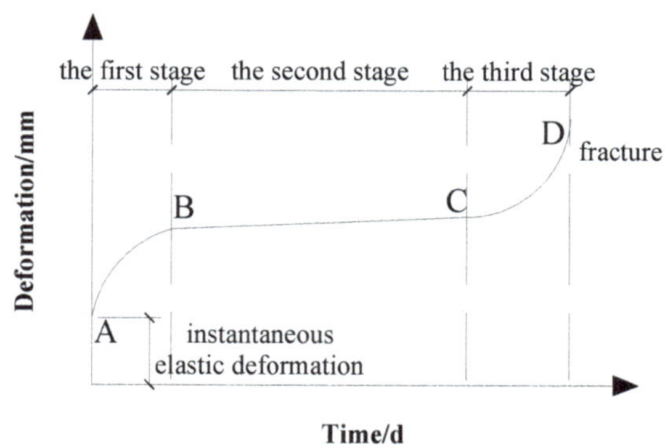


**Figure 10.** Diagram of the typical creep curve.

During the long-term loading test, the total deflection of the continuous beam is divided into three parts: the first part is the height of anti-arch $f_A$ of the beam when the prestressing is applied, the second is the instantaneous deflection of the beam when the hanging weight provides an instantaneous load, and the third is the additional deflection $f_C$ caused by the creep of wood under continuous long-term loading. It is worth mentioning that the deflection of the first and second parts are identified as short-term deflection $f_S$, and the deflection of the third part was identified as long-term deflection $f_L$. The sum of the three parts was the total deflection, as shown in Figure 11. The downward deflection is specified to be positive and the upward deflection to be negative. Due to the initial prestress was applied, the beams is reversed and the initial value of mid-span deflection was negative. The short-term, long-term, and total deflection values of beams of A and B in the test are shown in Table 4.

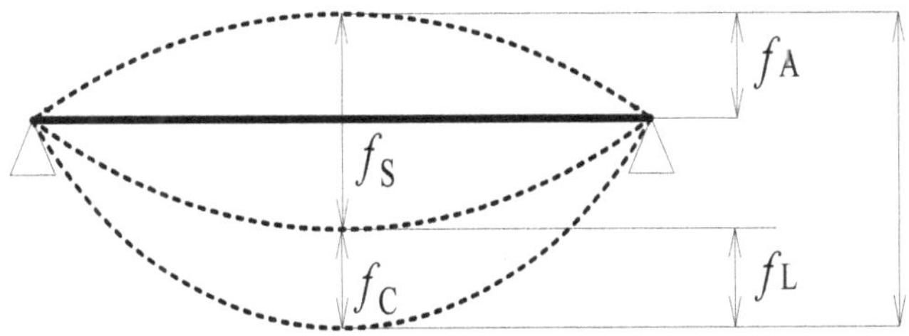


**Figure 11.** Deformation diagram of the prestressed glulam continuous beam during loading.

**Table 4.** Midspan deflection value of prestressed glulam continuous beams.

| Beam Number | | Short-Term Deflection $f_S$ (mm) | Long-Term Deflection $f_L$ (mm) | Total Deflection $f_T$ (mm) |
|---|---|---|---|---|
| $LA_1$ | Left span | −26.61 | 8.70 | −17.91 |
| | Right span | −25.35 | 8.42 | −16.93 |
| $LA_2$ | Left span | −20.03 | 6.00 | −14.03 |
| | Right span | −20.09 | 5.64 | −14.45 |
| $LA_3$ | Left span | −10.18 | 4.33 | −5.85 |
| | Right span | −10.28 | 4.56 | −5.72 |
| $LB_1$ | Left span | 7.51 | 4.93 | 12.44 |
| | Right span | 7.60 | 4.87 | 12.47 |
| $LB_2$ | Left span | −20.03 | 6.00 | −14.03 |
| | Right span | −20.09 | 5.64 | −14.45 |
| $LB_3$ | Left span | −29.21 | 8.97 | −20.24 |
| | Right span | −28.87 | 9.49 | −19.38 |

To explore the creep law of prestressed glulam continuous beams under long-term loading, the time-varying curve of midspan deflection of group A and B are drawn, as shown in Figure 12.

It can be seen from Figures 10 and 12 that the time-varying curve of midspan deflection of the continuous beam is close to the typical creep curve. For group A, the midspan long-term deflection of the beams had reached 60.8% to 85.0% of the total deflection on the 60th day, and the deflection at the 120th day had reached 96.1% to 99.8%; for group B, the midspan long-term deflection of the beams had accomplished 66.1% to 80.1% of the total deflection on the 60th day, and the deflection at the 120th day had accomplished 95.7% to 99.0%. It could be seen that the beam completed the primary creep faster and the creep rate decreased with time, which was consistent with the characteristic of the creep development of glulam. It is worth noting that LB1 was not applied prestressing. Therefore, the beam had a downward initial deflection after loading, and the initial deflection value was positive. In other cases, due to the application of different prestressing, the beams produced an upward anti-arch with a negative initial deflection value.

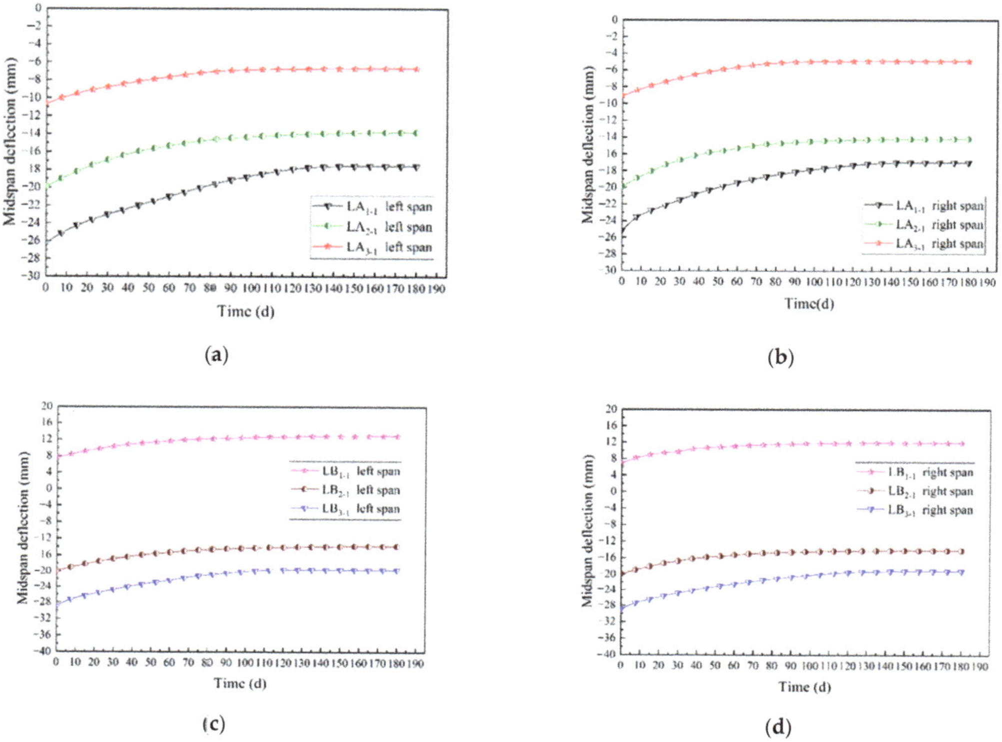


**Figure 12.** Time-varying curves of midspan deflection: (**a**) Left span of group A (different steel wires quantities); (**b**) right span of group A (different steel wires quantities); (**c**) left span of group B (different prestress values); (**d**) right span of group B (different prestress values).

To analyze the influences of the number of prestressed steel wires and the prestress level on the long-term creep rate of the continuous beams in the mid-span, the short-term deflection was ignored, which was assumed that the deflection of the beam was zero after instantaneous loading, then the time-varying curve of the creep deformation is gained and shown in Figure 13.

It can be seen from Figure 13 that for the beams in group A, compared with the continuous beam with two steel wires, the long-term deflection average of the beams with four and six steel wires decreased by 32.0% and 48.0%, respectively. It showed that the long-term deflection gradually decreased with the increase in the steel wires. Therefore, increasing the steel wire could effectively reduce the long-term deflection of the prestressed glulam continuous beam. For the beams in group B, compared with the continuous beam without prestressing, the long-term deflection average of the beams with 7 kN and 14 kN prestress values increased by 18.75% and 88.40%, respectively. It could be seen that the long-term deflection grew gradually with the increase in the applied prestress level. Because the anti-arch value of the beam increased with the applied prestress level, resulting in more stress on the beam, the creep deformation of the glulam was more significant.

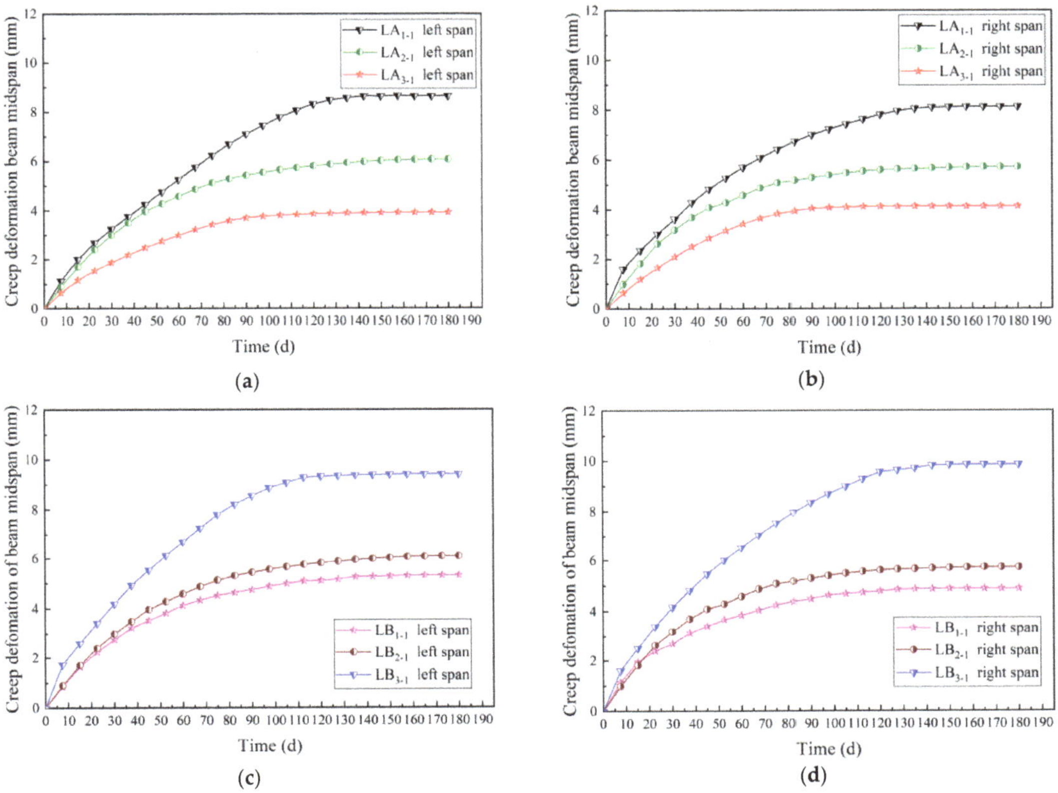


**Figure 13.** Time-varying curves of the creep deformation: (**a**) left span of group A (different steel wires quantities); (**b**) right span of group A (different steel wires quantities); (**c**) left span of group B (different prestress values); (**d**) right span of group B (different prestress values).

For group A, due to the number of steel wires increased, the growth rate of long-term deflection gradually slowed down. Increasing the number of steel wires can effectively restrain the long-term deflection change rate of the continuous beams; for group B, with the increase in applied prestress level, the growth rate of long-term deflection sped up. Therefore, increasing the prestress value could make the creep of glulam beams more obvious, promote the change rate of the long-term deflection of the continuous beams, and cause the deformation to develop rapidly.

### *3.5. Theoretical Analysis*

#### 3.5.1. Establishment of Creep Model

In order to research the creep law of the test beams with time, the existing long-term test data is applied to establish creep model. According to previous experience [30], the prediction models for wood creep are as follows:

$$y = a\left(1 - e^{f(t)}\right) \tag{1}$$

Equation (1) shows the Exponential Equation [43]. Where $y$ is the strain of material or deformation of component, $t$ is the time when the member occurs creep deformation, $a$ is a test parameter which is obtained by fitting the curve, and $f(t)$ is a function related to $t$.

$$y = a + bx^c \tag{2}$$

Equation (2) shows the power-law equation [44,45]. Where $y$ is the strain of material or deformation of component, $x$ is the time when the member occurs creep deformation, and $a$, $b$, and $c$ are the test parameters that are obtained by fitting the curve.

$$y = a + bt - ct^2 \tag{3}$$

Equation (3) shows the Polynomial Equation [46]. Where $y$ is the strain of material or deformation of component, $t$ is the time when the member occurs creep deformation, and $a$, $b$, and $c$ are the test parameters that are obtained by fitting the curve.

Based on the three models, the mid-span deflection variation of the test beam $LB_1$ under long-term loading was simulated, and the results are shown in Figure 14.

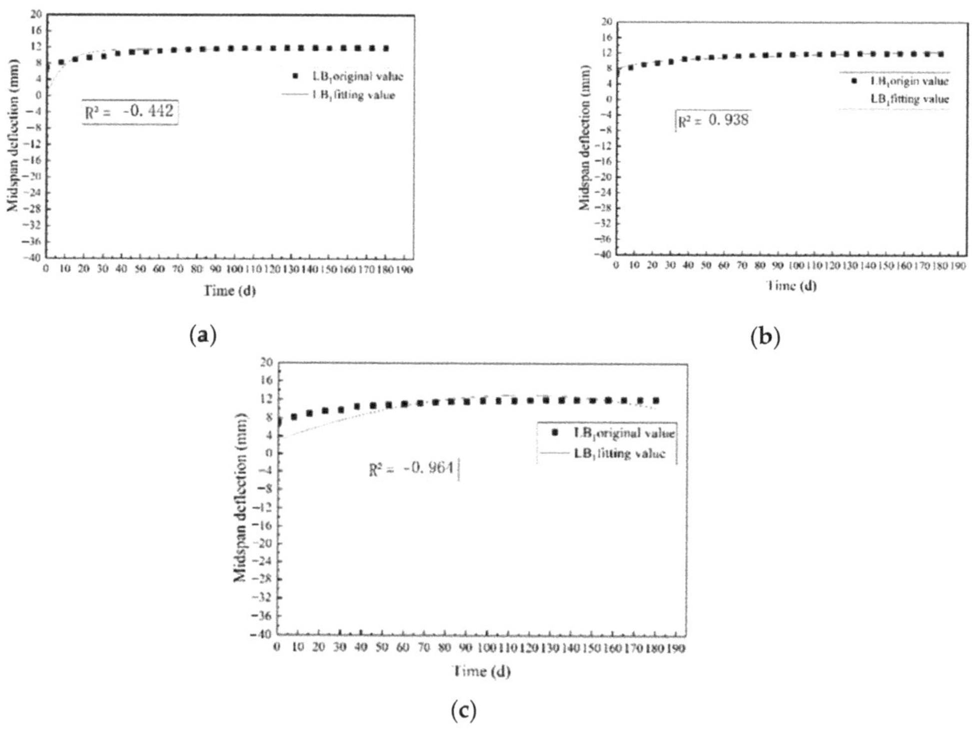


**Figure 14.** Midspan deflection fitting curve of $LB_1$ beam under three models: (**a**) exponential equation fitting curve; (**b**) power-law equation fitting curve; (**c**) polynomial equation fitting curve.

For the fitted curves, $R^2$ reflects the fitting degree of the regression equation, and expresses the relationship between the dependent variable and all independent variables, it means the percentage of the dependent variable whose variability can be explained by the regression equation. The fit result is better with the increase in the $R^2$ value, $R^2$ is represented by Equation (4):

$$R^2 = 1 - \frac{RSS}{TSS} \tag{4}$$

where $R^2$ is the coefficient of determination, RSS is the sum of squared differences between the fitted data and the original data, and TSS is the sum of squared differences between the original data and the mean value.

Research shows that the most commonly used empirical model is the power-law model [46,47]. It can be seen from Figure 14 that the determination coefficients $R^2$ of the creep model of the power-law equation, exponential equation, and polynomial equation were 0.938, −0.442, and −0.964, respectively. The comparison results show that the power-law equation could effectively predict the creep process, and the model is simpler and more effective than the other two. Therefore, the power-law equation was selected to fit the midspan deflection of the test beams, the fitting parameters *a*, *b*, *c*, and the determination coefficient $R^2$ were shown in Table 5.

**Table 5.** Result of fitting parameters.

| Beam Number | *a* | *b* | *c* | $R^2$ |
|---|---|---|---|---|
| $LA_1$ | −29.24 | 3.04 | 0.28 | 0.96 |
| $LA_2/LB_2$ | −26.49 | 6.18 | 0.34 | 0.96 |
| $LA_3$ | −14.75 | 2.82 | 0.22 | 0.94 |
| $LB_1$ | 4.04 | 12.77 | 0.09 | 0.97 |
| $LB_3$ | −30.71 | 3.11 | 0.42 | 0.98 |

Note: $LA_2$ and $LB_2$ in the table are the same beams, but they are in different groups, so different subscript characters are labeled.

It can be seen from Table 5 that all the $R^2$ values obtained by fitting the test beams are greater than 0.9, indicating that the fitting accuracy of the mid-span deflection results and the fitting curve is high, and the creep behavior of the prestressed glulam continuous beams was shown preferably.

3.5.2. Creep Deformation Coefficient $\theta$ and the Total Deflection

Three types of allowable value and calculation formulas were given for the deflection of flexural members in Eurcode5 [17]. Three types of allowable values contain instantaneous deflection $\omega_{inst}$, final total deflection $\omega_{fin}$, and final net deflection $\omega_{net,\,fin}$. The calculation of the total deformation of the flexural members is also specified clearly in American standards NDS-1997 [48]. According to the above standards, the creep deformation coefficient $\theta$ is defined as the ratio of the additional deflection produced by creep and the elastic deflection produced by loading in this paper. It is expressed by Equation (5):

$$\theta = \frac{f_C}{f_S} = \frac{f_L}{f_S} \tag{5}$$

where $\theta$ is the creep deformation coefficient, $f_c$ is the additional deflection value produced by the creep of the test beam, $f_s$ is the elastic deflection produced by the load (the short-term deflection value of the prestressed glulam continuous beam), and $f_L$ is the long-term deflection value of the prestressed glulam continuous beam.

The total deflection of flexural members under normal service conditions includes the elastic deflection caused by loading and the creep deflection caused by the creep of wood under continuous load. The elastic deflection $f_S$ of the flexural members could be calculated according to structural mechanics, while the creep deflection could be calculated by multiplying the deformation $f_Q$ by the creep deformation coefficient $\theta$. The calculation formula was shown in Equation (6):

$$f_T = f_S + \theta f_Q \tag{6}$$

where $f_T$ is the total deflection of the test beam, $f_Q$ is the deformation due to quasi-permanent combination of loads.

Because the service life of the building is generally 50 years, the power-law model is used to predict the total deflection value of the beams for 50 years in this paper, and the creep deformation coefficient of the test beams is calculated by Equation (5), the results are shown in Table 6. Table 6 shows the value range of creep deformation coefficient was 1.28–2.29 when the steel wires on the beam were from 2 to 6, and the $\theta$ value decreased with the steel wires increased; the value range of creep deformation coefficient was 1.69–2.17 when the prestress was from 0 to 14 kN, and the $\theta$ value decreased with the prestress increased. Therefore, the creep deformation coefficient $\theta$ was determined to be 1.28–2.29 in the range of familiar steel wire quantities and applied prestress.

**Table 6.** Creep deformation coefficient for 50 years.

| Project | Group A | | | Group B | | |
|---|---|---|---|---|---|---|
| | $LA_1$ | $LA_2$ | $LA_3$ | $LB_1$ | $LB_2$ | $LB_3$ |
| Short-term deflection $f_s$ (mm) | 6.12 | 5.53 | 3.54 | 7.01 | 5.53 | 5.44 |
| 50-year total deflection $f_T$ (mm) | 20.15 | 15.38 | 8.08 | 22.20 | 15.38 | 14.63 |
| Creep deformation coefficient $\theta$ | 2.29 | 1.78 | 1.28 | 2.17 | 1.78 | 1.69 |

## 4. Conclusions

- When the number of the prestressed steel wires was 2, 4, and 6, the total stress of steel wires decreased by 60.39%, 66.49%, and 69.61%, respectively, from the beginning to the end of loading; when the prestress value was 0, 7, and 14 kN, the total stress of steel wires decreased by 61.60%, 66.53%, and 68.74%, respectively, from the beginning to the end of loading. This is because by increasing the number of prestressed steel wires and the prestress value, the external load on the continuous beam increased, then the section stress of the glulam also increased, and the wood creep was more obvious, and the total stress of the steel wires decreased more.
- Compared with the continuous beam with two steel wires, the long-term deflection average of the beams with four and six steel wires decreased by 32.0% and 48.0%, respectively; compared with the continuous beam without prestressing, the long-term deflection average of the beams with 7 kN and 14 kN prestress value increased by 18.75% and 88.40%, respectively, which shows that increasing the number of steel wires could effectively restrain the long-term deflection change rate of the continuous beams, and increasing the prestress value made the creep of glulam beams more obvious and promoted the long-term deflection change rate of the continuous beams, the deformation developed faster.
- Based on the comparison between three common creep models of exponential equation, power-law equation, and polynomial equation, it found that the power-law equation was the most suitable for simulating the creep process of the prestressed glulam continuous beams, and the fitting curve had higher accuracy. According to the experiment and numerical fitting results, at familiar steel wire quantities and prestress levels, the creep deformation coefficient of the test beams within the design service life of 50 years was determined to be 1.28–2.29.
- The calculation formula of long-term deflection and creep deformation coefficient of test beams were given by long-term experiment and model fitting. Later, the accuracy of fitting can be verified by longer longer-term experiments.
- The long-term performance of prestressed glulam continuous beams was mainly studied from the perspective of experiment in this paper. In the future, its long-term constitutive relation can be exploited by finite element analysis.

**Author Contributions:** Conceptualization, N.G.; methodology, N.G. and S.Z.; software, S.Z.; validation, L.M.; formal analysis, S.Z., Y.Z. (Yan Zhao); data curation, S.Z., Y.Z. (Yunan Zhang); writing—original draft preparation, S.Z.; writing—review and editing, N.G., L.M. and Y.Z. (Yunan Zhang); funding acquisition, N.G. and Y.Z. (Yan Zhao). All authors have read and agreed to the published version of the manuscript.

**Funding:** The research work presented in this paper is supported by the Fundamental Research Funds for the Central Universities (2572021AW11), the Nature Science Foundation of Heilongjiang Province (LH2020E009), and the Natural Science Foundation of Fujian Province (2020J01402).

**Institutional Review Board Statement:** Not applicable.

**Informed Consent Statement:** Not applicable.

**Data Availability Statement:** The data presented in this study are available on request from the corresponding author.

**Conflicts of Interest:** The authors declare no conflict of interest.

## References

1. Du, C.Y.; Yu, C.H. Green, Space, Construction—Interpretation of Modern Western Timber Structure. *New Archit.* **2005**, *5*, 19–23.
2. He, M.J.; He, G.R.; Liang, F.; Li, Z. Development History of Chinese Wood Structures in the Past 20 Years. *Build. Struct.* **2019**, *49*, 83–90.
3. Liu, W.Q.; Yang, H.F. Research Progress of Modern Wood Structure. *J. Build. Struct.* **2019**, *40*, 16–43.
4. Shao, J.S.; Xue, W.C.; Liu, W.Q.; Wang, L. The Summarization of Research and Application of Wood Structure Strengthened with FRP. *Compos. Sci. Eng.* **2015**, *04*, 91–95.
5. Liu, W.; Yang, H. Experimental Study on Flexural Behavior of Engineered Wood Beams. *J. Build. Struct.* **2008**, *29*, 90–95.
6. Cao, L.; Zhang, Z.; Zeng, D.; He, G.; Cai, J. Experimental Research on Fatigue Behavior of Larch Glulam Beams. *J. Build. Struct.* **2016**, *37*, 27–35.
7. Zhang, J.; Wang, W.C.; Qiu, R.G.; Shen, H.; Xu, Q.F.; Gao, S. Experimental Study on Short-Term Flexural Behavior of Internal Prestressed Glulam Beams. *China Civ. Eng. J.* **2019**, *52*, 23–34.
8. Yang, H.F.; Ju, D.D.; Liu, W.Q.; Lu, W.D. Prestressed Glulam Beams Reinforced with CFRP Bars. *Constr. Build. Mater.* **2016**, *109*, 73–83. [CrossRef]
9. Mei, L.D.; Guo, N.; Li, L.; Zuo, H.L.; Zhao, Y. Study on Flexural Performance of Prestressed Glulam Continuous Beams under Control Influence. *J. Wood Sci.* **2021**, *67*, 48. [CrossRef]
10. Guo, N.; Wang, Y.J.; Zuo, H.L. Study of Short-Term Flexural Behavior for Glue-Bamboo and Lumber Beams under Different Pre-Stressed States. *Civ. Eng. J.* **2017**, *26*, 128–142. [CrossRef]
11. Guo, N.; Wang, W.B.; Zuo, H.L. Flexural Property of String Beam of Pre-Stressed Glulam Based on Influence of Regulation and Control. *Struct. Durab. Health Monit.* **2019**, *13*, 143–179. [CrossRef]
12. Guo, N.; Wu, M.T.; Li, L.; Li, G.D.; Zhao, Y. Bending Performance of Prestressed Continuous Glulam Beams. *Adv. Civ. Eng.* **2021**, *2021*, 5512350. [CrossRef]
13. Fonseca, E.M.M.; Leite, P.A.S. Fire Design of Wood Connections Loaded in Shear. *MATTER Int. J. Sci. Technol.* **2019**, *5*, 68–84. [CrossRef]
14. Peng, L.; Hadjisophocleous, G.; Mehaffey, J.; Mohammad, M. On the Fire Performance of Double-Shear Timber Connections. *Fire Saf. Sci.* **2011**, *10*, 1207–1218. [CrossRef]
15. Frangi, A.; Erchinger, C.; Fontana, M. Experimental Fire Analysis of Steel-to-Timber Connections Using Dowels and Nails. *Fire Mater.* **2009**, *34*, 1–19. [CrossRef]
16. Fonseca, E.M.M.; Leite, P.A.S.; Silva, L. Wood Connections Under Fire Conditions Protected with Gypsum Plasterboard Types A and F. In *Advances in Fire Safety Engineering*; Piloto, P.A.G., Rodrigues, J.P., Silva, V.P., Eds.; Lecture Notes in Civil Engineering; Springer International Publishing: Cham, Switzerland, 2020; Volume 1, pp. 93–106. ISBN 978-3-030-36239-3.
17. *BSEN 1995-1-1*; Eurocode 5: Design of Timber Structures. Part 1-1: General Common Rules and Rules for Buildings. British Standards Institution: London, UK, 2004.
18. Youngquist, J.A.; Product, P. *Wood Handbook: Wood as an Engineering Material*; General Technical Report FPL-GTR-190, Centennial Edition; Department of Agriculture, Forest Service, Forest Products Laboratory: Madison, WI, USA, 2010.
19. Fonseca, E.M.M.; Leite, P.A.S.; Silva, L.D.S.; Silva, V.S.B.; Lopes, H.M. Parametric Study of Three Types of Timber Connections with Metal Fasteners Using Eurocode 5. *Appl. Sci.* **2022**, *12*, 1701. [CrossRef]
20. Hunt, D.G. The Prediction of Long-Time Viscoelastic Creep from Short-Time Data. *Wood Sci. Technol.* **2004**, *38*, 479–492. [CrossRef]
21. Yazdani, N.; Johnson, E.; Duwadi, S. Creep Effect in Structural Composite Lumber for Bridge Applications. *J. Bridge Eng.* **2004**, *9*, 87–94. [CrossRef]
22. Guo, N.; Ren, J.; Li, L.; Zhao, Y.; Wu, M.T. Long-Term Loading Test of Reinforced Glulam Beam. *J. Renew. Mater.* **2022**, *10*, 183–201. [CrossRef]
23. Guo, N.; Yang, C.; Li, L.; Li, G.D.; Zhao, Y. Experimental Study on Flexural Performance of Regulated Reinforced Glulam Beam after Long-Term Loading. *Sustainability* **2021**, *13*, 5556. [CrossRef]
24. *GB/T 50329-2012*; Standard for Test Methods of Timber Structures. China Architecture & Building Press: Beijing, China, 2012.
25. *GB/T 1938-2009*; Method of Testing in Tensile Strength Parallel to Grain of Wood. Standardization Administration of the People's Republic of China: Beijing, China, 2009.
26. *GB/T 228.1-2010*; Metallic Materials—Tensile Testing—Part 1: Method of Test at Room Temperature. Standards Press of China: Beijing, China, 2010.
27. *GB 50005-2017*; Standard for Design of Timber Structures. China Architecture & Building Press: Beijing, China, 2017.
28. *O86-14*; Engineering Design in Wood. CSA Group: Toronto, ON, Canada, 2017.
29. *GB 50009-2012*; Load Code for the Design of Building Structures. China Architecture & Building Press: Beijing, China, 2012.
30. Yue, K.; Zhang, W.; Xia, Y.; Lu, X.N. A review on creep of wood material. *For. Grassl. Mach.* **2008**, 48–51.

31. *LY/T 1975-2011*; Standard for Evaluation of Duration of Load and Creep Effects of Wood and Wood-Based Products. Standards Press of China: Beijing, China, 2011.
32. Xu, Y.L.; Hua, Y.K. Creep Behavior and Bending Properties of Poplar LVL in Different Assembly Types. *Chin. J. Wood Sci. Technol.* **2002**, *16*, 10–12.
33. Lu, W.D.; Song, E.W.; Yue, K.; Liu, W.Q. Experimental Study on Creep Behavior of FRP-Reinforced Glulam Beam. *J. Build. Mater.* **2013**, *16*, 294–297.
34. Yang, T.Q.; Zhang, X.C.; Gang, Q.G. Temporal Characteristics of Loading for Creep Buckling of Plates. *Acta Mech. Sin.* **2000**, *32*, 319–325.
35. Zhang, F.W.; Chen, X. Experimental Study on Mechanical Behavior of Timber Beam Strengthened with Steel Plate during 1200-Day Sustained Loading. *J. Build. Struct.* **2017**, *38*, 78–85.
36. Dinwoodie, J.; Pierce, C.B.; Paxton, B.H. Creep in Chipboard. Part 4: The Influence of Temperature and Moisture Content on the Creep Behaviour of Range of Boards at Asingle Stress Levels. *Wood Sci. Technol.* **1984**, *18*, 205–224. [CrossRef]
37. Hariharan, K.; Dubey, P.; Jain, J. Time Dependent Ductility Improvement of Stainless Steel SS 316 Using Stress Relaxation. *Mater. Sci. Eng. A* **2016**, *673*, 250–256. [CrossRef]
38. Pommier, H.; Busso, E.P.; Morgeneyer, T.F.; Pineau, A. Intergranular Damage during Stress Relaxation in AISI 316L-Type Austenitic Stainless Steels: Effect of Carbon, Nitrogen and Phosphorus Contents. *Acta Mater.* **2016**, *103*, 893–908. [CrossRef]
39. Seo, M.-K.; Na, K.W.; Han, S.H.; Park, S.-H.; Ha, S.-D. Inhibitory Effect of Ethanol and Thiamine Dilaurylsulfate against Loosely, Intermediately, and Tightly Attached Mesophilic Aerobic Bacteria, Coliforms, and Salmonella Typhimurium in Chicken Skin. *Poult. Sci.* **2020**, *99*, 1571–1580. [CrossRef]
40. Cho, J.; Lee, C.-H. FE Analysis of Residual Stress Relaxation in a Girth-Welded Duplex Stainless Steel Pipe under Cyclic Loading. *Int. J. Fatigue* **2016**, *82*, 462–473. [CrossRef]
41. Sá, E.R.; Rodrigues, S.F.; Aranas, C.; Siciliano, F.; Reis, G.S.; Cabrera-Marrero, J.-M.; Silva, E.S. Softening-Precipitation Interaction in a Nb-and N-Bearing Austenitic Stainless Steel under Stress Relaxation. *J. Mater. Res. Technol.* **2020**, *9*, 7807–7816. [CrossRef]
42. Nguyen, N.-V.; Vu, Q.-A.; Kim, S.-E. An Experimental Study on Stress Relaxation Behaviour of High Strength Steel Wire: Microstructural Evolution and Degradation of Mechanical Properties. *Constr. Build. Mater.* **2020**, *261*, 119926. [CrossRef]
43. Jozsef, B.; Benjamin, A.J. *Mechanics of Wood and Wood Composites*; Springer: Berlin/Heidelberg, Germany, 1982.
44. Clouser, W.S. *Creep of Small Wood Beams under Constant Bending Load*; Department of Agriculture, Forest Service, Forest Products Laboratory: Madison, WI, USA, 1959.
45. Findley, W.N. Stress Relaxation and Combined Stress Creep of Plastics. *SPE J.* **1960**, *16*, 192–198.
46. Holzer, S.M.; Loferski, J.R.; Dillard, D.A. A Review of Creep in Wood: Concepts Relevant to Develop Long-Term Behavior Predictions For Wood Structures. *Wood Fiber Sci.* **1989**, *21*, 376–392.
47. Morlier, P. *Creep in Timber Structures: Report of RILEM Technical Committee 112-TSC*; E & FN Spon Ltd.: London, UK, 1994.
48. *NDS-1997*; National Design Specification for Wood Construction. American Forest & Paper Association: Washington, DC, USA, 2005.

materials

MDPI

Article

# Analysis of Four Types of Anchorage Devices for Prestressed Glulam Beam and Experimental Research

Mingfei Li [1,†], Mingtao Wu [1,†], Nan Guo [1,*], Lidan Mei [1] and Yan Zhao [2]

1 College of Civil Engineering, Northeast Forestry University, Harbin 150000, China; a475167970@163.com (M.L.); wmt0830@nefu.edu.cn (M.W.); mld@nefu.edu.cn (L.M.)
2 College of Civil Engineering and Architecture, Wuyi University, Wuyishan, Nanping 354300, China; zhaoyan_hit@163.com
* Correspondence: snowguonan@163.com
† These authors contributed equally to this work, and should be considered as co-first author.

**Abstract:** An anchorage device is an integral part of the prestressed Glulam beams. Therefore, its rationality and practicability have significant effects on the mechanical performance of the prestressed beams. To investigate the impact of the anchorage devices on the bearing capacity and stiffness of the prestressed beams, this paper compared and analyzed four kinds of anchors in detail through the finite element software. The results showed that when the initial mid-span deflection was 5 mm, 10 mm, and 15 mm, the bearing capacity of prestressed beams with four anchorage devices was 80.37–177.24%, 93.56–182.51%, and 95.62–194.60% higher than that of ordinary Glulam beam, respectively. When the initial mid-span top prestresses were 1 MPa, 1.5 MPa, and 2 MPa, the bearing capacity of prestressed beams with four anchorage devices was 101.71–172.57%, 105.85–175.88%, and 109.64–180.87% higher than that of ordinary Glulam beam, respectively. In addition, based on the simulation results, the prestressed beam with the external anchorage had the highest bearing capacity and stiffness. The deformation capacity of the beam with boot anchorage was the largest. The stress distribution of the beam installed under beam anchorage was the most uniform, and the beam with slotted anchorage was easy to cause stress concentration at the notch. Finally, based on the outstanding performance of the external anchorage, it was selected to carry out one experiment, and the experimental result showed that the simulation could predict the damage model and load–deflection relationship of the prestressed beams well.

**Keywords:** anchorage device; prestressed Glulam beam; finite element method; stiffness; bearing capacity



**Citation:** Li, M.; Wu, M.; Guo, N.; Mei, L.; Zhao, Y. Analysis of Four Types of Anchorage Devices for Prestressed Glulam Beam and Experimental Research. *Materials* **2021**, *14*, 6494. https://doi.org/10.3390/ma14216494

Academic Editor: Michele Bacciocchi

Received: 9 October 2021
Accepted: 25 October 2021
Published: 29 October 2021



## 1. Introduction

As one kind of natural green-building material, timber has many advantages, such as no pollution, low energy consumption, and degradability, which meets requirements for green sustainable development. Therefore, timber-structure buildings get more attention from researchers [1–7]. Compared with traditional wood materials, the Glulam beam has become an essential part of modern wood structure. It solves the disadvantages of low lumber recovery and complicated processing. It has the advantages of diverse cross-sections, component shapes, fewer centralized defects, higher strength, and more reasonable stress distribution [8–10]. However, the Glulam beam often suffers tensile failure due to lower tensile strength, resulting in the compressive strength not being fully utilized. It is a common method to use a variety of reinforced polymers to strengthen Glulam beams. The mechanical properties and failure modes of Glulam beams and basalt fiber reinforced polymer (BFRP) reinforced Glulam beams were compared and analyzed by Chen et al. [11]. Li et al. [12] investigated the influence of different bonding methods of flax fiber reinforced polymer (FFRP) on the flexural performance of Glulam beams. Wang et al. [13] researched the effect of glass fiber reinforced plastic (GFRP) on the bending capacity of Glulam beams.

Glišović and Johnsson carried out a series of experiments to study the impact of GFRP on the stiffness, bearing capacity, and ductility of Glulam beams [14,15]. The ultimate bearing capacity of Caribbean pine Glulam beams strengthened with fiber reinforced polymer (FRP) and without FRP were compared and analyzed [16]. Meanwhile, some researchers built the finite element model of near-surface bonded carbon fiber polymer wood beam to study the related influencing factors [17–19].

Due to the successful application of prestressing technology in concrete structures, many scholars also began to introduce prestressing technology into wood structures. Some scholars analyzed the flexural ultimate bearing capacity and flexural stiffness of four groups of different linear prestressed reinforced beams [20]. Yang et al. [21] carried out four-point bending tests of carbon fiber reinforced plastic (CFRP) reinforced Glulam beams to make a profound study of the theoretical models of the flexural capacity of unreinforced beams, reinforced beams, and prestressed reinforced beams. Lin et al. [22] demonstrated that the prestress could improve more bending capacity of Glulam beams compared with reinforcement. Zhang et al. [23] concluded that compared with traditional Glulam beams, the prestressed Glulam beams with line and curve steel bars had higher bending capacity. A new type of prestressed laminated beam is proposed, which is prestressed by rotating bolts [24–27]. Yang et al. found that prestressed beams' ultimate capacity, stiffness, and ductility were higher than traditional Glulam beams. The prestress loss of prestressed glued wood beam reinforced FRP plate was calculated and analyzed, and then the theoretical analysis was verified by the measured value of reverse arch deflection [28]. Mei [29] proposed a novel prestressed Glulam continuous beam and analyzed the effects of prestressing control on the failure mode, ultimate bearing capacity, and load-deformation relationship of continuous laminated beams. In addition, Mario V. and Alessandra J. [30] dealing with problems inherent to MEMS devices, highlights the procedural process of identifying ghost solutions starting from numerical solutions.

The existing studies have enriched the flexural properties of prestressed material reinforced Glulam beams, but there is a gap in the research on prestressing anchorage devices. Therefore, this paper established finite element models of four different types of anchorage devices (external anchorage device, boot anchorage device, under beam anchorage device, and slotted anchorage device) by using the finite element software ABAQUS. Through the analysis of load-deflection curves and bending stress distribution, the effects of four different types of anchorage devices on the flexural capacity and stiffness of prestressed beams under different prestress and pre-deflection were explored. Finally, based on the finite element analysis results, the four-point bending test selected and studied the anchorage device with the best flexural property. The test measured values were compared and verified with the simulation values.

## 2. Finite Element Experiment

### 2.1. Finite Element Modeling

As shown in Figure 1, the prestressed beam is composed of a Glulam beam, anchorage device, reinforcement, and bolt. Figure 2 presents the diagram of screw bolt composition. All dimensions are in mm, similarly hereinafter.

The Glulam was modeled as anisotropic, and all anchorage devices, screw bolt compositions, and steel bars were modeled as isotropic. First, we built and assembled the parts of the model. Next, the load was applied after the bolt was extended. This simulation included the related prestress and stress state. Subsequently, the bending behavior of prestressed beams was modeled in the quasi-static time-independent domain.

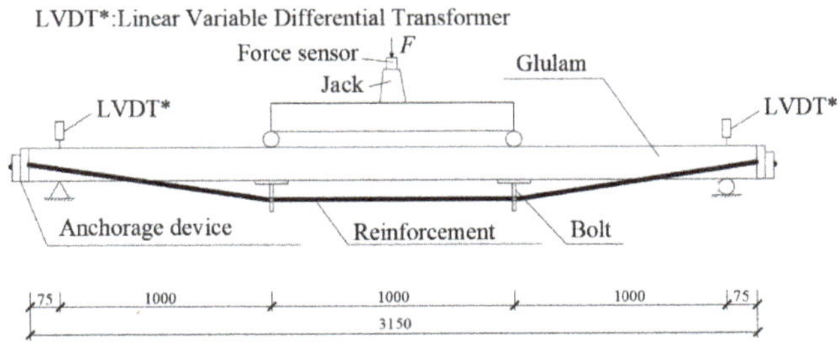


**Figure 1.** Prestressed beam.

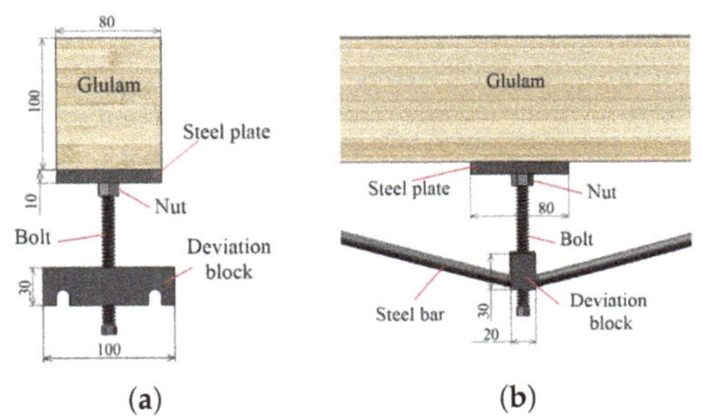


**Figure 2.** Diagram of screw bolt composition (dimensions are presented in (mm)). (**a**) side view and (**b**) main view.

There are four types of anchorage devices: external anchorage device, boot anchorage device, under beam anchorage device, and slotted anchorage device, as shown in Figure 3. For convenience, they are named mode I, mode II, mode III and mode IV.

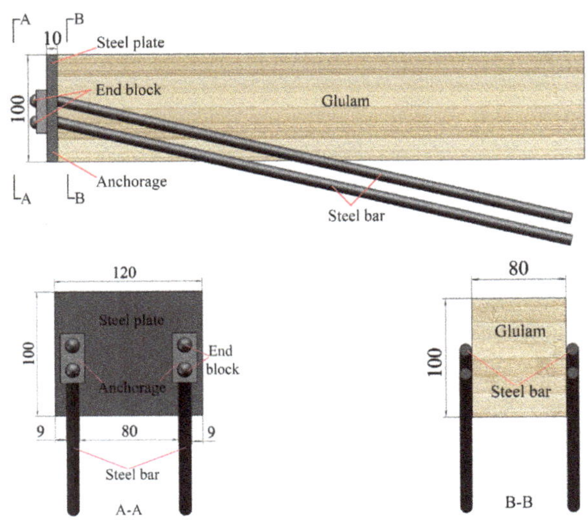


(**a**)

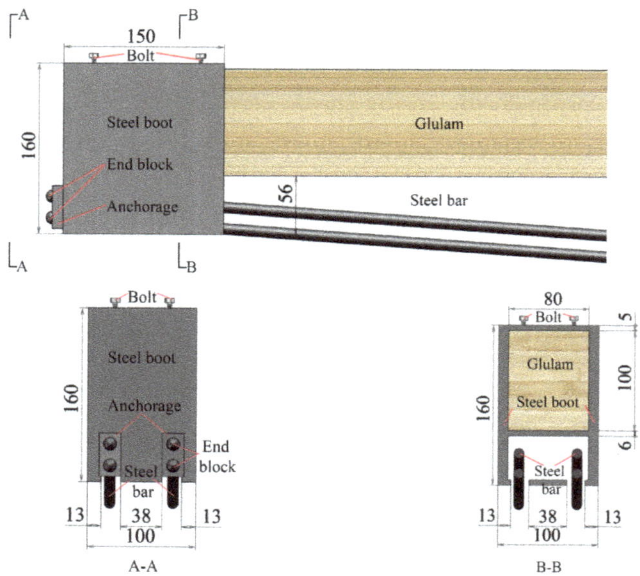
A
B
150
Bolt
Steel boot
Glulam
160
End block
Anchorage
56
Steel bar
A
B
Bolt
Steel boot
Anchorage
160
End block
Steel
bar
13
38
13
100
A-A
80
Bolt
5
Glulam
100
Steel boot
160
6
Steel
bar
13
38
13
100
B-B

(**b**)

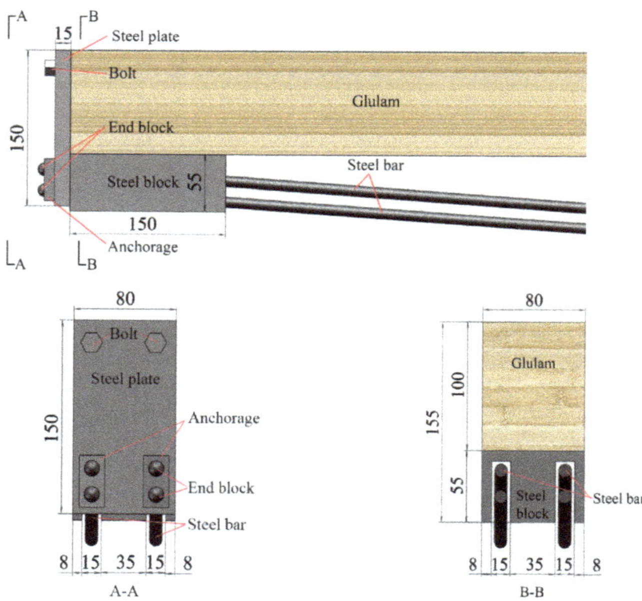
A
15
B
Steel plate
Bolt
Glulam
150
End block
Steel block
55
Steel bar
150
Anchorage
A
B
80
Bolt
Steel plate
150
Anchorage
End block
Steel bar
8 15 35 15 8
A-A
80
Glulam
100
155
55
Steel block
Steel bar
8 15 35 15 8
B-B

(**c**)

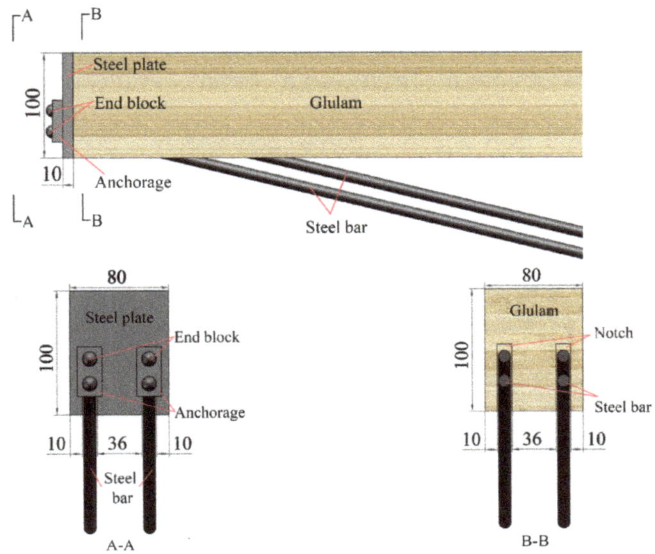


(**d**)

**Figure 3.** Four types of anchorage devices (dimensions are presented in (mm)). (**a**) mode I: external anchorage device, (**b**) mode II: boot anchorage device, (**c**) mode III: under beam anchorage device and (**d**) mode IV: slotted anchorage device.

### 2.2. Constitutive Model

The linear orthotropic constitutive equation can be expressed as follows:

$$D = \begin{bmatrix} D_{1111} & D_{1122} & D_{1133} & 0 & 0 & 0 \\ & D_{2222} & D_{2233} & 0 & 0 & 0 \\ & & D_{3333} & 0 & 0 & 0 \\ & & & D_{1212} & 0 & 0 \\ & sym & & & D_{1313} & 0 \\ & & & & & D_{2323} \end{bmatrix} \quad (1)$$

where

$$D_{1111} = E_1(1 - v_{23}v_{32})r \quad (2)$$

$$D_{2222} = E_2(1 - v_{13}v_{31})r \quad (3)$$

$$D_{3333} = E_3(1 - v_{12}v_{21})r \quad (4)$$

$$D_{1122} = E_1(v_{21} + v_{31}v_{23})r = E_2(v_{12} + v_{32}v_{13})r \quad (5)$$

$$D_{1133} = E_1(v_{31} + v_{21}v_{32})r = E_3(v_{13} + v_{12}v_{23})r \quad (6)$$

$$D_{2233} = E_2(v_{32} + v_{12}v_{31})r = E_3(v_{23} + v_{21}v_{13})r \tag{7}$$

$$D_{1212} = G_{12} \tag{8}$$

$$D_{1313} = G_{13} \tag{9}$$

$$D_{2323} = G_{23} \tag{10}$$

$$r = \frac{1}{1 - v_{12}v_{21} - v_{23}v_{32} - v_{31}v_{13} - 2v_{21}v_{32}v_{13}} \tag{11}$$

where 1, 2, 3 are, respectively, the longitudinal (*L*), radial (*R*), and tangential directions (*T*) of Glulam (as shown in Figure 4). Therefore, $E_i$ is Young's module in one direction (e.g., $E_1$ is Young's in the longitudinal direction), $G_{ij}$ is the shear module in the certain plane; and $v_{ij}$ is the Poisson's ratio ($i, j$ = 1, 2, 3). The cross-section of the Glulam beam is shown in Figure 5.

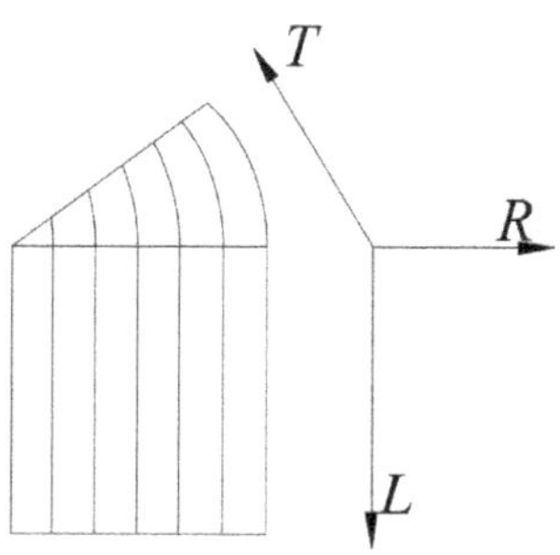


**Figure 4.** Three-dimensional image of Glulam.

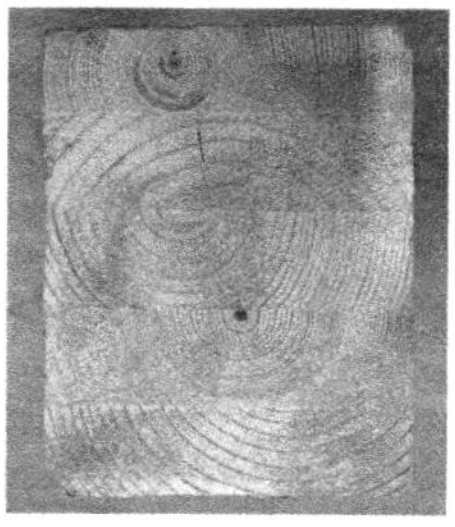

**Figure 5.** Cross-section of Glulam beam.

The above values must satisfy the requirements below:

$$E_i, G_{ij}, v_{ij} > 0 \tag{12}$$

$$1 - v_{12}v_{21} - v_{23}v_{32} - v_{31}v_{13} - 2v_{21}v_{32}v_{13} > 0 \tag{13}$$

### *2.3. Material Properties*

Material compressive properties of Glulam used in the modeling were from the experimental results, as shown in Figure 6. Based on standards [31,32] and related studies [33], the compressive plastic stress was 35 $N/mm^2$. The orientations definition of Glulam is shown in Figure 7, and the corresponding elastic properties of Glulam are listed in Table 1. The tensile strength and the elastic module of the steel bar were 1720 N/mm$^2$ and 2.0 × 10$^5$ N/mm$^2$, respectively. Furthermore, the radius of the reinforcement was 4.95 mm. On the other hand, the yield strength and the elastic module of other steel members were 345 N/mm$^2$ and 2.0 × 10$^5$ N/mm$^2$, respectively.

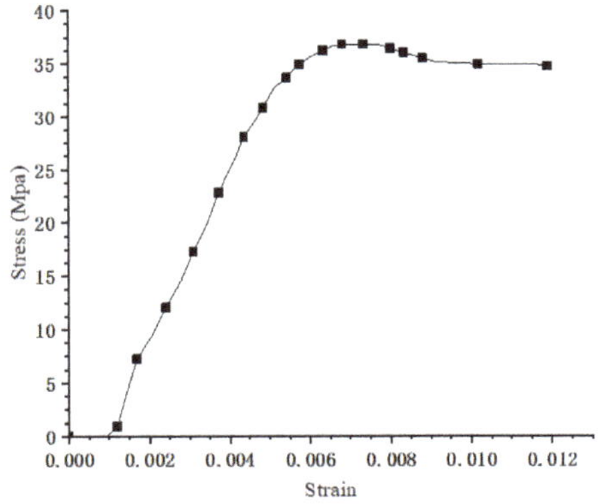


**Figure 6.** Material compressive properties of Glulam.

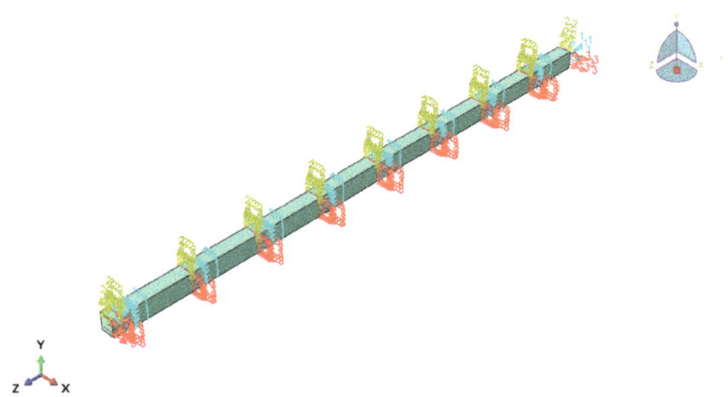


**Figure 7.** Orientations definition of Glulam.

**Table 1.** Corresponding elastic properties of Glulam.

| $D_{1111}$ | $D_{2222}$ | $D_{3333}$ | $D_{1212}$ | $D_{1313}$ | $D_{1122}$ | $D_{1133}$ | $D_{2323}$ | $D_{2233}$ |
|---|---|---|---|---|---|---|---|---|
| 8654.084 | 488.532 | 488.532 | 564.574 | 564.574 | 221.982 | 221.982 | 150.553 | 182.109 |

## *2.4. Finite Element Simulation*

Figure 8 shows the prestressed beams models with four types of various anchorage devices. The surface-to-surface contact was employed to model the interaction between the Glulam and the steel devices, including steel plates and anchorage devices. The former and the latter were defined as the slave surface and master surface, respectively. As there should not be any delamination between anchorage devices and the beams, the laminations were tied. Top prestress and upward pre-deflection were applied to the beams by extending the bolt.

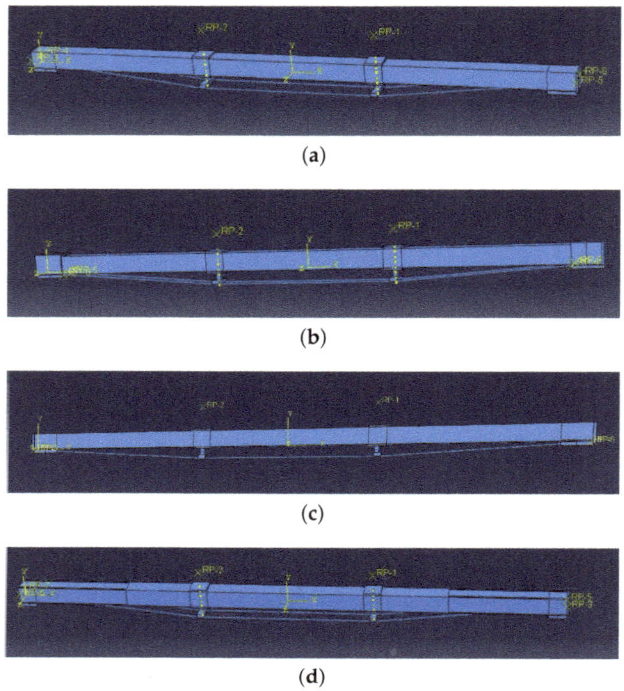

Figure 8. Four prestressed beams models. (a) mode I: external anchorage device, (b) mode II: boot anchorage device, (c) mode III: under beam anchorage device and (d) mode IV: slotted anchorage device.

Three-dimensional finite element models were developed to simulate the prestressing state and the loading state after applying the prestress of the prestressed Glulam beams. The Glulam beam was 3150 mm in length, 150 mm in depth, 100 mm in width, and 3000 mm in span length. The mesh size was 20 mm, except for the steel bars, the mesh property of other parts was C3D8R, the steel bars' mesh property was B31, its size was 50 mm. At the final stage, the flexural tests of prestressed Glulam beams were simulated, following modeling of the prestressing process and then the loading process.

## 3. Results and Discussion

### 3.1. Load-Deflection Curves

The load-deflection curves include two stages, the elastic stage and the plastic stage. At the elastic stage, with load increasing, deflection increases linearly. At the plastic stage, compression plasticity could be better utilized due to the existence of steel bars. Therefore, stages with relatively low slopes could be found in the curves. The upward direction is defined as negative, and the opposite direction is positive.

Keeping consistent top pre-deflection, control the initial mid-span deflection was 5 mm, 10 mm, and 15 mm, respectively. As is shown in Figure 9, as the initial mid-span deflection increased, the bending capacities of all prestressed beams had certainly improved, and initial deflection is defined as negative deflection. Meanwhile, the stiffness showed significant improvement. Compared with the same size of Glulam beams, the capacity of all prestressed Glulam beams was improved. Within the scope of regulation, the prestressed beams with mode I anchorage device keep the largest bending capacity under the same top deflection. Compared with the other three forms of the prestressed beam, with the initial top deflection increasing, the bending capacity of the prestress beam gets the most significant improvement, but that was still lower than other forms of prestressed beams. At the elastic stage, the stiffness of the beam with mode II anchorage device was the highest, but at a later stage, its stiffness showed a pronounced decline, this kind of beam shows more plasticity. This is because the iron bolt can improve the stiffness initially. Still, with the load increasing, the steel bar could not produce relatively high stress, and the stress concentration could be found in the interface between the iron bolt and beam end.

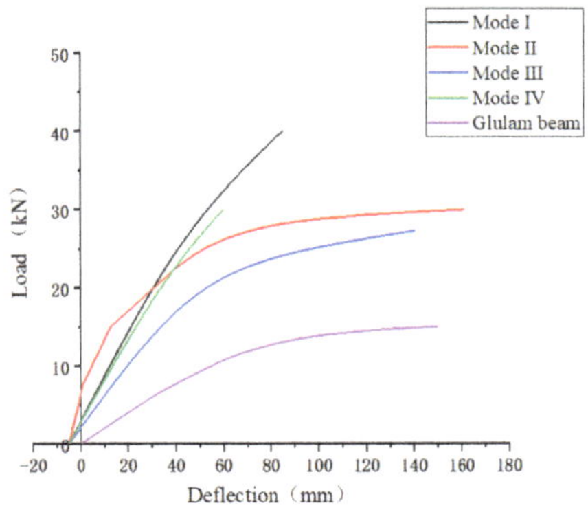


(**a**)

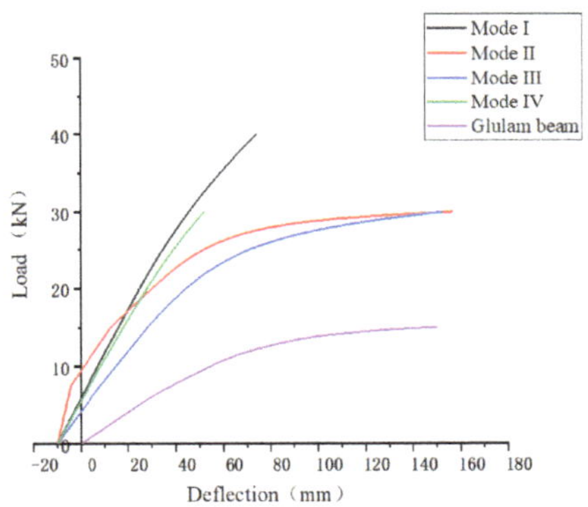


(**b**)

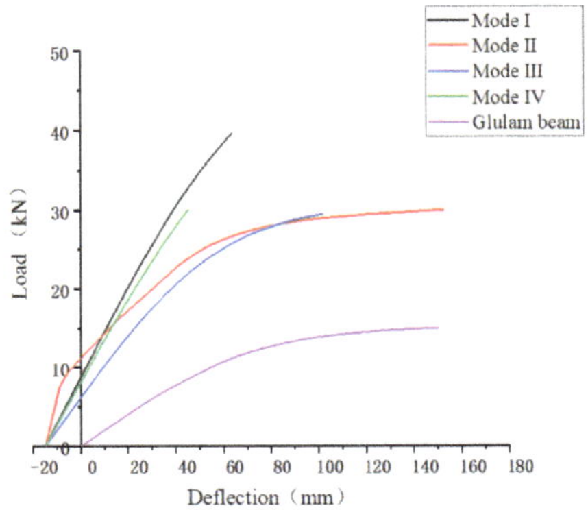


(**c**)

**Figure 9.** Load-deflection curves with various initial mid-span deflection. (**a**) Five mm initial mid-span deflection, (**b**) 10 mm initial mid-span deflection and (**c**) 15 mm initial mid-span deflection.

Next, 1 MPa, 1.5 MPa, and 2 MPa were selected to be initial mid-span top prestress. The load-deflection curves are shown in Figure 10, initial deflection is defined as negative deflection. Compared with the same size Glulam beam without prestressing, the capacity of all prestressed Glulam beams was significantly improved. With the increase of the top mid-span stress, the bending capacity of the prestressed beams does not show noticeable improvement, but their stiffness illustrates certain augment. The bending capacity of the prestressed beam with mode I anchorage device remains the priority, but with the increase of the top prestress, it does not show significant improvement.

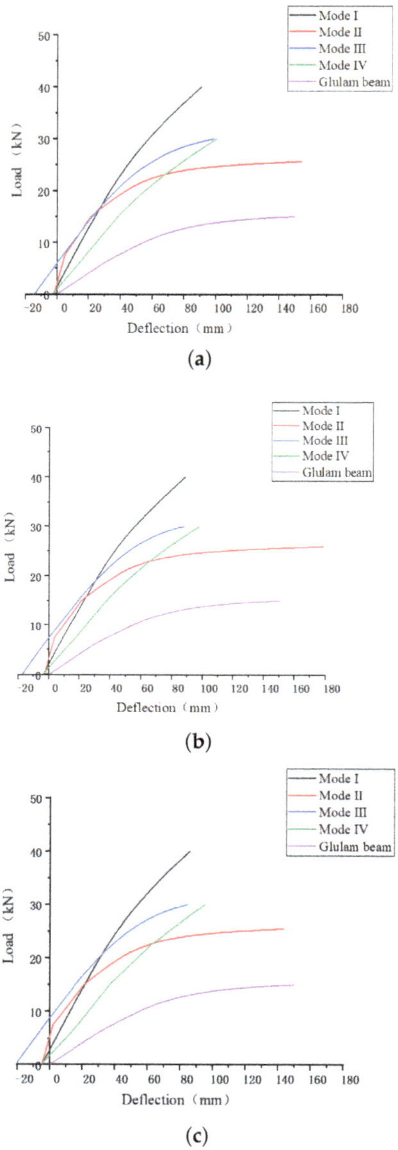


**Figure 10.** Load-deflection curves with various initial mid-span top prestress. (**a**) One MPa initial mid-span top prestress, (**b**) 1.5 MPa initial mid-span top prestress and (**c**) 2 MPa initial mid-span top prestress.

On the other hand, the prestressed beam with a mode III anchorage device is easily influenced by the increase of the prestress, and its bending performance and stiffness get the most noticeable improvement. Initially, the stiffness of the prestressed beam with mode IV was the lowest, but when the deformation of the Glulam beam is relatively large, the beam would be in contact with the reinforcement. Therefore, the ability of collaborative work would be improved in the later stage, which will not result in a significant reduction in its stiffness.

### 3.2. The Bending Stress Distribution

The bending stress distributions of Glulam beams with 30 kN load are shown in Figure 11. There are the bending stress distribution figures of half-span Glulam beams because of the symmetry. Figure 11a shows the stress distribution of the prestressed beam with mode I anchorage device. Between the two loading points, the stress distribution is uniform. Meanwhile, the stress value is not high, and there is no stress concentration in the whole beam. Figure 11b shows the bending stress distribution of the boot anchorage prestressed beam. The stress distribution between the loading point and midspan is uniform. However, the compression stress between the bending section is higher than in other areas. However, the highest stress is found near the bolt anchorage because of the bolt effect, which causes the stress concentration and stiffness to decline. The compression stress of the prestressed beam with mode III anchorage device between the bending section is uniform, as shown in Figure 11c, but the stress was much high. Figure 11d presents the bending stress distribution of the slotted anchorage prestressed beam. The bending stress near the mid-span is not high. However, as expected, the bending stress near the notch area is very high due to stress concentration caused by the notch.

(**a**)

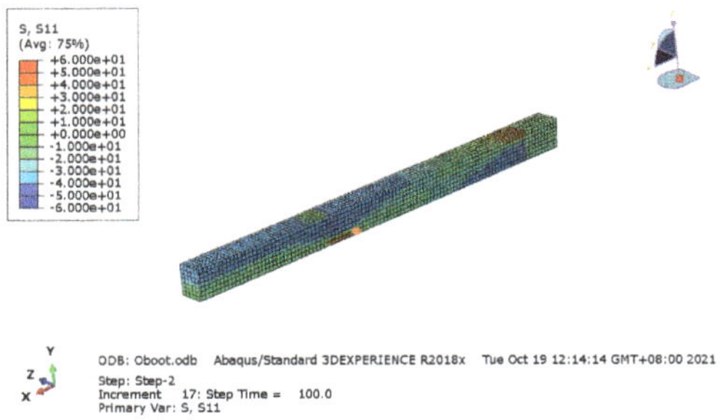


(**b**)

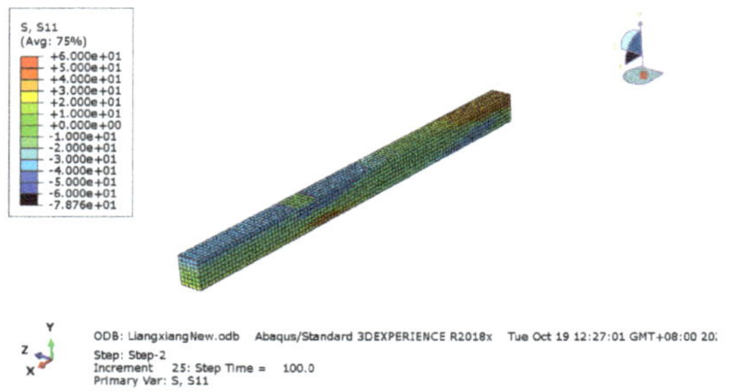


(**c**)

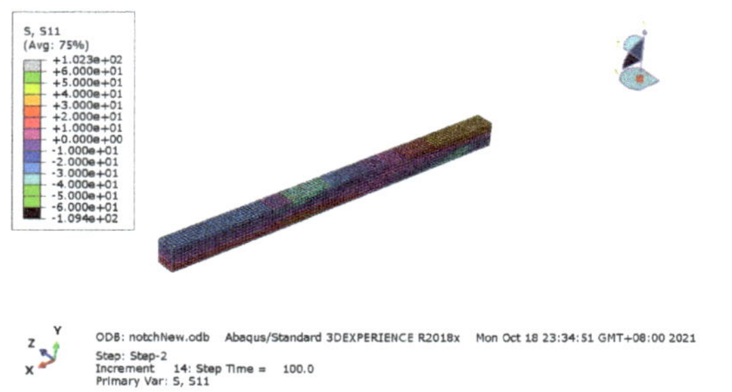


(**d**)

**Figure 11.** Bending stress distributions of Glulam beams with 30 kN load. (**a**) mode I: external anchorage device, (**b**) mode II: boot anchorage device, (**c**) mode III: under beam anchorage device and (**d**) mode IV: slotted anchorage device.

### *3.3. Impact of Initial Upward Pre-Deflection*

When initial deflection was 5 mm, 10 mm, and 15 mm, respectively, the bearing capacity of prestressed beams and Glulam beam is shown in Table 2. Compared with the Glulam beam, the bearing capacity of all prestressed beams is significantly increased by 80.37–194.60%. As the initial deflection increases, bearing capacity is raised to a certain extent. Compared with the prestressed beam with a top deflection of 5 mm, the bearing capacity of the prestressed beam with top deflection of 10 mm and 15 mm is increased by 0–7.31% and 0–8.46%, respectively, and the bearing capacity of mode III is the largest; the stiffness of prestressed beams with top deflection of 10 mm and 15 mm is increased by 1.87–13.30% and 3.46–24.67%, respectively, among which the stiffness of mode IV is the largest. With top deflection increasing, the capacity and stiffness of mode I are always the highest.

### *3.4. Impact of the Initial Top Pre-Stress*

When initial top stress was 1 MPa, 1.5 MPa, and 2 MPa, the bearing capacity of prestressed beams and Glulam beam is shown in Table 3. Compared with the Glulam beam, the bearing capacity of prestressed beams was increased by 101.71–180.87%. With the top stress increasing, bearing capacity does not significantly increase, but the stiffness shows noticeable improvement. For example, compared with the prestressed beams with top stress of 1 MPa, the bending performance of prestressed beams with the stress of 1.5 MPa and 2 MPa increased by 0.02–2.05% and 0.49–3.93%, respectively; the stiffness of prestressed beams with the stress of 1.5 MPa and 2 MPa increased 1.19–8.51% and 1.33–11.77%, separately.

**Table 2.** Bearing capacity with various initial deflection.

| Deflection (mm) | Anchoring Device Type | Yield Point Load (kN) | Deflection (mm) | Improvement Rate (%) |
|---|---|---|---|---|
| - | - | 11.34 | 64.94 | - |
| 5 | Mode I | 31.45 | 57.38 | 177.24% |
| | Mode II | 25.31 | 53.78 | 123.13% |
| | Mode III | 20.46 | 55.16 | 80.37% |
| | Mode IV | 30.00 | 59.85 | 164.47% |
| 10 | Mode I | 32.05 | 50.60 | 182.51% |
| | Mode II | 25.20 | 52.77 | 122.11% |
| | Mode III | 21.96 | 52.66 | 93.56% |
| | Mode IV | 30.00 | 51.89 | 164.47% |
| 15 | Mode I | 33.42 | 46.70 | 194.60% |
| | Mode II | 25.30 | 51.92 | 123.04% |
| | Mode III | 22.19 | 46.12 | 128.55% |
| | Mode IV | 30.00 | 45.09 | 164.47% |

**Table 3.** Bearing capacity with several top prestress.

| Initial Top Pre-Stress (MPa) | Anchoring Device Type | Yield Point Load (kN) | Deflection (mm) | Improvement Rate (%) |
|---|---|---|---|---|
| - | - | 11.34 | 64.94 | - |
| 1 | Mode I | 30.92 | 60.46 | 172.57% |
| | Mode II | 22.88 | 64.83 | 101.71% |
| | Mode III | 27.16 | 69.35 | 139.43% |
| | Mode IV | 26.97 | 85.54 | 137.75% |
| 1.5 | Mode I | 31.29 | 59.74 | 175.88% |
| | Mode II | 23.35 | 63.18 | 105.85% |
| | Mode III | 27.18 | 63.45 | 139.64% |
| | Mode IV | 26.97 | 83.42 | 137.80% |
| 2 | Mode I | 31.86 | 56.94 | 180.87% |
| | Mode II | 23.78 | 63.97 | 109.64% |
| | Mode III | 27.42 | 61.19 | 141.71% |
| | Mode IV | 27.10 | 81.55 | 138.91% |

## 4. Experimental Research

Based on the FE result and engineering feasibility, the prestressed beam with an external anchorage device (Mode I) could keep better-bending capacity and stiffness. Therefore, it was chosen to do experimental research.

### *4.1. Mechanical Properties of Materials*

The raw material was spruce-pine-fir (SPF) from Canada. The dried spruce plates were formed into a rectangular shape with Phenolic adhesives and cured at a temperature of 25 °C. The grade of the timber was $I_c$. Based on standards [31,32] and the former study, compressive and tensile tests were carried out in the laboratory of Northeast Forestry University. The experimental data are shown in Table 4.

**Table 4.** Experimental data of timber.

| Type | Peak Compressive ($f_{cu}$) | Peak Tensile ($f_{tu}$) |
|---|---|---|
| Strength (MPa)/COV(%) | 35/4.63% | 81.32/6.42% |
| Elastic modulus (MPa)/COV(%) | 10,350.2/6.14% | 10,350.2/6.18% |

All steel specimens were the low-relaxation prestressed steel wire. The nominal yield strength of the steel bar was 1570 MPa, and its tensile strength and elastic modulus were 1640.51 MPa and $1.92 \times 10^5$ MPa, respectively, with a coefficient of variation of 1.71% and 0.35%.

### 4.2. Testing Procedure

The test contained two dimensions of 3150 mm × 80 mm × 100 mm prestressed laminated beams. Specimens fixed with 7 mm diameter low relaxation prestressed steel wires. Specific parameters are shown in Table 5.

**Table 5.** Parameters of specimens.

| Specimen | Size (mm × mm × mm) | Reinforcement Ratio (%) | Pre-Force (kN) |
|---|---|---|---|
| B1<br>B2 | 3150 × 80 × 100 | 19.24 % | 9 |

### 4.3. Bending Test

A Four-point bending test was adopted by the Chinese national standards [32,34]. The photo and schematic diagram of the bending test are shown in Figures 12 and 13. Based on a related study [35], the distance between two supports was 3000 mm, and a shear span ( distance from the support reaction force to the nearest applied concentrated load) was 1000 mm. A linear variable displacement transducer (LVDT) with a range of 50 mm was attached to the top side of each specimen at support points to measure the deflections. Two LVDTs and one LVDT with a range of 150 mm were attached to the bottom side of each beam at two loading points and the mid-span point to obtain corresponding vertical deflections, respectively. Five strain gauges were bonded to the loading-point lateral surface with equal spacing to get the strain located at the loading point. Meanwhile, the strain situated at the top and bottom of the half-span beam was gauged by the two 100 mm × 3 mm strain gauges separately. The 2 mm × 3 mm strain gauges were pasted to the steel wire surfaces at an L/2 distance from the beams' support points.

**Figure 12.** Photo of test.

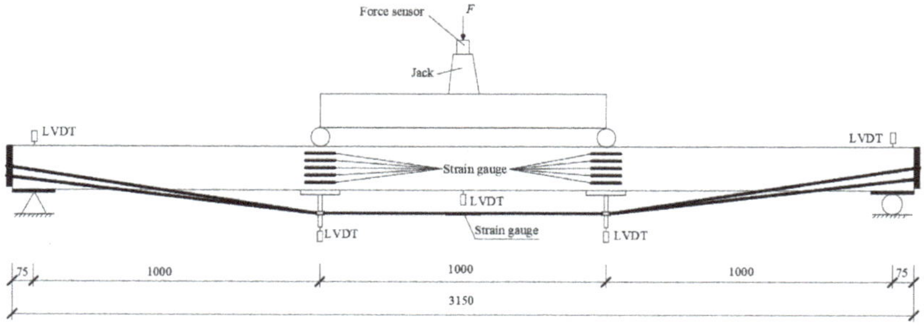


**Figure 13.** Schematic diagram of the test.

After assembling the prestressed laminated beam, turning the nut counterclockwise causes the screw to rotate. The steering block moves downward to stretch the reinforcement and simultaneously applies to top prestress and upward pre-deflection to the laminated beams. The prestress value can be controlled by adjusting the distance from the deviation block to the beam bottom.

### 4.4. Failure Modes

The failure mode of the B1 is shown in Figure 14. The bottom fibers began to break initially when the load was about 85% of the ultimate load due to the knot, as Figure 14a. With load increasing, Figure 14b shows the wrinkle at the top surface near the left loading point of the beam. At this time, the load is 94.61% of the ultimate load. As the load continues to increase, there is a split above the knot, as shown in Figure 14c. Finally, the beam failed due to tension damage at the bottom of the beam, as Figure 14d.

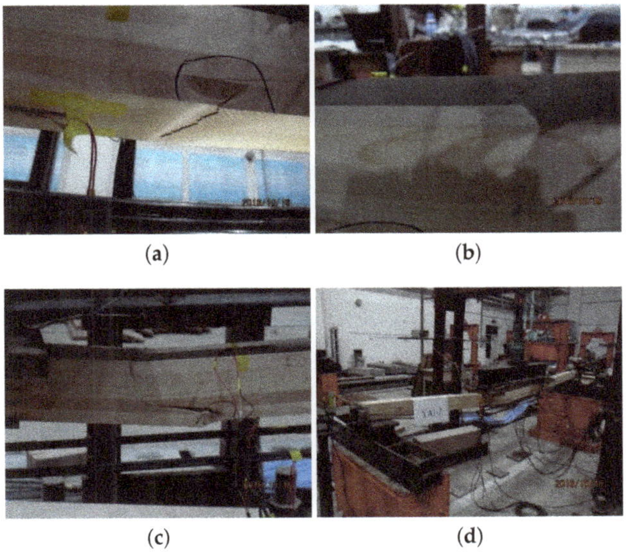

**Figure 14.** B1 failure mode. (**a**) bottom fibers break, (**b**) top surface wrinkle, (**c**) split above the knot and (**d**) beam failed.

Figure 15 showed the failure mode of the B2. Initially, there were no apparent changes before the fold shown at the top of the beam, as shown in Figure 15a. As the increase of the load, the fold extends along the section of the beam, as Figure 15b. Figure 15c presented that when the load reached 32 kN, the fold developed to the bottom with reduced bearing capacity. Finally, the beam is suddenly destroyed during the loading stage, as shown in Figure 15d.

This kind of failure mode was consistent with the simulation result. As shown in Figure 16, the tested beam was damaged caused by the top compression damage, the compressive strength of the top area can be fully utilized, and obvious folds could be found in this area.

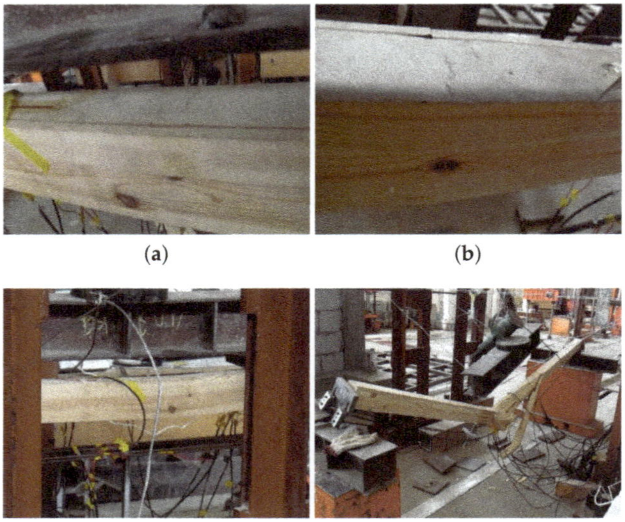

(a) (b) (c) (d)

**Figure 15.** B2 failure mode. (**a**) Top fold, (**b**) fold extends along the section, (**c**) fold extended to the bottom and (**d**) destroyed suddenly.

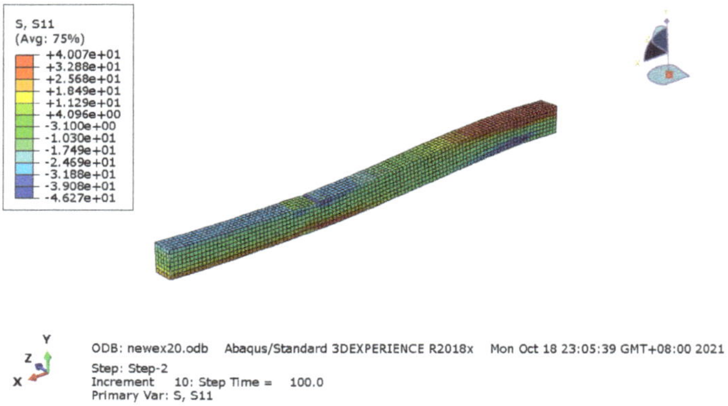


**Figure 16.** Stress distribution.

### 4.5. Load-Deflection Curves and Bearing Capacity

Figure 17 shows the load-deflection curves of tested beams. Before the peak point, there is no obvious turning point, and the curves are similar to the line, which is identical to the curves of the simulation result. Initial upward deflection is defined as a negative deflection. By comparison, the simulation can predict the load-deflection curves of the tested beams. As the initial defect of the timber was not taken into consideration, the prediction results capacity was higher than the results from the test.

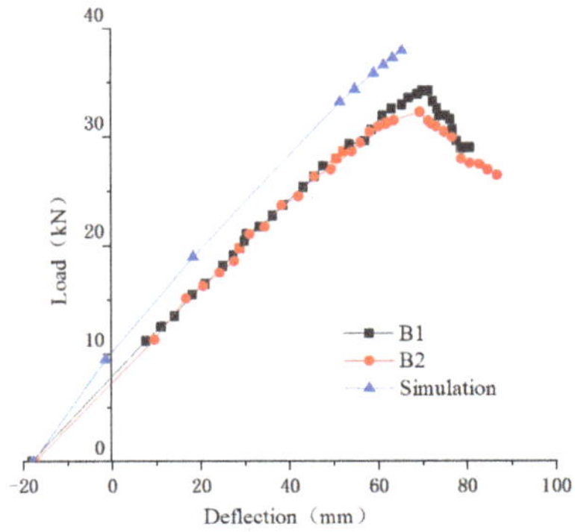


**Figure 17.** Load–deflection curves of tested beams and simulation.

As shown in Table 6, the bearing capacity of the mode I beam is close to the simulation results. However, compared with the simulation results, the stiffness is relatively low due to the difference between initial defects and raw materials of laminated materials.

**Table 6.** Data comparison of two beams.

| Specimen | A (mm$^2$) | Reinforcement Ratio | Pre-Force (kN) | Bearing Capacity (kN) | Deflection (mm) | Failure Mode |
|---|---|---|---|---|---|---|
| B1 | 100 × 80 | 19.24% | 9 | 34.38 | 69.75 | tensile damage |
| B2 | | | | 32.33 | 69.25 | compression damage |

## 5. Conclusions

- Compared with the Glulam beam, prestressed beams with four types of anchoring devices could be obviously improved. When the prestress is 1 MPa, 1.5 MPa and 2 MPa, the bearing capacity is increased by 101.71–172.57%, 105.85–175.88%, and 109.64–180.87%, respectively; when the top deflection is 5 mm, 10 mm and 15 mm, the bending performance of the prestress beams is improved by 80.37–177.24%, 93.56–182.51%, and 95.62–194.60%;
- Under different prestress and pre-deflection values, the stiffness and bearing capacity of prestressed beam with the mode I is the highest; but the bearing capacity of prestressed beams with the mode III gets the hugest improvement with the initial top stress increasing. Compared with 1 MPa prestressed beams, the bearing capacity of 1.5 MPa and 2 MPa prestressed beams is increased by 8.51% and 11.77%, respectively; compared with the beam with pre-deflection of 5 mm, the bearing capacity of the beam with pre-deflection of 10 mm and 15 mm is increased by 13.30% and 24.67%, respectively;
- Compared with the stress distribution of the other three kinds of prestressed beams, the beam with mode I is most uniform. Therefore the properties of the raw material could be fully utilized;
- Experimental data show the failure mode of the prestressed tested beams is the brittle failure. Compared between experimental results and simulation results, the simulation could better predict the basic linearity of the load-displacement curve, ultimate bending capacity, and the failure mode.

**Author Contributions:** Conceptualization, M.L. and M.W.; methodology, M.L. and M.W.; formal analysis, Y.Z.; investigation, M.L., M.W. and L.M.; writing—original draft preparation, M.L. and M.W.; writing—review and editing, L.M. and N.G.; visualization, Y.Z.; supervision, N.G.; funding acquisition, N.G. All authors have read and agreed to the published version of the manuscript.

**Funding:** This research was funded by the Fundamental Research Funds for the Central Universities (2572021AW11) , the Natural Science Foundation of Heilongjiang Province (LH2019E005), the Natural

Science Foundation of Heilongjiang Province (LH2020E009), and the Natural Science Foundation of Fujian Province (2020J01402).

**Institutional Review Board Statement:** Not applicable.

**Informed Consent Statement:** Not applicable.

**Data Availability Statement:** The data presented in this study are available on request from the corresponding author.

**Conflicts of Interest:** The authors declare no conflict of interest.

## References

1. Canada Wood. Structural types and advantages of modern wood structure buildings. *Int. Wood Ind.* **2020**, *50*, 10–11.
2. Jochen, K.; Staffan, S. Probabilistic representation of the duration of load effects in timber structures. *Eng. Struct.* **2011**, *33*, 462–467.
3. Haiyan, Z.; Qian, L. The Application of the Wooden Structure to the Canadian School Buildings. *New Archit.* **2020**, *6*, 110–113.
4. Yue, C. American medium and high-rise wood structure buildings. *Constr. Sci. Technol.* **2019**, *17*, 31–34.
5. Chengmou, F.; Shengzhe, N. *Modern Wood Structure*; Harbin Institute of Technology Press: Harbin, China, 2007; Volume 10, pp. 1–5.
6. Josef, K. *Multi-Story Timber Structure*; Birkhäuser: Basel, Switzerland, 2008; pp. 188–200.
7. Ehrhart, T.; Steiger, R.; Lehmann, M.; Frangi, A. European Beech (*Fagus sylvatical* L.) Glued Laminated Timber: Lamination Strength Grading, Production and Mechanical Properties. *Eur. J. Wood Wood Prod.* **2020**, *78*, 971–984. [CrossRef]
8. Zhonghua, H. A Study on Joint Finger Technique for Laminations in the Structural Larch Glulam. Master's Thesis, Chinese Academy of Forestry Sciences, Beijing, China, 2012.
9. Liqing, L. Development of the glued-laminated timber structure housing. *China Build. Mater. Sci. Technol.* **2008**, *17*, 63–66.
10. Xu, S. *Experimental Research on Flexural Performance of Reinforced Glulam Beams*; Northeast Forestry University: Harbin, China, 2016.
11. Aijun, C.; Guojun, C.; Guosheng, C.; Wang, X.; Peng, R. Experimental study on the mechanical behavior of Glulam timber beams reinforced with BFRP. *J. Cent. S. Univ. For. Technol.* **2019**, *39*, 107–113.
12. Yishi, L. *Experimental Study on Flexural Behavior of Flax Fiber Reinforced Polymer Reinforced Glulam Beams*; Northeast Forestry University: Harbin, China, 2020.
13. Feibin, W.; Xiaolin, Y. Study on Bending Behavior of Glulam Beams Reinforced with Bonded-in GFRP Rods. *China For. Prod. Ind.* **2017**, *44*, 18–23.
14. Ivan, G.; Boško, S. *Flexural Reinforcement of Glulam Beams with CFRP Plates*; Springer: Dordrecht, The Netherlands, 2016; Volume 49, pp. 2841–2855.
15. Johnsson, H.; Blanksvärd, T. Glulam members are strengthened by carbon fiber reinforcement. *Mater. Struct.* **2007**, *40*, 142–149. [CrossRef]
16. Juliano, F.; Antonio, A. Glulam beams reinforced with FRP externally-bonded: Theoretical and experimental evaluation. *Mater. Struct.* **2011**, *44*, 1431–1440.
17. Khelifa, M.; Celzard, A. Numerical analysis of flexural strengthening of timber beams reinforced with CFRP strips. *Compos. Struct.* **2014**, *111*, 393–400. [CrossRef]
18. Fava, G.; Carvelli, V. Pull-out strength of glued-in FRP plates bonded in Glulam. *Constr. Build. Mater.* **2013**, *43*, 362–371. [CrossRef]
19. Valipour, H.; Crews, K . Efficient finite element modeling of timber beams strengthened with bonded fiber-reinforced polymers. *Constr. Build. Mater.* **2011**, *25*, 3291–3300. [CrossRef]
20. Jin, Z.; Weichang, W. Experimental study on short-term flexural behavior of internal prestressed Glulam beams. *China Civ. Eng. J.* **2019**, *52*, 23–34.
21. Huifeng, Y.; Dongdong, J. Prestressed Glulam beams reinforced with CFRP bars. *Constr. Build. Mater.* **2016**, *109*, 73–83.
22. Cheng, L.; Huifeng, Y. Experimental Study on the Flexural Behavior of Prestressed Glulam Beams. *Struct. Eng.* **2014**, *30*, 160–164.
23. Jin, Z.; Hao, S. Short-Term Flexural Behavior of Prestressed Glulam Beams Reinforced with Curved Tendons. *J. Struct. Eng.* **2020**, *146*, 04020086.
24. Yang, W.; Shaocong, Y. Experimental and theoretical investigation of steel-reinforced bamboo scriber beams. *Eng. Struct.* **2020**, *223*, 111179.
25. McConnell, E.; McPolin, E. Post-tensioning of Glulam timber with steel tendons. *Constr. Build. Mater.* **2014**, *73*, 426–433. [CrossRef]
26. Negrao, J. Preliminary study on wire prestressing methods for timber pieces reinforcement. *Constr. Build. Mater.* **2014**, *102*, 109–110. [CrossRef]
27. Ansari, B.; Guan, Z. Structural behavior of glued laminated timber beams prestressed by compressed wood. *Constr. Build. Mater.* **2011**, *29*, 24–32. [CrossRef]
28. Yubin, Z.; Wen, L. Prestress loss of prestressed Glulam beam reinforced with FRP. *World Earthq. Eng.* **2019**, *35*, 99–107.
29. Lidan, M.; Nan, G. Study on flexural performance of prestressed Glulam continuous beams under control influence. *J. Wood Sci.* **2021**, *67*,48.
30. Mario, V.; Alessandra, J. A Semi-Linear Elliptic Model for a Circular Membrane MEMS Device Considering the Effect of the Fringing Field. *Sensors* **2021**, *21*, 5237.

31. GB/T 1935-2009. *Methods of Testing in Compressive Strength Parallel to Grain of Wood*; Standardization Administration of China: Beijing, China, 2009.
32. GB/T 50329-2012. *Standard for Test Methods of Timber Structures*; Standardization Administration of China: Beijing, China, 2012.
33. Mingtao, W.; Nan, G. Bending Performance of Prestressed Continuous Glulam Beams. *Adv. Civ. Eng.* **2021**, *2021*, 5512350.
34. GB/T 50005-2017. *Standard for Design of Timber Structure*; Standardization Administration of China: Beijing, China, 2017.
35. Lidan, M.; Nan, G. Influence of the Force Arm on the Flexural Performance of Prestressed Glulam Beams. *Adv. Civ. Eng.* **2021**, *2021*, 8831406.

MDPI AG
Grosspeteranlage 5
4052 Basel
Switzerland
Tel.: +41 61 683 77 34

*Materials* Editorial Office
E-mail: materials@mdpi.com
www.mdpi.com/journal/materials